Universitext

Universitext

(continued after the index)

Falko Lorenz

Algebra

Volume II:
Fields with Structure, Algebras and Advanced Topics

With the collaboration of the translator, Silvio Levy

 Springer

Falko Lorenz
FB Mathematik Institute
University Münster
Münster, 48149
Germany
lorenz@math.uni-muenster.de

ISBN: 978-0-387-72487-4   e-ISBN: 978-0-387-72488-1

Library of Congress Control Number: 2005932557

Mathematics Subject Classification (2000): 11-01,12-01,13-01

springer.com

*"Because certainty is desirable in didactic discourse
— the pupil wishes to have nothing uncertain deliv-
ered to him — the teacher cannot let any problem
stand, circling it from a distance, so to speak. Things
must be determined at once (staked out, as the Dutch
say[1]), and so one believes for a while that one owns
the unknown territory, until another person rips out
the stakes again and immediately sets them down,
nearer or farther as the case may be, once again."*

J. W. Goethe, in *Werke* (Weimar 1893), part II, vol. 11
("Science in General"), p. 133.

Foreword

In this second volume of *Algebra*, I have followed the same expository guidelines
laid out in the preface to the first volume, with the difference that now pedagogic
considerations can take a secondary role in favor of a more mature viewpoint on the
content.

I imagine the reader of this second volume to be a student who already has a
working knowledge of algebra and is eager to extend and deepen this knowledge
in one direction or another. Thus, in sections that can to a large extent be studied
independently of the rest, I have made a broader choice of presentation.

There was good reason, in my opinion, to let the material in the first volume
be guided by an emphasis on fields. Thus it was natural to present in this second
volume certain classes of fields having additional structure. Among these we deal
first with ordered fields, in part to arouse interest in the area of real algebra, which
is given short shrift in most current textbooks (though it was much esteemed in the
nineteenth century and gained new momentum in the 1920s through the work of
Artin and Schreier). It also seemed worth broaching certain aspects of the theory of
quadratic forms.

Next, special attention is devoted to the theory of valued fields. Local fields
represent today, over a hundred years after their discovery by Hensel, a completely
standard prerequisite in many areas of mathematics.

Besides making an effort not to treat superficially any area once selected for
coverage, I also aimed for some diversity. Thus I decided not to stay within the
confines of field theory proper, but rather to include another major theory, that of
semisimple algebras. In this context it seemed a matter of course to discuss the
rudiments of finite group representations as well. This path to the subject, offered
here instead of the more direct one opened up by Schur, is well worth the trouble,

---

[1] Goethe writes the past participle of the Dutch verb *bepalen*, bridging the two meanings:
literally 'to plant stakes', but in its normal usage 'to prescribe, determine, fix, set in place.'

especially if one is interested also in questions of rationality in representation theory, as they are treated at the end of the book.

Undoubtedly this volume contains more material than can be covered in one undergraduate semester. Some sections are perhaps suitable for introductory graduate seminars. Several topics absent from average textbooks are included here, as they fit naturally with our treatment: among them we mention the Witt calculus, Tsen rank theory, and local class field theory.

My warm thanks go to all who helped in the creation of this book: the students in my course, for their invigorating interest; my faculty colleagues, for much good advice and for prodding me on with their frequent inquiries about when the book would be ready; Florian Pop, for a conversation in Heidelberg, which persuaded me to include the topic of local class field theory; Hans Daldrop, Burkhardt Dorn and Hubert Schulze Relau, for their critical reading of large portions of the manuscript, and the latter also for his careful work on the index; Bernadette Bourscheid for her efficient preparation of the original typescript; the publishers of the first German edition (1990), BI-Wissenschaftsverlag, for their renewed cooperation, and particularly the editor Hermann Engesser, for understanding and patient advice.

In the preparation of the second German edition (1997) I again benefited from suggestions, praise and criticism from colleagues, including S. Böge, B. Huppert, J. Neukirch, P. Roquette and K. Wingberg, and from the involvement of students — not only from and my course but also from elsewhere — whose watchful reading led to improvements. Special thanks go to Susanne Bosse for her professional resetting of the text in LaTeX.

Now I am pleased to see this second volume of my work being made available in English as well. I'm very thankful to Springer New York and its mathematics editor, Mark Spencer, for his support and good advice. The translation, like that of the first volume, was done by Silvio Levy, and once again he has suggested helpful improvements to the exposition. I shall look back upon this fruitful collaboration with fondness and appreciation.

Münster, September 2007   Falko Lorenz

Contents

20

Ordered Fields and Real Fields

1. As a first class of fields with additional structure we now turn to ordered fields.

Definition 1. Let K be a field. Given an order relation $\leq$ on the set K, we say that K — or, more formally, the pair $(K, \leq)$ — is an *ordered field* if

(1) $\leq$ is a *total order*.

(2) $a \leq b$ implies $a + c \leq b + c$.

(3) $a \leq b$ and $0 \leq c$ imply $ac \leq bc$.

In this situation we also say that $\leq$ is a *field order* — or, if no confusion can arise, just an *order* — on K.

If $\leq$ is an order on a set, we write $a < b$ if $a \leq b$ and $a \neq b$. Often instead of $a \leq b$ or $a < b$ we write $b \geq a$ or $b > a$. Because of (2) we have in an ordered field

$$a \leq b \implies b - a \geq 0.$$

Thus the order $\leq$ of an ordered field K is entirely determined by the set

$$P = \{a \in K \mid a \geq 0\},$$

called the *positive set* (or *set of positive elements*) of $\leq$. We have

(4) $P + P \subseteq P$ and $PP \subseteq P$;

(5) $P \cap -P = \{0\}$;

(6) $P \cup -P = K$.

Conversely, if P is a subset of a field K satisfying properties (4), (5) and (6), the relation $\leq$ defined by

$$a \leq b \iff b - a \in P$$

makes K into an ordered field (with positive set P). For this reason we sometimes also call such a subset P of K an *order* of K.

Let K be an ordered field. Then

$$a^2 > 0 \quad \text{for every } a \in K^\times.$$

Thus all sums of squares of elements in $K^\times$ are also strictly positive:

$$a_1^2 + a_2^2 + \cdots + a_n^2 > 0 \quad \text{for every } a_1, \ldots, a_n \in K^\times.$$

It follows that any ordered field has characteristic 0.

Definition 2. For K an *arbitrary* field, we denote by

$$SQ(K)$$

the set of all sums of squares in K, that is, the set of all elements of the form $a_1^2 + \cdots + a_n^2$ for $a_i \in K$, $n \in \mathbb{N}$.

Squares and their sums play an important role in the theory of ordered fields, as we will see. First we point out some simple formal properties of $SQ(K)$:

(7) $SQ(K) + SQ(K) \subseteq SQ(K), \quad SQ(K)SQ(K) \subseteq SQ(K).$

Note the analogy between (7) and (4).

Definition 3. A subset T of a field K is called a (*quadratic*) *preorder* on K if

(8) $T + T \subseteq T, \quad TT \subseteq T, \quad$ and

(9) $K^2 \subseteq T.$

Here K^2 denotes the set $K^{\times 2} \cup \{0\}$ of all squares of elements in K. In particular, then, 0 lies in T and 1 lies in $T^\times := T \smallsetminus \{0\}$.

For any field K, the set $SQ(K)$ of sums of squares is a quadratic preorder; it is contained in any quadratic preorder of K, and hence is the minimal quadratic preorder on K.

F1. *Let T be a quadratic preorder of a field K. There is equivalence between*:

(i) $T \cap -T = \{0\}$;

(ii) $T^\times + T^\times \subseteq T^\times$;

(iii) $-1 \notin T$.

If char $K \neq 2$ *these conditions are also equivalent to*

(iv) $T \neq K$.

Proof. (i) $\Leftrightarrow$ (ii) and (i) $\Rightarrow$ (iii) are clear. Suppose (i) does not hold, so there exists $a \neq 0$ in K such that $a \in T \cap -T$. Since a and $-a$ lie in T, so does their product $-a^2$. But because of (9) we have $(a^{-1})^2 \in T$, so $-1 = -a^2 a^{-2} \in T$, so (iii) is contradicted. The implication (iii) $\Rightarrow$ (iv) is clear.

Now suppose, in contradiction to (iii), that $-1 \in T$. If char $K \neq 2$, any element a of K can be written as a difference of two squares (say, those of $\frac{1}{2}(a \pm 1)$). There follows $T = K$, the negation of (iv). $\square$

Remark 1. If char $K = 2$, the quadratic preorders T of K are precisely the intermediate fields of the extension K/K^2.

Remark 2. Let T be a quadratic preorder of K with property (i). The set K has a *partial order* $\leq$ given by

$$a \leq b \iff b - a \in T,$$

and properties (2) and (3) in Definition 1 hold. Moreover $a^2 \geq 0$ for all $a \in K$.

We now investigate under what conditions a quadratic preorder of a field K can be extended to an order of the field K. We first show:

Lemma 1. *Suppose given a quadratic preorder T of a field K with $-1 \notin T$. Let a be an element of K such that $a \notin -T$. Then*

$$T' := T + aT$$

is a quadratic preorder of K, also satisfying $-1 \notin T'$. (Note that $T \subseteq T'$ and $a \in T'$.)

Proof. Obviously we have $T' + T' \subseteq T'$, $T'T' \subseteq T'$, and $K^2 \subseteq T \subseteq T'$, so T' is a quadratic preorder. Assume that $-1 \in T'$, meaning that there are elements $b, c \in T$ such that $-1 = b + ac$. Then c is nonzero since $-1 \notin T$. There follows

$$-a = \frac{1+b}{c} = \frac{1}{c^2}(1+b)c \in T,$$

contradicting the assumption that $-a \notin T$. $\qquad\square$

Theorem 1. *Suppose a quadratic preorder T on a field K does not contain -1. Then T is the intersection of all field orders P of K that contain T:*

$$(10) \qquad\qquad T = \bigcap_{T \subseteq P} P.$$

In particular, for every quadratic preorder T not containing -1 there is an order P of K such that $T \subseteq P$.

Proof. By Zorn's Lemma, T lies in a maximal preorder P of K such that $-1 \notin P$. We claim that P is actually an order of K. By F1, we just have to show that

$$P \cup -P = K.$$

Let a be an element of K such that $a \notin -P$. We must show that $a \in P$. By Lemma 1, the set $T' = P + aP$ is a preorder of K, and we have $-1 \notin T'$. Because P is maximal this implies that $P + aP = P$, and thus $a \in P$.

Now let b be any element of K not contained in T. We still must show that there is an order P of K such that $T \subseteq P$ and $b \notin P$. We apply Lemma 1 to T, with $a := -b$. Then $T' = T - bT$ is a preorder of K not containing -1. From the part of the theorem already shown we conclude that there is an order P of K containing T'; then $-b$ lies in P and therefore $b \notin P$, since $b \neq 0$. $\qquad\square$

Definition 4. A field K is called *formally real*, or just a *real field*, if -1 is not a sum of squares in K:

$$-1 \notin SQ(K).$$

Any ordered field is obviously of this type. Conversely, any real field admits at least one order:

Theorem 2 (Artin–Schreier). *If K is a field, the following conditions are equivalent*:

(i) K *is formally real, that is, -1 is not a sum of squares in K.*

(ii) K *can be made into an ordered field.*

More generally:

Theorem 3. *Let K be a field of characteristic different from 2. Given $b \in K$, there is equivalence between*:

(i) b *being a sum of squares in K*;

(ii) b *being a totally positive element, that is, positive for any field order on K.*

In particular, if K does not admit an order, every element of K is a sum of squares; if, on the other hand, K is real, it admits an order.

Proof. We consider the quadratic preorder $T = SQ(K)$ and assume that $b \notin T$. Then $T \neq K$, whence, by F1, $-1 \notin T$. By Theorem 1, K has some order P such that $b \notin P$. This proves the implication (ii) $\Rightarrow$ (i). The converse is clear.   □

Theorem 4. *Let $(K, \leq)$ be an ordered field and E/K a field extension. The following statements are equivalent*:

(i) *The order $\leq$ of K can be extended to an order on E.*

(ii) -1 *is **not** a sum of elements of the form $a\alpha^2$ with $a \geq 0$ in K and α in E.*

(iii) *Given elements $a_1, \ldots, a_m$ in K with each $a_i > 0$, the quadratic form*

$$a_1 X_1^2 + \cdots + a_m X_m^2$$

*is **anisotropic over** E, that is, it only admits the trivial zero $(0, \ldots, 0)$ in E^m.*

Condition (iii) is a simple reformulation of (ii). The equivalence between (i) and (ii) follows immediately from a more general fact:

Theorem 5. *Let $(K, \leq)$ be an ordered field. If E/K is a field extension and β is an element of E, there is equivalence between*

(i) β *being of the form*

$$\beta = \sum_i a_i \alpha_i^2$$

with each $a_i \in K$ positive and each α_i in E;

(ii) β *being positive in any order of the field E that extends the order $\leq$ on K.*

Proof. The implication (i) $\Rightarrow$ (ii) is trivial. Let T be the set of all sums of elements of the form $a\alpha^2$ with $a \geq 0$ in K and α in E. Clearly T is a quadratic preorder of E. Assume that (i) does not hold, so that $\beta \notin T$. Then $T \neq E$, which by F1 implies that $-1 \notin T$. By Theorem 1 there is then an order P of E such that $P \supseteq T$ and $\beta \notin P$. Therefore (ii) does not hold. $\square$

F2. *Let* $(K, \leq)$ *be an ordered field. If* E/K *is a finite extension of **odd degree**, there is an extension of* $\leq$ *to an order of the field* E.

This follows from Theorem 4 and the next statement:

F3. *If* E/K *is a field extension of **odd degree**, any quadratic form*

$$a_1 X_1^2 + \cdots + a_m X_m^2$$

that is anisotropic over K *is also anisotropic over* E.

Proof. We use induction on the degree $n = E : K$. We can thus assume that $E = K(\beta)$ for some $\beta \in E$. Let $f(X)$ be the minimal polynomial of β over K. Suppose that

$$a_1 \alpha_1^2 + \cdots + a_m \alpha_m^2 = 0$$

with $\alpha_1, \ldots, \alpha_m$ in E, not all 0. We can assume without loss of generality that $\alpha_1 = 1 = a_1$. Since $E = K(\beta)$, there exist polynomials $g_2(X), \ldots, g_m(X) \in K[X]$ of degree at most $n - 1$ such that

$$1 + \sum_{i=2}^m a_i g_i(X)^2 \equiv 0 \bmod f(X).$$

Therefore there exists a polynomial $h(X) \in K[X]$ such that

$$(11) \qquad 1 + \sum_{i=2}^m a_i g_i(X)^2 = h(X) f(X).$$

Now, the sum on the left-hand side is a polynomial of even, strictly positive degree, since by assumption the form $X_1^2 + a_2 X_2^2 + \cdots + a_m X_m^2$ is anisotropic over K. On the other hand, its degree is at most $2(n-1) = 2n-2$. Consequently it follows from (11) that $h(X)$ has odd degree, at most $n - 2$. Thus $h(X)$ too has an *irreducible* divisor $f_1(X)$ of *odd degree less than* n.

Consider the extension $K(\beta_1)$, where β_1 is a root of f_1. Then $E_1 : K$ is odd and less than n. If we replace X in (11) by β_1, we obtain

$$1 + \sum_{i=2}^m a_i g_i(\beta_1)^2 = 0,$$

contradicting the induction assumption. $\square$

F4. *Let* $(K, \leq)$ *be an ordered field. In an algebraic closure* C *of* K, *let* E *be the subfield of* C *obtained from* K *by adjoining the square roots of all **positive** elements of* K. *Then* $\leq$ *can be extended to an order of the field* E.

Proof. We again resort to Theorem 4. Suppose there is some relation

$$-1 = \sum_{i=1}^{m} a_i \alpha_i^2$$

with $a_i \in K$ positive and α_i in E. The $\alpha_1, \ldots, \alpha_m$ belong to a subfield $K(w_1, \ldots, w_r)$ of E, where $w_i^2 \in K$ and $w_i^2 \geq 0$; let the w_i be chosen so that r is as small as possible. Then

(12) $$w_r \notin K(w_1, \ldots, w_{r-1}).$$

Now $\alpha_i = x_i + y_i w_r$ with $x_i, y_i \in K(w_1, \ldots, w_{r-1})$ for $1 \leq i \leq m$. There follows

$$-1 = \sum_{i=1}^{m} a_i (x_i^2 + y_i^2 w_r^2) + 2 \left(\sum_{i=1}^{m} a_i x_i y_i \right) w_r.$$

From (12) we obtain

$$-1 = \sum_{i=1}^{m} a_i (x_i^2 + y_i^2 w_r^2) = \sum_{i=1}^{m} a_i x_i^2 + \sum_{i=1}^{m} (a_i w_r^2) y_i^2,$$

contradicting the minimality of r. $\qquad\square$

Definition 5. A real field R is called *real-closed* if the only *algebraic* extension E/R with E *real* is the trivial one, $E = R$.

F5. *Let* R *be a real-closed field.*

(a) *Every polynomial of odd degree over* R *has a root in* R.

(b) R *admits exactly one order.*

(c) *The set* R^2 *of all squares in* R *is an order of* R.

Proof. (a) Let $f \in R[X]$ have odd degree, and assume without loss of generality that it is irreducible. Take the extension $E = R(\alpha)$, where α is a root of f. By F2 (and Theorem 2) we know that E is *real*. It follows that $E = R$; that is, $\alpha \in R$.

(b) and (c): Since R is real, it admits an order P, by Theorem 2. Take $\alpha \in P$ and form the extension $E = R(\beta)$, where β is a square root of α. By F4, E is real. Hence E coincides with R and α is a square in R. We conclude that $P \subseteq R^2$, and hence that $P = R^2$, as needed. $\qquad\square$

Remark. Let R be a *real-closed* field and α a positive element of R. By F5, α has a unique positive square root in R. We denote it by $\sqrt{\alpha}$.

F6. *Let K be a real field. There exists an **algebraic** extension R/K of K such that R is a **real-closed** field. Such a field R is called a **real closure** of the real field K.*

Proof. Take an algebraic closure C of K and consider all the *real* subfields of C containing K. By Zorn's Lemma there is a maximal such subfield, and it is necessarily real-closed. □

If R and R' are real algebraic closures of a real field K, it is not at all the case that R/K and R'/K must be isomorphic. Indeed, part (I) of the next theorem says that to get a counterexample we merely need to take a field with at least two distinct orders. The subfield $K = \mathbb{Q}(\sqrt{2})$ of $\mathbb{R}$ is obviously such an example (and indeed it admits exactly two orders; see §20.2).

Theorem 6. *Let $(K, \leq)$ be an ordered field.*

(I) *There exists a **real-closed** extension R of K such that the (unique) order on R extends the given order $\leq$ on K. Such an R is called a **real closure** of the ordered field $(K, \leq)$.*

(II) *If R_1 and R_2 are real closures of $(K, \leq)$, the extensions R_1/K and R_2/K are isomorphic, and indeed there is exactly one K-isomorphism $R_1 \to R_2$; moreover, this isomorphism preserves order.*

Proof. In an algebraic closure C of K, let E be the subfield of C obtained from K by adjoining the square roots of all positive elements of $(K, \leq)$. By F4, E is a *real* field. By F6, then, E has a real closure R (which we can regard as a subfield of C). Since every $a \geq 0$ in K is a square in E and thus also in R, the element a must also be positive in the order determined by R. This proves assertion (I).

The proof of part (II), concerning uniqueness, is not so immediate, and we postpone it until page 12. □

Theorem 7 (Euler–Lagrange). *Let $(R, \leq)$ be an ordered field with the following properties:*

(a) *Every polynomial of odd degree over R has a root in R.*

(b) *Every positive element in R is a square in R.*

*Then the field $C = R(i)$ obtained from R by adjoining a square root i of -1 is **algebraically closed**. Consequently R itself is **real-closed**.*

Remark. Since the field $\mathbb{R}$ of real numbers is ordered and satisfies (a) and (b), the theorem implies the algebraic closedness of $\mathbb{C} = \mathbb{R}(i)$, the field of complex numbers (Fundamental theorem of algebra).

Proof of Theorem 7. Let E/C be a finite field extension. We must show that $E = C$. By passing to a normal closure if needed, we can assume that E/R is Galois. Let H be a Sylow 2-subgroup of the Galois group $G(E/R)$ of E/R and let R' be the corresponding fixed field. By the choice of H the degree $R' : R$ is *odd*; then assumption (a) immediately implies that $R' = R$. Therefore $G(E/R)$ is a 2-group, and so $G(E/C)$ likewise.

Now assume for a contradiction that $E \neq C$. Then $G(E/C) \neq 1$, and $G(E/C)$ has as a 2-group (by F8 in Chapter 10 of Volume I) a subgroup of index 2. Consequently there is an intermediate field F of E/C such that $F : C = 2$. Then $F = C(w)$ with $w^2 \in C$ but $w \notin C$. This is impossible, for the following reason (see also §20.5): all elements of C have a square root in C. Indeed, if $z = a + bi$ with $a, b \in R$ is an arbitrary element of C, the formula

$$(13) \qquad w = \sqrt{\frac{|z|+a}{2}} + i\varepsilon \sqrt{\frac{|z|-a}{2}},$$

with $\varepsilon = 1$ if $b \geq 0$ and $\varepsilon = -1$ if $b < 0$, expresses a square root of z in C. (Here we have used the abbreviation $|z| = \sqrt{a^2 + b^2} \in R$.) Because $|a| \leq |z|$ the square roots that appear in (13) do exist in R, and as claimed we have

$$w^2 = \frac{|z|+a}{2} - \frac{|z|-a}{2} + 2i\varepsilon\sqrt{\frac{|z|^2-a^2}{4}} = a + i\varepsilon|b| = a + ib = z. \qquad \square$$

The theorem just proved has a remarkable complement:

Theorem 8 (Artin). *Let C be an algebraically closed field. If K is a subfield of C such that $C : K < \infty$ and $C \neq K$, then $C = K(i)$ with $i^2 + 1 = 0$, and K is a real-closed field.*

Proof. (1) C is perfect, since it is algebraically closed. Since C/K is a finite extension, this implies that K is also perfect; for if $p := \operatorname{char} K > 0$, we have $C : K^p = C^p : K^p \leq C : K$, hence $K = K^p$ (see F14 and F19 in Chapter 7). Hence C/K is *Galois*.

(2) Let i be an element of C such that $i^2 + 1 = 0$. Set $K' = K(i)$ and assume that $K' \neq C$. Take a prime q dividing the order of the Galois group G of C/K'. By Sylow's First Theorem, G has an element of order q, so there is an intermediate field L of C/K' such that

$$C : L = q.$$

(a) If $q = \operatorname{char} K$, we immediately get, by §14.4(b) in Volume I, a contradiction with the fact that C is algebraically closed. Thus

$$\operatorname{char} K \neq q.$$

Here is another, more direct proof of the same fact. Suppose $\operatorname{char} K = q$. Consider the map $\wp : C \to C$ defined by $\wp(x) = x^q - x$ (Chapter 14, remarks after Theorem 3). If $S = S_{C/L}$ is the trace of the cyclic extension C/L of degree q, we obviously have $S\wp(x) = S(x^q - x) = S(x^q) - Sx = S(x)^q - Sx = \wp(Sx)$, hence

$$(14) \qquad S\wp = \wp S.$$

Since C is algebraically closed, $\wp$ is surjective; therefore S too is surjective, since C/L is separable. From (14) then we get $\wp(L) = L$. But this is impossible, because by Theorem 3 in Chapter 14, C is obtained from L by adjunction of the roots of a polynomial of the form $X^q - X - a$ over L, and there is no such root in L.

(b) Since char $K \neq q$, the field C contains a primitive q-th root of unity ζ. Consider the degree $L(\zeta) : L$; on the one hand it is at most $q - 1$, and on another it divides $C : L = q$. Therefore ζ lies in L. From Theorem 1 in Chapter 14 it follows that C is the splitting field a polynomial of the form $X^q - \gamma$ over L, where $\gamma \in L$ is *not* a q-th power in L. Since $C : L = q$ and C is algebraically closed, the polynomial $X^{q^2} - \gamma$ *cannot be irreducible* over L. Therefore by Theorem 2 of Chapter 14 the only remaining possibility is that $q = 2$ and γ is of the form $-4\lambda^4$ for some $\lambda \in L$. But since $i \in L$, this same γ *is* a square in L — a contradiction!

Thus we have shown that in fact $C = K(i)$, and also that K cannot have characteristic 2.

(3) Since $C \neq K$, we have $i \notin K$. Moreover C/K is Galois. Let a and b be any elements of K. Since $C = K(i)$ is algebraically closed, there is a square root of $a + ib$ in C; that is, there are elements x, y in K such that

$$(x + iy)^2 = a + ib.$$

Apply the norm map $N_{C/K} : C \to K$ to conclude that $(x^2 + y^2)^2 = a^2 + b^2$. Thus $a^2 + b^2$ is a square *in K*. It follows that every sum of squares in K is itself a square in K: in symbols, $\mathrm{SQ}(K) = K^2$. Because -1 does not lie in K^2, it follows that K is a *real* field. The extension $C = K(i)$ of K is algebraically closed and not real; thus K is real-closed, and we have proved Artin's Theorem. □

Remark. To avoid getting a false impression about what Theorem 8 says, note that the field $\mathbb{C}$, for instance, has subfields K satisfying $\mathbb{C} : K = 2$ but $K \neq \mathbb{R}$; see §18.3(iv) in the appendix of Volume I.

2. As we have seen, *quadratic forms* arise naturally in the theory of formally real fields; here we address a further interesting relationship between ordered fields and quadratic forms, which goes back to Sylvester and turns out to be a very useful tool in several respects.

Let (K, P) be an ordered field and R a real-closed extension of K whose order induces the given order P on K. We denote by sgn_P the corresponding sign map; that is, for $a \in K$ we set $\mathrm{sgn}_P(a) = 1, -1$ or 0 according to whether $a > 0$, $a < 0$ or $a = 0$.

Given a quadratic form b with coefficients in K, we now define the *signature* $\mathrm{sgn}_P(b)$ of b with respect to the order P of K. As is well-known, b is equivalent to a diagonal form:

(15) $$b \simeq [b_1, \ldots, b_n], \quad \text{with each } b_i \in K.$$

(see LA II, pp. 37 ff). Denote by $b \otimes R$ the quadratic form over R determined by b. Then set

(16) $$\mathrm{sgn}_P(b) = \sum_{i=1}^{n} \mathrm{sgn}_P(b_i) = \mathrm{sgn}(b \otimes R).$$

In view of *Sylvester's Theorem*, this expression is well defined (see LA II, pp. 63 ff., for example).

Theorem 9 (Sylvester). *Let (K, P) be an ordered field and R a real-closed extension of K whose order induces the given order P on K. Consider for a given polynomial $f \neq 0$ in $K[X]$ the K-algebra $A = K[X]/f$, and the symmetric bilinear form (quadratic form) induced on A by the trace $S_{A/K}$ of A:*

(17) $$s_{f/K} : (x, y) \mapsto S_{A/K}(xy).$$

Let C be an algebraic closure of R. Then:

(a) *The number of distinct roots of f in R equals the signature of $s_{f/K}$:*

(18) $$\left| \{ \alpha \in R \mid f(\alpha) = 0 \} \right| = \mathrm{sgn}_P(s_{f/K}).$$

In particular, $\mathrm{sgn}_P(s_{f/K}) \geq 0$.

(b) *One-half the number of roots of f in C that don't lie in R is equal to the inertia index of $s_{f/K}$ (LA II, p. 66). In particular, all the roots of f lie in R if and only if $s_{f/K}$ is positive semidefinite.*

Proof. For simplicity we give the proof only for the case where f has no multiple roots (but see §20.11).

By Euler–Lagrange (Theorem 7) we have $C = R(i)$ with $i^2 + 1 = 0$. Therefore if

$$f(X) = f_1(X) f_2(X) \ldots f_r(X)$$

is the prime factorization of f in $R[X]$, we have

$$\deg f_i \leq 2 \quad \text{for } 1 \leq i \leq r.$$

In view of the natural isomorphism

$$R[X]/f \simeq K[X]/f \otimes R,$$

we obviously have $s_{f/R} = s_{f/K} \otimes R$. Therefore

$$\mathrm{sgn}_P(s_{f/K}) = \mathrm{sgn}(s_{f/R}).$$

By the Chinese Remainder Theorem there is a canonical isomorphism

$$R[X]/f \simeq R[X]/f_1 \times \cdots \times R[X]/f_r$$

of R-algebras. Thus $s_{f/R}$ is equivalent to the orthogonal sum of the $s_{f_i/R}$. Therefore

$$\mathrm{sgn}(s_{f/R}) = \sum_{j=1}^{r} \mathrm{sgn}(s_{f_i/R}),$$

and a similar relation holds for the inertia index.

Now, we have $R[X]/f_i \simeq R$ or C, according to whether f_i has degree 1 or 2; in the first case $s_{f_i/R}$ is equivalent to the form x^2 and in the second to $x^2 - y^2$. Altogether, then, the signature of $s_{f/K}$ equals the number of f_i's of degree 1, and the inertia index of $s_{f/K}$ is the number of f_i's of degree 2. This is what we needed to prove. $\square$

Theorem 9 in itself already represents a remarkable fact, but it turns out that one can also draw important consequences form it. For example, the uniqueness part of Theorem 6 can be derived from Theorem 9. We first show:

F7. *Let $(E, \leq)$ be an ordered field and K a subfield of E, with E/K algebraic. We regard K as an ordered field with the order P induced on K by $\leq$. If R is a real-closed field and*

$$\sigma : K \to R$$

*is an **order-preserving** homomorphism from K to R, then σ has a **unique** extension*

$$\rho : E \to R$$

*to an **order-preserving** homomorphism from E to R.*

Proof. (1) First assume that $E : K < \infty$. Then $E = K(\alpha)$, for some $\alpha \in E$. Set $f = \mathrm{MiPo}_K(\alpha)$ and let R_1 be a real closure of the ordered field $(E, \leq)$ and thus also of (K, P). Using Theorem 9 we see that, since

$$\mathrm{sgn}_{\sigma P}(s_{\sigma f / \sigma K}) = \mathrm{sgn}_P(s_{f/K}),$$

σf has a root β in R. Hence σ extends to a field homomorphism

$$\rho : E = K(\alpha) \to R.$$

Let $\rho_1, \ldots, \rho_n$ be *all* the extensions of σ into homomorphisms from E to R; we claim that at least one of them preserves order. Otherwise there is for each i some element $\gamma_i > 0$ in E such that $\rho_i(\gamma_i) < 0$ in R. From the earlier part of the proof we know that σ extends to a homomorphism $\rho : E(\sqrt{\gamma_1}, \ldots, \sqrt{\gamma_n}) \to R$, whose restriction to E must be one some ρ_i. But then $\rho_i(\gamma_i) = \rho(\gamma_i) = \rho(\sqrt{\gamma_i})^2 > 0$, contrary to our assumption.

(2) Let E/K be any algebraic extension. By Zorn's Lemma σ has a *maximal* extension

$$\sigma' : K' \to R$$

to an order-preserving homomorphism from an intermediate field K' of E/K with values in R. The extension E/K' is algebraic because E/K is, so by part (1) we immediately get $K' = E$.

For an arbitrary alpha in E such that f = MiPo... -> Given an alpha in K and denoting its

(3) There remains to show the uniqueness of ρ. Given $\alpha \in E$, with minimal polynomial over K denoted by $f = \mathrm{MiPo}_K(\alpha)$, let

$$\alpha_1 < \alpha_2 < \cdots < \alpha_r$$

be all the roots of f in E, in order. Likewise, let

$$\alpha'_1 < \alpha'_2 < \cdots < \alpha'_s$$

be the roots of σf in $E' := \rho E$. Clearly, $s = r$. There is a unique k such that $\alpha = \alpha_k$. Since ρ preserves order, we must have $\rho \alpha = \alpha'_k$. Thus ρ is uniquely determined. $\square$

Proof of the uniqueness part of Theorem 6. Let R_1 and R_2 be real closures of $(K, \leq)$. Applying F7 with $E = R_1$, $R = R_2$ and $\sigma = \mathrm{id}_K$, we obtain a unique order-preserving K-homomorphism $\rho: R_1 \to R_2$. Now, ρR_1 is real-closed because R_1 is; thus $\rho R_1 = R_2$ (since $R_2/\rho R_1$ is algebraic), and we see that ρ is an isomorphism from R_1 to R_2.

Next, let $\tau: R_1 \to R_2$ be any K-homomorphism. For each strictly positive $\alpha \in R_1$, we have

$$\tau(\alpha) = \tau(\sqrt{\alpha})^2 > 0;$$

thus τ must preserve order and hence must equal ρ.   □

Here is a second fundamental fact whose proof can be built on Theorem 9:

Theorem 10. *Let K be a **real-closed** field and let $E = K(x_1, \ldots, x_m)$ be a finitely generated extension of K. If E is **real**, there exists a homomorphism*

$$K[x_1, \ldots, x_m] \to K$$

of K-algebras.

Proof. (1) We first reduce to the case that the extension has transcendence degree 1. Let E' be an intermediate field of E/K with $\mathrm{TrDeg}(E/E') = 1$ (we can assume $\mathrm{TrDeg}(E/K) > 1$, otherwise $E = K$ and there is nothing to prove). Let R be a real closure of the real field E, and R' an algebraic closure of E' in R. Using Theorem 7 one easily sees that R' is *real-closed*. If our theorem's assertion holds for extensions of transcendence degree 1, there is a homomorphism

$$\varphi: R'[x_1, \ldots, x_m] \to R'$$

of R'-algebras. But

$$\mathrm{TrDeg}(K(\varphi x_1, \ldots, \varphi x_m)/K) \leq \mathrm{TrDeg}(R'/K) = \mathrm{TrDeg}(E'/K) = \mathrm{TrDeg}(E/K) - 1.$$

By induction we can thus assume the existence of a homomorphism of K-algebras $K[\varphi x_1, \ldots, \varphi x_m] \to K$. Composing with the restriction of φ to $K[x_1, \ldots, x_m]$, we achieve the required reduction.

(2) Now let $E = K(x, y_1, \ldots, y_r)$, where x is transcendental over K and the y_i are algebraic over $K(x)$. We seek a homomorphism of K-algebras

$$K[x, y_1, \ldots, y_r] \to K.$$

By the primitive element theorem, there exists $y \in E$ such that $E = K(x, y) = K(x)[y]$, and moreover y can be assumed *integral* over $K[x]$. Then the y_i satisfy

$$y_i = \frac{g_i(x, y)}{h(x)} \quad \text{for } 1 \leq i \leq r,$$

where the $g_i(X, Y)$ are polynomials in $K[X, Y]$ and $h \in K[X]$ is nonzero. Any homomorphism of K-algebras

(19) $$\varphi: K[x, y] \to K$$

that satisfies $h(\varphi x) \neq 0$ defines (uniquely) a homomorphism of K-algebras

$$K[x, y_1, \ldots, y_r] \to K.$$

So if we show there are infinitely many K-algebra homomorphisms $\varphi : K[x, y] \to K$, we are done, since only finitely many of them fail the condition $h(\varphi x) \neq 0$ (h has finitely many roots in K, and if φx is such a root, only finitely many values of φy are possible). Let

$$f = f(x, Y) = Y^n + c_1(x)Y^{n-1} + \cdots + c_n(x)$$

be the minimal polynomial of y over $K(x)$. Because y is integral over $K[x]$, all the $c_i(x)$ lie in $K[x]$. Take $a \in K$. We must look for roots of the polynomial

$$f_a(Y) := f(a, Y) \in K[Y]$$

in K. For if b is such a root, there is a unique K-algebra homomorphism (19) such that $\varphi x = a$ and $\varphi y = b$. (To see this, note that the kernel of the substitution homomorphism $K[x, Y] \to K[x, y]$ defined by $x \mapsto x$, $Y \mapsto y$ is the principal ideal of $K[x, Y]$ generated by $f(x, Y)$.) So the matter boils down to proving that there are infinitely many $a \in K$ for which

(20) $\qquad\qquad f_a(b) = f(a, b) = 0 \quad$ for some $b \in K$.

(3) By assumption the field $E = K(x, y)$ is *real*, and so admits an order $\leq$. Let R be a real closure of $(E, \leq)$. We know that the polynomial

$$f = f(x, Y) \in K[x][Y]$$

has a root in R, namely the element y in $E \subseteq R$. Therefore, by Theorem 9,

(21) $\qquad\qquad\qquad \operatorname{sgn}(s_{f/K(x)}) > 0.$

Here of course we have given $K(x)$ the order induced by the order $\leq$ of E. We now wish to show that, for infinitely many $a \in K$ as well,

(22) $\qquad\qquad\qquad \operatorname{sgn}(s_{f_a/K}) > 0.$

For each such a there is then, by Theorem 9, some $b \in K$ satisfying (20), and that will prove the theorem.

The quadratic form $s_{f/K[x]}$ is equivalent to a diagonal form

(23) $\qquad\qquad [h_1(x), \ldots, h_n(x)], \quad$ with $h_i(x) \in K[x].$

Lemma 2 below asserts the existence of infinitely many $a \in K$ such that

(24) $\qquad\qquad \operatorname{sgn} h_i(x) = \operatorname{sgn} h_i(a) \quad$ for all $1 \leq i \leq n.$

For all such values of a, the quadratic form

(25) $\qquad\qquad\qquad [h_1(a), \ldots, h_n(a)]$

has the same signature over K as (23). But it is easy to check that for *almost all* $a \in K$ the quadratic Form $s_{f_a/K}$ is equivalent to the form (25). Consequently, (21) implies that (22) is satisfied for infinitely many $a \in K$. $\qquad\qquad\square$

Lemma 2. *Let K be a **real-closed** field, and let $h_1(x), \ldots, h_n(x)$ be polynomials in one variable x over K. Let $\leq$ be an order of the field $K(x)$ and sgn the sign map it determines. There are infinitely many elements a in K satisfying condition* (24).

Proof. First let $h(x) \in K[x]$ be any polynomial. Since K is real-closed, the prime factorization of h has the form

$$h(x) = u(x - c_1) \ldots (x - c_r)q_1(x) \ldots q_s(x),$$

where the $q_1(x), \ldots, q_s(x)$ are normalized quadratic polynomials. Let $q(x)$ be one of the $q_i(x)$. Then

$$q(x) = x^2 + bx + c = (x + b/2)^2 + (c - b^2/4).$$

Since $q(x)$ is irreducible, $c - b^2/4$ is strictly positive. Therefore we have not only $q(x) > 0$ but also $q(a) > 0$ for any $a \in K$. Consequently,

$$\operatorname{sgn} h(x) = \operatorname{sgn} u \prod_{i=1}^{r} \operatorname{sgn}(x - c_i),$$

$$\operatorname{sgn} h(a) = \operatorname{sgn} u \prod_{i=1}^{r} \operatorname{sgn}(a - c_i) \quad \text{for all } a \in K.$$

Therefore the assertion of the lemma only needs to be shown for a finite number of linear polynomials

$$x - c_1, \, x - c_2, \, \ldots, \, x - c_m,$$

where we may assume in addition, without loss of generality, that the c_i are all distinct. Now rename the elements $c_1, \ldots, c_m, x$ of $K(x)$, in the order determined by $\leq$, as follows:

(26) $$t_1 < t_2 < \cdots < t_{m+1},$$

where x stands in i-th place, say. Now the desired assertion is clear, because we can replace $x = t_i$ in this chain of inequalities by infinitely many different elements $a \in K$, without perturbing the order. $\qquad \square$

21

Hilbert's Seventeenth Problem
and the Real Nullstellensatz

1. The theory of real fields, whose fundamentals we covered in the last chapter, arose from *Hilbert's Seventeenth Problem*, one of the challenges he posed in his celebrated address to the International Congress of Mathematicians in 1900:

   *Let $f \in \mathbb{R}[X_1, \ldots, X_n]$ be a polynomial in n variables over the field $\mathbb{R}$ of real numbers. Suppose that f is **positive definite**; that is, $f(a_1, \ldots, a_n) \geq 0$ for all $a_1, \ldots, a_n \in \mathbb{R}$. Is f a **sum of squares of rational functions** in $\mathbb{R}(X_1, \ldots, X_n)$?*
(Hilbert himself had shown that, already for $n = 2$, we must allow rational functions and not just polynomials.)

   It was in order to approach this problem that Émil Artin and Otto Schreier developed the theory of formally real fields, and indeed by building on this theory Artin was able to solve the problem. He proved, more generally:

Theorem 1 (Artin). *Let K be a real field admitting a unique order, and let R be a real closure of K. If $f \in K[X_1, \ldots, X_n]$ is a polynomial in n variables over K such that*

$$(1) \qquad\qquad f(a_1, \ldots, a_n) \geq 0 \quad \text{for all } a_1, \ldots, a_n \text{ in } R,$$

then f is a sum of squares of rational functions in $K(X_1, \ldots, X_n)$.

Proof. Suppose that f is *not* a sum of squares in $K(X_1, \ldots, X_n)$. By Theorem 3 in Chapter 20, there is an order $\leq$ of the field

$$F := K(X_1, \ldots, X_n)$$

such that

$$(2) \qquad\qquad f = f(X_1, \ldots, X_n) < 0.$$

We need only show that (2) implies that R contains elements $a_1, \ldots, a_n$ such that

$$f(a_1, \ldots, a_n) < 0.$$

Let R_F be a real closure of the ordered field $(F, \leq)$. In R_F we have $f < 0$, so there exists $w \in R_F$ such that

$$w^2 = -f.$$

Consider the algebraic closure R_0 of K in R_F. It is easy to see (cf. Theorem 7 in Chapter 20) that R_0 is real-closed and therefore is a real closure of the *ordered* field K (seeing that K admits only one order). By the uniqueness part of Theorem 6 in the last chapter, therefore, we are entitled to assume that $R = R_0$. Now we apply Theorem 10 of Chapter 20 to the extension $R(X_1, \ldots, X_n, w)$ of R, which is real since it is a subfield of R_F. We obtain a homomorphism of R-algebras

$$\varphi : R[X_1, \ldots, X_n, w, 1/w] \to R.$$

If $a_1, \ldots, a_n$ are the images of $X_1, \ldots, X_n$ under φ, we do in fact obtain

$$f(a_1, \ldots, a_n) = \varphi(f) = -\varphi(w)^2 < 0,$$

because $\varphi(w)$ cannot vanish (consider $\varphi(w)\varphi(1/w)$).   □

Remark 1. For $K = \mathbb{Q}$ one can replace assumption (1) in Theorem 1 by:

(1′) $f(a_1, \ldots, a_n) \geq 0$ for all $a_1, \ldots, a_n$ *in K*.

This is because $K = \mathbb{Q}$ is dense in $\mathbb{R}$ (and *a fortiori* in the field R of real algebraic numbers), so by continuity condition (1) is satisfied if (1′) is.

Remark 2. What happens to Theorem 1 when we relax the condition that the real field K has a single order, and take as our starting point any ordered field (K, P), but making R be its real closure? The answer is simple: Assumption (1) then implies only that

(3) $f \in \text{SQ}_P(K(X_1, \ldots, X_n))$,

where $\text{SQ}_P(K(X_1, \ldots, X_n))$ denotes the set of all finite sums of the form

(4) $$\sum_i p_i f_i^2 \quad \text{with } p_i \in P, \ f_i \in K(X_1, \ldots, X_n).$$

The proof works much as the one above; we just have to invoke Theorem 5 of Chapter 20, instead of Theorem 3, at the beginning. The results below, too, can be modified correspondingly. Moreover if K is a subfield of $\mathbb{R}$ it is clear that assumption (1′) already implies the statement (3).

Now we would like to formulate Theorem 1 somewhat more generally, and clothe it in geometric garb. Let K be a real field and R a fixed real closure of K. We recall the terminology of Chapter 19 in vol. I, the role of the extension C/K (which was arbitrary in that chapter) being played here by the extension R/K. Thus, for an ideal $\mathfrak{a}$ in $K[X_1, \ldots, X_n]$, we denote by

$$\mathcal{N}(\mathfrak{a}) = \mathcal{N}_R(\mathfrak{a}) = \{(a_1, \ldots, a_n) \in R^n \mid f(a_1, \ldots, a_n) = 0 \text{ for all } f \in \mathfrak{a}\}$$

the set of zeros of a in R^n. Such a subset V of R^n is called an *algebraic K-set* of R^n. The *ideal of V* is defined as

$$\mathscr{I}(V) = \mathscr{I}_K(V) = \{g \in K[X_1, \ldots, X_n] \mid g(a) = 0 \quad \text{for all } a \in V\},$$

and the quotient algebra

$$K[V] = K[X_1, \ldots, X_n]/\mathscr{I}(V)$$

is the *affine coordinate ring* of V. The elements of $K[V]$ can be regarded as functions on V with values in R. Being a homomorphic image of $K[X_1, \ldots, X_n]$, the ring $K[V]$ is generated by n elements:

(5) $$K[V] = K[x_1, \ldots, x_n],$$

and so is an *affine K-algebra*, as discussed in Definition 4 of Chapter 19. If the algebraic K-set V is *irreducible*, $K[V]$ is an integral domain (F3 in Chapter 19); in this case we denote by

$$K(V) := \text{Frac } K[V]$$

the fraction field of $K[V]$ and we call $K(V)$ the *field of rational functions on V*. We say that f in $K(V)$ is *defined at* $a \in V$ if there is a representation $f = g/h$, with $g, h \in K[V]$, such that $h(a) \neq 0$. In this case the expression

$$f(a) := g(a)/h(a)$$

yields a well defined value for f in a. We say that $f \in K(V)$ is *positive definite* if, whenever f is defined at $a \in V$, we have $f(a) \geq 0$. If in fact $f(a) > 0$ at all such points a, we call f *strictly positive definite*.

We can now formulate a generalization of Theorem 1:

Theorem 2. *Let K be a real field admitting a unique order, and let R be a real closure of K. Suppose given an affine K-variety V in R^n, that is, an irreducible algebraic K-set V of R^n. Then we have, for all $f \in K(V)$,*

(6) $$f \text{ is positive definite} \Rightarrow f \in \text{SQ}(K(V)),$$

Proof. In essence we can proceed as in the proof of Theorem 1. Let

$$f = g/h \quad \text{with } g, h \in K[V], \ h \neq 0.$$

Assume that $f \notin \text{SQ}(K(V))$. Then there is an order $\leq$ of the field $F := K(V)$ such that $f < 0$. Let R_F be a real closure of the ordered field $(F, \leq)$; we can choose R_F in such a way that it contains the given real closure R of K as a subfield.

Let $K[V] = K[x_1, \ldots, x_n]$ be as in (5). Since $f < 0$ we have $\sqrt{-f} \in R_F$. Now apply Theorem 10 of Chapter 20 to the extension $R(x_1, \ldots, x_n, \sqrt{-f})$ of R, which is *real* because it is a subfield of R_F. We obtain a homomorphism of R-algebras

$$\varphi : R[x_1, \ldots, x_n, \sqrt{-f}, 1/gh] \to R.$$

Now if $a_1, \ldots, a_n$ are the images of $x_1, \ldots, x_n$, we first see that $a := (a_1, \ldots, a_n)$ is a element of V. Next we have $g(a), h(a) \neq 0$, so f is defined at $a \in V$, with $f(a) = g(a)/h(a) \neq 0$. But $-f$ is mapped by φ to a square in R, so

$$f(a) = g(a)/h(a) = \varphi(f) < 0,$$

which proves the theorem.   □

Remarks. (i) For $V = \mathbb{R}^n$ the converse of (6) also holds, by continuity.

(ii) In general, however, the converse of (6) is false. Here is a counterexample: For $K = \mathbb{R}$, let V be the algebraic K-set in K^2 defined by the *prime* polynomial

$$f(X, Y) = Y^2 - (X^3 - X^2)$$

from $K[X, Y]$. We wish to show that V is a K-variety, that is, *irreducible*. (Caution: Because f is irreducible, the algebraic K-set $\mathcal{N}_{\mathbb{C}}(f)$ of $\mathbb{C}^2$ is certainly irreducible, but this does not imply that $V = \mathcal{N}_{\mathbb{R}}(f)$ must also be). Let $g(X, Y) \in K[X, Y]$ be a polynomial that vanishes on V. Division with remainder gives

(7) $$g(X, Y) = h(X, Y) f(X, Y) + u(X) Y + v(X),$$

with $h(X, Y)$ in $K[X, Y]$ and $u(X), v(X)$ in $K[X]$. Now, it is easy to check that all points of V apart from $(0, 0)$ are accounted for by the parametrization

(8) $$x = t^2 + 1, \quad y = t(t^2 + 1),$$

with $t \in K$. If we substitute (8) in (7), we get

$$t(t^2 + 1) u(t^2 + 1) + v(t^2 + 1) = 0 \quad \text{for all } t \in K.$$

By looking at degrees we see that $u(X) = v(X) = 0$. All told, then, we get

$$\mathcal{I}_K(V) = (f),$$

showing that $\mathcal{I}_K(V)$ is a prime ideal, so V is indeed irreducible. We have

$$K[V] = K[x, y], \quad \text{with } y^2 = x^3 - x^2,$$

and x is nonzero in $K[V]$. Therefore

$$x - 1 = y^2/x^2$$

is a square in $K(V)$, but the (polynomial!) function $x - 1$ takes the value $-1 < 0$ at the point $(0, 0)$.

It is worth remarking that this phenomenon, surprising at first sight, is connected with the fact that $(0, 0)$ is a *singular* point of V: we have

$$\frac{\partial f}{\partial X}(0, 0) = 0 = \frac{\partial f}{\partial Y}(0, 0).$$

2. Can the Hilbert Nullstellensatz be modified to work in real fields? As a first counterpart of its "geometric form" (Theorem 2 of Chapter 19), we have the following statement, where we call a prime ideal of a commutative ring with unity *real* if the fraction field of its quotient ring is a real field.

Theorem 3. *Let K be a real field admitting a unique order, and let R be a real closure of K. If a **prime ideal** $\mathfrak{p}$ of a polynomial ring $K[X_1, \ldots, X_n]$ in n variables over K is **real**, then*

$$\mathcal{N}_R(\mathfrak{p}) \neq \varnothing;$$

in other words, there exists $(a_1, \ldots, a_n) \in R^n$ at which all elements of $\mathfrak{p}$ vanish.

Proof. Let $x_1, \ldots, x_n$ be the images of $X_1, \ldots, X_n$ under the quotient map defined by $\mathfrak{p}$. Then

$$(9) \qquad K[X_1, \ldots, X_n]/\mathfrak{p} = K[x_1, \ldots, x_n].$$

Since $\mathfrak{p}$ is assumed prime, $K[x_1, \ldots, x_n]$ is an integral domain. Since $\mathfrak{p}$ is also *real*, the fraction field $F := K(x_1, \ldots, x_n)$ is a real field. Let R_F be a real closure of F, which we can assume to contain the given real closure R of K. Now apply Theorem 10 of Chapter 20 to the extension $R(x_1, \ldots, x_n)$ of R (which is real because it is a subfield of R_F); we obtain a homomorphism of R-algebras

$$R[x_1, \ldots, x_n] \twoheadrightarrow R.$$

If $a_1, \ldots, a_n$ are the images of $x_1, \ldots, x_n$, the element $a = (a_1, \ldots, a_n) \in R^n$ obviously lies in $\mathcal{N}_R(\mathfrak{p})$. $\qquad\square$

The theorem just proved can be sharpened easily:

Theorem 4. *Let the assumptions and the notation be as in Theorem 3, and let $K[x_1, \ldots, x_n]$ denote, as in (9), the quotient algebra arising from $\mathfrak{p}$, with fraction field $F = K(x_1, \ldots, x_n)$. Let $\leq$ be an order of the field F. For each element g of $K[x_1, \ldots, x_n]$ satisfying $g < 0$, there exists some $a \in \mathcal{N}_R(\mathfrak{p})$ such that*

$$(10) \qquad\qquad g(a) < 0.$$

Proof. We merely have to choose R_F as a real closure of the *ordered* field $(F, \leq)$ and apply Theorem 10 of Chapter 20 to the intermediate field

$$R(x_1, \ldots, x_n, \sqrt{-g}\,)$$

of R_F/R. We obtain a homomorphism of R-algebras

$$R[x_1, \ldots, x_n, \sqrt{-g}, 1/g] \twoheadrightarrow R.$$

Again letting $a_1, \ldots, a_n$ be the images of $x_1, \ldots, x_n$, the tuple $a = (a_1, \ldots, a_n)$ is an element of $\mathcal{N}_R(\mathfrak{p})$ satisfying (10). $\qquad\square$

Next, in order to establish a suitable real counterpart for the algebraic Hilbert Nullstellensatz $\mathscr{IN}(\mathfrak{a}) = \sqrt{\mathfrak{a}}$ (Theorem 1 of Chapter 19), we must make some preparations.

Let A be any commutative ring such that $1 \neq 0$.

As in the case of a field, we denote by

$$SQ(A) = \left\{ \sum_i a_i^2 \mid a_i \in A \right\}$$

the set of sums of squares in A, and we define a *quadratic preorder* of A to be any additively and multiplicatively closed subset T of A containing all squares in A:

(11) $$T + T \subseteq T, \quad TT \subseteq T, \quad SQ(A) \subseteq T.$$

If T is a quadratic preorder of A and $a \in A$, then $T + aT$ is a quadratic preorder of A containing a. As a generalization of Lemma 1 of the preceding chapter we have:

Lemma 1. *Let T be a quadratic preorder of A such that $-1 \notin T$. Let a, b be elements of A satisfying $ab \in T$. Then at least one of $T + aT$ and $T - bT$ is a quadratic preorder of A not containing -1.*

Proof. Assume, contrary to the conclusion, that there exist $t_1, t_2, t_3, t_4 \in T$ such that

$$-1 = t_1 + at_2 = t_3 - bt_4.$$

Multiplying $-at_2 = 1 + t_1$ by $bt_4 = 1 + t_3$ we obtain $-abt_2t_4 = 1 + t_5$, with $t_5 \in T$. Therefore $-1 = t_5 + abt_2t_4 \in T$, contradicting the lemma's assumption.   □

Lemma 2. *Among all quadratic preorders of A that do not contain -1, let T be a maximal element. Then*

(a) $T \cup -T = A$ *and*

(b) $T \cap -T$ *is a **prime ideal** of A.*

(In particular, if $A = K$ is a *field*, T is an *order* of K.)

Proof. (a) Take $a \in A$. Apply Lemma 1 with $a = b$; because T is maximal, we get $a \in T$ or $-a \in T$. (b) By part (a), $T \cap -T$ is an ideal of A. Lemma 1 and part (a) imply that this ideal is prime.   □

Lemma 3. *Let T_0 be a quadratic preorder of A such that $-1 \notin T_0$. There exists a quadratic preorder T of A with the following properties:*

$$T_0 \subseteq T; \quad T \cup -T = A; \quad T \cap -T \text{ is a prime ideal of } A.$$

Proof. This follows immediately from Lemma 2 using Zorn's Lemma.   □

Lemma 4 (Prestel). *Let f be an element of A such that*

(12) $$tf \neq 1 + s \quad \text{for all } s, t \in \mathrm{SQ}(A).$$

There exist a prime ideal $\mathfrak{p}$ of A and an order $\leq$ of the fraction field of $A/\mathfrak{p}$ satisfying

(13) $$\bar{f} \leq 0,$$

where $\bar{f}$ denotes the image of f in $A/\mathfrak{p}$.

Proof. Condition (12) means that the quadratic preorder $T_0 = \mathrm{SQ}(A) - f\,\mathrm{SQ}(A)$ does not contain -1. The assertion then follows easily from Lemma 3: If T is as in the conclusion of Lemma 3, consider the image $\bar{T}$ of T in $\bar{A} = A/\mathfrak{p}$, where $\mathfrak{p}$ is the prime ideal $T \cap -T$. Then $\bar{T}$ is a quadratic preorder of $\bar{A}$ such that $\bar{T} \cup -\bar{T} = \bar{A}$ and $\bar{T} \cap -\bar{T} = \{0\}$. Therefore $\bar{T}$ can be extended in an obvious way to an order $\bar{P}$ of $\mathrm{Frac}\,\bar{A}$, and since $-f \in T_0 \subseteq T$ we have $-\bar{f} \in \bar{P}$, which is (13). $\square$

Using Lemma 4 we can prove the following complement to Theorem 2:

Theorem 5. *Let the assumptions and the notation be as in Theorem 2. If a polynomial function $f \in K[V]$ is strictly positive definite, that is, if*

$$f(a) > 0 \quad \text{for all } a \in V,$$

there are sums of squares s and t in the ring $K[V]$ such that f can be represented as

$$f = \frac{1+s}{t}.$$

Proof. Suppose not. Since $K(V)$ is *real*, we can apply Lemma 4 to $A = K[V]$ and $f \in A$. We obtain a prime ideal $\mathfrak{P}$ of $K[X_1, \ldots, X_n]$ such that $\mathfrak{P} \supseteq \mathscr{I}(V)$, and an order $\leq$ of the fraction field F of $K[X_1, \ldots, X_n]/\mathfrak{P}$ such that $g \leq 0$, where g denotes the image of f $K[X_1, \ldots, X_n]/\mathfrak{P}$. But by Theorem 4 there is then an $a \in \mathscr{N}(\mathfrak{P}) \subseteq V$ such that $g(a) = f(a) \leq 0$. Contradiction! $\square$

Theorem 6 (Dubois Nullstellensatz). *Let K be a real field admitting a unique order, and let R be a real closure of K. Suppose given an ideal $\mathfrak{a}$ of the polynomial ring $K[X_1, \ldots, X_n]$ in n variables over K. Define the **real radical** $r(\mathfrak{a})$ of $\mathfrak{a}$ as the set of all $f \in K[X_1, \ldots, X_n]$ for which there exists a natural number m (depending on f) and a sum of squares s in $K[X_1, \ldots, X_n]$ satisfying*

(14) $$f^{2m} + s \in \mathfrak{a}.$$

The ideal of the zero set $\mathscr{N}(\mathfrak{a})$ of $\mathfrak{a}$ in R^n satisfies

(15) $$\mathscr{I}\mathscr{N}(\mathfrak{a}) = r(\mathfrak{a}).$$

Proof. (1) Set $W = \mathscr{N}(\mathfrak{a})$. If (14) is satisfied, the polynomial $f^{2m} + s$ vanishes on W. Since all its values lie in R, f must also vanish on W; that is, $f \in \mathscr{I}(W)$. Thus the inclusion $\supseteq$ in (15) is proved.

(2) Now assume that $W = \mathcal{N}(\mathfrak{a}) = \varnothing$. We then have to prove that there is a sum of squares s of $K[X_1, \ldots, X_n]$ such that

$$(16) \qquad\qquad\qquad\qquad 1 + s \in \mathfrak{a}.$$

Let $\mathfrak{a}$ be generated by $f_1, \ldots, f_r$ (see Theorem 3 in Chapter 19). Consider the K-variety $V = R^n$, and in $K[V] = K[X_1, \ldots, X_n]$ the function $f := f_1^2 + \cdots + f_r^2 \in \mathfrak{a}$. Since $\mathcal{N}(\mathfrak{a}) = \varnothing$, we have $f(a) > 0$ for all $a \in R^n = V$. An application of Theorem 5 then yields sums of squares s, t in $K[V] = K[X_1, \ldots, X_n]$ such that $1 + s = tf$. Now (16) follows, as desired.

(3) Having dealt with the case $\mathcal{N}(\mathfrak{a}) = \varnothing$, we now tackle the general case using the *Rabinovich trick* (see the proof of Theorem 2 in Chapter 19). Let f be a polynomial in $K[X_1, \ldots, X_n]$ that vanishes on the zero set $\mathcal{N}_R(\mathfrak{a})$ of a certain ideal $\mathfrak{a}$ of $K[X_1, \ldots, X_n]$. In the polynomial ring $K[X_1, \ldots, X_n, X_{n+1}]$ in $n+1$ variables over K, consider the ideal

$$\mathfrak{A} = (\mathfrak{a}, 1 - X_{n+1} f).$$

Since the zero set of $\mathfrak{A}$ in R^{n+1} is obviously *empty*, part (2) of the proof shows that $1 \in r(\mathfrak{A})$. Thus there holds a relation of the form

$$1 + s = \sum_i h_i g_i + h(1 - X_{n+1} f),$$

for certain polynomials $g_i \in \mathfrak{a}$ and $s, h_i, h \in K[X_1, \ldots, X_n, X_{n+1}]$, where in addition s is a sum of squares. The substitution $X_{n+1} \mapsto 1/f$ then gives

$$1 + s(X_1, \ldots, X_n, 1/f) = \sum_i h_i(X_1, \ldots, X_n, 1/f) g_i(X_1, \ldots, X_n).$$

Multiplication by an appropriate (even) power f^{2m} yields

$$f^{2m} + \tilde{s}(X_1, \ldots, X_n) = \sum_i \tilde{h}_i(X_1, \ldots, X_n) g_i(X_1, \ldots, X_n),$$

for certain polynomials $\tilde{h}_i, \tilde{s} \in K[X_1, \ldots, X_n]$, where moreover $\tilde{s}$ is a sum of squares in $K[X_1, \ldots, X_n]$. Since $g_i \in \mathfrak{a}$ this implies $f \in r(\mathfrak{a})$, proving the theorem. $\qquad\square$

Remark. From (15) we get, in particular, that the real radical $r(\mathfrak{a})$ of an ideal $\mathfrak{a}$ is also an *ideal* of $K[X_1, \ldots, X_n]$. In fact, Theorem 6 can easily be shown to imply that $r(\mathfrak{a})$ is the intersection of *all real prime ideals* $\mathfrak{p}$ of $K[X_1, \ldots, X_n]$ containing $\mathfrak{a}$. (See §21.4 in the Appendix. Also see §21.5 for another proof of Theorem 6.)

3. We now wish to extend Artin's Theorem to certain *semialgebraic* sets, that is, sets that can be described by finitely many equalities and inequalities. First we introduce one more piece of notation: Let A be a commutative ring with $1 \neq 0$. If

M is any subset of A, let $\mathrm{SQ}_M(A)$ denote the *quadratic preorder on A generated by* M. If $M = \{d_1, \ldots, d_r\}$ is finite, we of course have

$$\mathrm{SQ}_M(A) = \left\{ \sum_\varepsilon q_\varepsilon d_1^{\varepsilon_1} \ldots d_r^{\varepsilon_r} \mid q_\varepsilon \in \mathrm{SQ}(A) \right\},$$

where each sum runs over all tuples $\varepsilon = (\varepsilon_1, \ldots, \varepsilon_r) \in \{0, 1\}^r$. We have the following generalization of Theorem 2:

Theorem 7. *Let K be a real field admitting a unique order, and let R be a real closure of K. Suppose given an affine K-variety W of R^n and finitely many distinct nonzero functions $d_1, \ldots, d_r$ in $K(W)$. If a function $f \in K(W)$ is positive definite on the set*

(17) $$\{b \in W \mid d_i(b) > 0 \text{ for } 1 \leq i \leq r\},$$

then f lies in the set $\mathrm{SQ}_{\{d_1,\ldots,d_r\}}(K(W))$ given by

(18) $$\left\{ \sum_\varepsilon q_\varepsilon d_1^{\varepsilon_1} \ldots d_r^{\varepsilon_r} \mid q_\varepsilon \in \mathrm{SQ}(K(W)) \right\},$$

where each sum runs over all tuples $\varepsilon = (\varepsilon_1, \ldots, \varepsilon_r) \in \{0, 1\}^r$.

Proof. Consider the extension

$$F := K(W)\left(\sqrt{d_1}, \ldots, \sqrt{d_r}\right)$$

arising from $K(W)$ by the adjunction of square roots of the $d_1, \ldots, d_r$. We claim that f is a sum of squares in F:

(19) $$f \in \mathrm{SQ}(F).$$

If this is proved, it is easy to check that f must indeed belong to the subset of $K(W)$ described by (18). So now assume for a contradiction that (19) does not hold. Then the field F has an order $\leq$ such that $f < 0$. Thus the real closure R_F of the ordered field $(F, \leq)$ contains a square root $\sqrt{-f}$ of $-f$, and as before we can assume that $R_F \supseteq R$. Set $K[W] = K[x_1, \ldots, x_n]$ and let h be the product of all numerators and denominators occurring in (fixed) representations of $f, d_1, \ldots, d_r$ as quotients of elements of $K[W]$. Apply Theorem 10 of Chapter 20 to the intermediate field

$$E := R\left(x_1, \ldots, x_n, \sqrt{d_1}, \ldots, \sqrt{d_r}, \sqrt{-f}\right)$$

of R_F/R. This yields a homomorphism of R-algebras

$$R\left[x_1, \ldots, x_n, \sqrt{d_1}, \ldots, \sqrt{d_r}, \sqrt{-f}, 1/h\right] \to R.$$

We thus obtain an element $a = (a_1, \ldots, a_n)$ of W such that

$$f(a) < 0 \text{ and } d_i(a) > 0 \quad \text{for } 1 \leq i \leq r.$$

But this contradicts the assumption that f takes positive values at all points of (17) where it is defined. $\qquad\square$

F1. *With K and R as in Theorem 7, let V and W be affine K-varieties in R^n and R^m, respectively. Assume that $K[W]$ is a subalgebra of $K[V]$. Then there exist $d_1,\ldots,d_r$ in $K[W]$ such that, for every function f in $K(W)$ that is positive definite as a function on V, we have*

(20) $$ f \in SQ_{\{d_1,\ldots,d_r\}}(K(W)). $$

Proof. Set $K[V] = K[x_1,\ldots,x_n]$ and $K[W] = K[y_1,\ldots,y_m]$. Because $K[W]$ is contained in $K[V]$, we have $y_i = s_i(x_1,\ldots,x_n)$, with polynomials $s_1,\ldots,s_m$ in $K[X_1,\ldots,X_n]$. Thus we get a map

$$ s : V \to W, $$

associating to each $a = (a_1,\ldots,a_n) \in V$ the element

$$ s(a) = \big(s_1(a),\ldots,s_m(a)\big), $$

which is plainly seen to lie in W. Now, there are certainly functions $d_1,\ldots,d_r$ in $K[W]$ such that, for every $b \in W$,

(21) $$ d_i(b) > 0 \quad \text{for } 1 \le i \le r \quad \Longrightarrow \quad b \in s(V) $$

— if nothing else, take $d_1 = -1$. But for an r-uple of functions $d_1,\ldots,d_r \in K[W]$ such that property (21) holds, Theorem 7 guarantees that any $f \in K(W)$ that is positive definite as a function on V lies in $SQ_{\{d_1,\ldots,d_r\}}(K(W))$. $\qquad\square$

Remark. Since one can always choose $d_1 = -1$ in F1, the statement of F1 is properly speaking vacuous; but in concrete situations things generally depend on being able to make a nontrivial choice of functions $d_1,\ldots,d_r$ with property (21). Here is an illustration:

Example. With K and R as above (in Theorem 7), take $V = R^n$. Thus

$$ K[V] = K[x_1,\ldots,x_n], $$

where $x_1,\ldots,x_n$ are algebraically independent. Over $K[V]$, consider the polynomial

$$ u(X) = \prod_{i=1}^{n}(X - x_i) = X^n - s_1 X^{n-1} + \cdots \pm s_n, $$

where the coefficients $s_1,\ldots,s_n$ are the *elementary symmetric functions* (page 174 in vol. I). We now set

$$ K[W] = K[s_1,\ldots,s_n]. $$

Let $f \in K(W)$. We assume that f is positive definite as a function on $V = R^n$; that is, assume that

$$ f(b) \ge 0 $$

for all $b = (b_1,\ldots,b_n)$ in $s(R^n) := \{b \in R^n \mid b_i = s_i(a) \text{ for some } a \in R^n\}$ at which f is defined. Clearly the following statements are equivalent:

(i) $b \in s(R^n)$.

(ii) All the roots of $u_b(X) := X^n - b_1 X^{n-1} + \cdots \pm b_n$ lie in R.

By *Sylvester's Theorem* (Theorem 9 in Chapter 20), statement (ii) is equivalent to

(iii) The quadratic form $s_{u_b/R}$ is positive semidefinite.

Let B be the matrix of the quadratic form $s_{u/K(W)}$ in the basis $\overline{1}, \overline{X}, \ldots, \overline{X}^{n-1}$ of $K(W)[X]/u$. Then

(22)
$$B = \left(\mathrm{Tr}(\overline{X}^{i+j})\right)_{0 \le i,j \le n-1} = \left(\sum_{k=1}^{n} x_k^{i+j}\right)_{0 \le i,j \le n-1}.$$

Denote by d_j the principal minor of B involving rows and columns $1, 2, \ldots, j$. Like all the coefficients of B, so are also

$$d_1 = n, \quad d_2 = (n-1)s_1^2 - 2n s_2, \quad \ldots, \quad d_n = \det B$$

polynomials in $s_1, \ldots, s_n$ (as elements of $K(W)$). Now if for some $b \in W$ the condition

(23)
$$d_i(b) > 0 \quad \text{for all } 2 \le i \le n$$

is satisfied, then $s_{u_b/R}$ is positive definite (see LA II, p. 69). Thus condition (23) implies statement (i), and we can apply Theorem 7. The result is synthesized in the next proposition: take into account that the elements f of $K(W) = K(s_1, \ldots, s_n)$ are precisely the *symmetric rational functions* in $K(x_1, \ldots, x_n)$ (Chapter 15, F3 in vol. I).

F2 (Procesi, Lorenz). *Let K and R be as in the statement of Theorem 7. Every positive definite symmetric function f in n variables $x_1, \ldots, x_n$ over K has the form*

$$f = \sum_{\varepsilon} q_\varepsilon d_2^{\varepsilon_2} \ldots d_n^{\varepsilon_n},$$

where the sum runs over all $\varepsilon = (\varepsilon_2, \ldots, \varepsilon_n) \in \{0, 1\}^{n-1}$, the coefficients q_ε are sums of squares of symmetric functions in n variables over K, and $d_1, d_2, \ldots, d_n$ are the principal minors of the matrix B of (22), each d_j involving rows and columns $1, \ldots, j$.

The next result complements Theorem 7 in the case of *strictly positive definite* polynomials.

Theorem 8. *Let K be a real field admitting a unique order, and let R be a real closure of K. Suppose given an affine K-variety W of R^n and finitely many distinct nonzero functions $d_1, \ldots, d_r$ in $K[W]$. If a (polynomial) function $f \in K[W]$ is **strictly positive definite** on the set*

(24)
$$\{b \in W \mid d_i(b) > 0 \text{ for } 1 \le i \le r\},$$

it must have the form

$$(25) \qquad f = \frac{1+s}{t} \quad \text{with } s, t \in SQ_M(K[W][1/d_1, \ldots, 1/d_r]),$$

where we have set $M = \{d_1, \ldots, d_r\}$.

Proof. Introduce the abbreviations

$$A := K[W][1/d_1, \ldots, 1/d_r] \quad \text{and} \quad T := SQ_M(A).$$

If -1 lies in T, so does $f - 1 = (f/2)^2 - (1 - f/2)^2$, and thus (25) can be fulfilled with $t = 1$. Therefore we can assume from now on that $-1 \notin T$. Suppose that (25) does not hold; that is,

$$tf \neq 1 + s \quad \text{for all } s, t \in T.$$

Another way to say this is that

$$-1 \notin T - fT.$$

Now, $T - fT$ is a quadratic preorder of A. We then use Lemma 3, according to which there exists a quadratic preorder P of A containing $T - fT$ and such that

$$P \cup -P = A \quad \text{and} \quad P \cap -P \text{ is a prime ideal of } A.$$

On the fraction field F of the integral domain $\bar{A} = A/P \cap -P$ there is an order $\leq$ induced by P. For this order we have

$$\bar{f} \leq 0 \quad \text{and} \quad \bar{d}_i > 0 \quad \text{for } 1 \leq i \leq r$$

(where the bar denotes passage to the quotient ring), because $-f \in P$, $d_1, \ldots, d_r \in P$ and $d_1, \ldots, d_r \in A^\times$. As before, choose a real closure R_F of F containing the given real closure R of K as a subfield. Write $K[W] = K[x_1, \ldots, x_n]$. An application of Theorem 10 of Chapter 20 to the intermediate field

$$R(\bar{x}_1, \ldots, \bar{x}_n, \sqrt{\bar{d}_1}, \ldots, \sqrt{\bar{d}_r}, \sqrt{-\bar{f}})$$

of R_F/R yields a homomorphism of R-algebras

$$R[\bar{x}_1, \ldots, \bar{x}_n, 1/\sqrt{\bar{d}_1}, \ldots, 1/\sqrt{\bar{d}_r}, \sqrt{-\bar{f}}] \to R.$$

Thus we obtain an element $a = (a_1, \ldots, a_n) \in R^n$ lying in W and satisfying

$$f(a) \leq 0 \quad \text{and} \quad d_i(a) > 0 \quad \text{for } 1 \leq i \leq r.$$

But this contradicts our assumption that f takes only strictly positive values at points of the set (24). $\square$

In view of the theorem just proved, we can state the following complement to the result about symmetric functions in F2:

F3. *Let K and R be as in the statement of Theorem 7, and let $f \in K[x_1, \ldots, x_n]$ be a **symmetric polynomial** in n variables over K. If f is **strictly positive definite**, it can be written as*

$$f = \frac{1 + s}{t},$$

where s and t have the form

$$\sum_{\mu} q_{\mu} d_2^{-\mu_2} \ldots d_n^{-\mu_n},$$

where the sum runs over all $(n-1)$-tuples $\mu = (\mu_2, \ldots, \mu_n)$ of integers such that

$$\mu_i \geq -1 \quad for \ 2 \leq i \leq n,$$

the q_{μ} denote sums of squares of symmetric polynomials in n variables over K, and $d_2, \ldots, d_n$ are defined as in F2.

Orders and Quadratic Forms

1. As we saw in Chapter 20, quadratic forms come very naturally into the theory of formally real fields. Clearly there is a close internal connection, and we now investigate it more systematically. In this chapter K will always denote a field with

$$(1) \qquad\qquad \mathrm{char}\, K \neq 2.$$

We will consider vector spaces over K, always finite-dimensional, and quadratic forms on such spaces, and recall some fundamental facts about them (for details see, for example, LA II, p. 31 ff). If q is a quadratic form on a vector space V over K, we sometimes call (V, q) a *quadratic space over* K; but when no misunderstanding can arise, we will usually leave V out of the notation and say that q is a quadratic space. We will also denote by q the unique *symmetric bilinear form* $V \times V \to K$ for which

$$(2) \qquad\qquad q(x, x) = q(x).$$

Every quadratic form q is equivalent (under a linear transformation $V \to K^{\dim V}$) to a diagonal form:

$$(3) \qquad\qquad q \simeq [a_1, \ldots, a_m].$$

In the sequel we will tacitly assume all quadratic forms to be *nondegenerate*. This means that the entries a_i in (3) all lie in $K^\times$. The *orthogonal sum* of two quadratic spaces (V, q) and (V', q') over K, denoted by $(V, q) \perp (V', q')$, is defined as $(V \oplus V', q \perp q')$ with

$$(4) \qquad\qquad (q \perp q')(x \oplus x') = q(x) + q'(x').$$

Likewise their *tensor product* $(V, q) \otimes (V', q')$ is defined as $(V \otimes V', q \otimes q')$ with $q \otimes q'$ the (unique) bilinear form satisfying

$$(5) \qquad\qquad (q \otimes q')(x \otimes x', \, y \otimes y') = q(x, y)\, q'(x', y')$$

for $x, y \in V$ and $x', y' \in V'$. With the standard identifications $K^m \oplus K^n = K^{m+n}$ and $K^m \otimes K^n = K^{mn}$, we therefore have

$$(6) \qquad [a_1, \ldots, a_m] \perp [b_1, \ldots, b_n] = [a_1, \ldots, a_m, b_1, \ldots, b_n],$$

$$(7) \qquad [a_1, \ldots, a_m] \otimes [b_1, \ldots, b_n] = [a_1 b_1, a_1 b_2, \ldots, a_m b_n].$$

The *k-fold orthogonal sum* of a quadratic space $q = (V, q)$ is denoted by $k \times q$.

Every q possesses an orthogonal decomposition

$$(8) \qquad q = q_0 \perp q_1$$

with an *anisotropic* component q_0 — meaning that no $v \neq 0$ in the domain of q_0 satisfies $q_0(v) = 0$ — and a *hyperbolic* component q_1, which by definition means $q_1 \simeq k \times H = k \times [1, -1]$. If some q' equivalent to q has a similar decomposition $q' = q_0' \perp q_1'$, *Witt's theorem* implies that $k = k'$ and $q_0 \simeq q_0'$. We call k the *Witt index* and q_0 a *core form* of q. Thus all the core forms of q are equivalent.

Definition 1. We call two quadratic spaces q and q' over K *similar*, or *Witt-equivalent*, and we write

$$(9) \qquad q \sim q',$$

if they have equivalent core forms. We denote by $\langle q \rangle$ the class of forms Witt-equivalent to q, and call it the *Witt class* of q. Then it makes sense to talk about the *set* of all Witt classes of quadratic forms over a given field K. This set is denoted by $W(K)$ and called the *Witt ring of K*, for reasons about to become clear. The Witt class of the diagonal form $[a_1, \ldots, a_n]$ is written

$$(10) \qquad \langle a_1, \ldots, a_n \rangle .$$

Now, $q \simeq q'$ implies $q \sim q'$; thus every element of $W(K)$ has the form (10). Since core forms are unique up to equivalence, q and q' are Witt-equivalent if and only if there exist integers $m, n \geq 0$ such that

$$(11) \qquad q \perp (m \times H) \simeq q' \perp (n \times H).$$

For every q we have

$$(12) \qquad q \otimes H \simeq q \perp (-q) \sim 0,$$

because for $q \simeq [a_1, \ldots, a_m]$, equations (7) and (6) imply that

$$q \otimes H \simeq [a_1, \ldots, a_m] \otimes [1, -1] \simeq [a_1, -a_1, \ldots, a_m, -a_m]$$
$$\simeq [a_1, -a_1] \perp \ldots \perp [a_m, -a_m] \simeq H \perp \ldots \perp H \sim 0.$$

From this we easily derive:

F1. *The rules* $\langle q \rangle + \langle q' \rangle = \langle q \perp q' \rangle$ *and* $\langle q \rangle \cdot \langle q' \rangle = \langle q \otimes q' \rangle$ *give rise to well defined operations* $+$ *and* $\cdot$ *on* $W(K)$, *which make* $W(K)$ *into a commutative ring with unity. The zero element of* $W(K)$ *is the Witt class of the hyperbolic plane* $H = [1, -1]$ *(and of the zero-dimensional space* 0*); the additive inverse of an element* $\langle q \rangle = \langle a_1, \ldots, a_n \rangle$ *is* $\langle -q \rangle = \langle -a_1, \ldots, -a_n \rangle$; *the multiplicative unity of* $W(K)$ *is the Witt class* $\langle 1 \rangle$.

Next, if P is an *order* on the field K, the *signature* $\mathrm{sgn}_P(q)$ *of a quadratic space* q *with respect to* P is defined as follows: Write $q \simeq [a_1, \ldots, a_m]$ and set

(13) $$\mathrm{sgn}_P(q) = \sum_{i=1}^{m} \mathrm{sgn}_P(a_i),$$

where sgn_P on the right denotes the sign function on K under P. The expression $\mathrm{sgn}_P(q)$ is well defined; see (16) in Chapter 20. Clearly

$$\mathrm{sgn}_P(q \perp q') = \mathrm{sgn}_P(q) + \mathrm{sgn}_P(q'),$$
$$\mathrm{sgn}_P(q \otimes q') = \mathrm{sgn}_P(q)\,\mathrm{sgn}_P(q'),$$
$$\mathrm{sgn}_P(H) = 0.$$

In this way sgn_P gives rise to a surjective homomorphism from the Witt ring $W(K)$ to the integers, for which we use the same notation:

(14) $$\mathrm{sgn}_P : W(K) \to \mathbb{Z}.$$

Theorem 1. *The correspondence* $P \mapsto \mathrm{sgn}_P$ *is a bijection between the set of all orders* P *of* K *and the set of all ring homomorphisms* $s : W(K) \to \mathbb{Z}$.

The content of Theorem 1 is part of a result that, in spite of its simplicity, was only formulated in 1970, in a paper of *J. Leicht and F. Lorenz,* and independently by *D. Harrison.* The result describes the *prime ideals* of the ring $W(K)$. First we establish that the correspondence $q \mapsto \dim q$ induces a well defined ring homomorphism $W(K) \to \mathbb{Z}/2$; its kernel

(15) $$I(K) = \{\langle q \rangle \mid \dim\, q \equiv 0 \bmod 2\}$$

is called the *fundamental ideal* of $W(K)$. By definition, $I(K)$ consists of the Witt classes of all *even-dimensional* (nondegenerate) quadratic spaces. The adjective "fundamental" is justified because $I(K)$ plays a very important role in the theory of quadratic forms, although we will say no more about it here.

Because $W(K)/I(K) \simeq \mathbb{Z}/2$, the fundamental ideal is *maximal.*

Now let $\mathfrak{p}$ be any *prime ideal* of $W(K)$. For every $a \in K^{\times}$ we have

(16) $$(1 + \langle a \rangle)(1 - \langle a \rangle) = 1 - \langle a^2 \rangle = 0 \quad \text{in } W(K).$$

One of the two factors on the left lies in $\mathfrak{p}$, so

(17) $$\langle a \rangle \equiv \pm 1 \bmod \mathfrak{p}.$$

The canonical homomorphism

(18) $\mathbb{Z} \to W(K)/\mathfrak{p}$

is therefore *surjective*, because $W(K)$ is generated by the classes $\langle a \rangle$. But $W(K)/\mathfrak{p}$ is an integral domain, so either

(19) $W(K)/\mathfrak{p} \simeq \mathbb{Z}$ or $W(K)/\mathfrak{p} \simeq \mathbb{Z}/p$

for some (well defined) prime p. In the first case $\mathfrak{p}$ is a *minimal prime ideal* of $W = W(K)$. Proof: for any prime ideal $\mathfrak{q} \subseteq \mathfrak{p}$ the natural epimorphism $W/\mathfrak{q} \to W/\mathfrak{p}$ has trivial kernel, because $W/\mathfrak{q}$ is also a homomorphic image of $\mathbb{Z}$.

Now let $\mathfrak{p}$ be a prime ideal such that $W/\mathfrak{p} \simeq \mathbb{Z}$, and let p be a prime number. There is a unique prime ideal $\mathfrak{p}_p$ above $\mathfrak{p}$ such that

(20) $W/\mathfrak{p}_p \simeq \mathbb{Z}/p;$

namely, $\mathfrak{p}_p = \mathfrak{p} + (p \times W)$.

If $\mathfrak{p}$ is a prime ideal such that $W(K)/\mathfrak{p} \simeq \mathbb{Z}/2$, we necessarily have $\mathfrak{p} = I(K)$, because the generators $\langle a \rangle$ of $W(K)$ are all units in $W(K)$, and so there can be only one ring homomorphism $W(K) \to \mathbb{Z}/2$.

Given these preliminaries, we now consider any prime ideal $\mathfrak{p}$ of $W(K)$ distinct from $I(K)$. We will show that the set

(21) $P := \{a \in K^{\times} \mid \langle a \rangle \equiv 1 \bmod \mathfrak{p}\} \cup \{0\}$

is an *order* on K. First note that $P \cup -P = K$, because any $a \in K^{\times}$ satisfies (17). Next, it is clear that $PP \subseteq P$. We now assert that $a, b \in P$ implies $a + b \in P$. Obviously we can assume that all three of these elements are nonzero. Now

(22) $\langle a \rangle + \langle b \rangle = \langle a, b \rangle = \langle a+b, ab(a+b) \rangle = \langle a + b \rangle (1 + \langle ab \rangle),$

where the second equality is checked by evaluating the quadratic form $[a, b]$ at the elements of the $[a, b]$-orthogonal basis $\{(1, 1), (-b, a)\}$. Therefore the assumption $\langle a \rangle \equiv \langle b \rangle \equiv 1 \bmod \mathfrak{p}$ implies

(23) $1 + 1 \equiv \langle a + b \rangle (1 + 1) \bmod \mathfrak{p}.$

But since $\mathfrak{p} \neq I(K)$ we have $1 + 1 \not\equiv 0 \bmod \mathfrak{p}$, so (23) says that $\langle a+b \rangle \equiv 1 \bmod \mathfrak{p}$, showing that $a + b \in P$.

Finally, it is clear that $-1 \notin P$, because our assumption that $\mathfrak{p} \neq I(K)$ implies that $1 \not\equiv -1 \bmod \mathfrak{p}$.

We have proved that for any prime ideal $\mathfrak{p}$ of $W(K)$ apart from $I(K)$, the set P defined by (21) is an order of K. Now let $\mathrm{sgn}_P : W(K) \to \mathbb{Z}$ be the corresponding signature map. For every $f \in W(K)$ we have

(24) $f \equiv \mathrm{sgn}_P(f) \times 1 \bmod \mathfrak{p},$

because, by definition, each generator $f = \langle a \rangle$ of $W(K)$ satisfies (24). Let q be the *kernel* of sgn_P. By (24), q is contained in p. If p is a *minimal* prime ideal, it must coincide with q. If instead p is not minimal, it follows from the equivalence $W(K)/\mathfrak{q} \simeq \mathbb{Z}$ and the preceding considerations that $\mathfrak{p} = \mathfrak{q}_p$, for some prime number $p \neq 2$. Moreover q is the only minimal prime ideal contained in p, because a minimal prime ideal $\mathfrak{q}' \subseteq \mathfrak{p}$ corresponds to an order of K, as seen above, and in view of (21) this order coincides with the one defined by p.

Putting it all together we obtain a characterization of the spectrum (set of prime ideals) of $W(K)$:

Theorem 2. (A) *If K is not real, the ideal $I(K)$ in (15) is the only prime ideal of $W(K)$.*

(B) *For the rest of the theorem's statement, assume K real. There are two types of prime ideals in $W(K)$:*

 (a) *Minimal prime ideals* p, *for which* $W(K)/\mathfrak{p} \simeq \mathbb{Z}$.

 (b) *Maximal prime ideals* p, *for which* $W(K)/\mathfrak{p} \simeq \mathbb{Z}/p$ *with p prime.*

Concerning (a), the minimal prime ideals are in one-to-one correspondence with the orders of K, as follows: If P is an order of K, the kernel of the corresponding signature map $\text{sgn}_P : W(K) \rightarrow \mathbb{Z}$ is a minimal prime ideal p of $W(K)$, and

$$(25) \qquad\qquad f \equiv \text{sgn}_P(f) \times 1 \bmod \mathfrak{p} \quad \text{for all } f \in W(K);$$

conversely, if p is a minimal prime ideal of $W(K)$, the subset P of W consisting of 0 and all elements $a \in K^\times$ such that $\langle a \rangle \equiv 1 \bmod \mathfrak{p}$ is an order of K satisfying (25).

Concerning (b), for every minimal prime ideal p of $W(K)$ and every prime p there is a unique prime ideal $\mathfrak{p}_p$ above p that satisfies $W(K)/\mathfrak{p}_p \simeq \mathbb{Z}/p$, namely, the maximal ideal

$$\mathfrak{p}_p = \mathfrak{p} + p \times W(K) = \{ f \mid \text{sgn}_P(f) \equiv 0 \bmod p \},$$

where P denotes the order corresponding to p. All the maximal ideals of $W(K)$ can be so expressed. If $p \neq 2$, each $\mathfrak{p}_p$ contains exactly one minimal prime ideal, namely p; whereas if $p = 2$, we have $\mathfrak{p}_2 = I(K)$ for all p.

Obviously, Theorem 1 is contained in Theorem 2. For if $s : W(K) \rightarrow \mathbb{Z}$ is any homomorphism of rings with unity, the ideal $\mathfrak{p} := \ker s$ is prime in $W(K)$ and satisfies $W(K)/\mathfrak{p} \simeq \mathbb{Z}$. By Theorem 2 there is an order P of K associated to p, and it satisfies (25). An application of s to (25) then yields $s(f) = \text{sgn}_P(f)$ for all $f \in W(K)$.

We take Theorem 1 as our cue for our next bit of terminology:

Definition 2. Any ring homomorphism $s : W(K) \rightarrow \mathbb{Z}$ is called a *signature* of the field K. We denote by $\text{Sign}(K)$ the set of all signatures of K.

For the sake of a precise description of the spectrum of $W(K)$, we formulated Theorem 2 in detail. As a digest of the theorem we explicitly state again:

Theorem 3. *A field K is real if and only if $\mathrm{Sign}(K) \neq \varnothing$. For a real field K the correspondences $P \mapsto \mathrm{sgn}_P$ and $s \mapsto \ker(s)$ are natural bijections between the set of orders of K, the set $\mathrm{Sign}(K)$ of signatures of K and the set of minimal prime ideals of $W(K)$.*

Theorem 3 allows us to deal with signatures of a field K instead of orders on K. We now demonstrate the possibilities of this reformulation with some examples. In place of Theorem 2 in Chapter 20, we get this counterpart:

$$K \text{ is real} \iff \mathrm{Sign}(K) \neq \varnothing.$$

Next, let L/K be a field extension. The correspondence casts a new light on the question of when does an order on K extend to an order on L. Let's call a signature t of L an *extension* of a signature s of K if the diagram

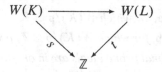

commutes, where the horizontal arrow represents the natural homomorphism $r_{L/K}$: $W(K) \to W(L)$ defined by the extension of the field of constants. One can then check easily that, if P and Q are orders on K and L, respectively, Q is an extension of P if and only if sgn_Q is an extension of sgn_P. Therefore we get another criterion for whether an order can be extended:

F2. *Let L/K be a field extension. An element $s \in \mathrm{Sign}(K)$ can be extended to an element $t \in \mathrm{Sign}(L)$ if and only if s vanishes on the kernel of $r_{L/K} : W(K) \to W(L)$.*

Proof. Necessity is clear; we need to prove sufficiency. Let B be the image of $r_{L/K}$ and let $\bar{s} : B \to \mathbb{Z}$ be the ring homomorphism that makes this diagram commute (it exists by assumption):

(26)
$$W(K) \xrightarrow{\ r\ } B \hookrightarrow W(L)$$
$$s \searrow \quad \swarrow \bar{s}$$
$$\mathbb{Z}$$

The kernel of $\bar{s}$ is a prime ideal $\mathfrak{q}$ of B such that $B/\mathfrak{q} \simeq \mathbb{Z}$. In view of (26) it follows from Theorem 2 that $\mathfrak{q}$ must be a *minimal* prime ideal of B. We now show that there exists a prime ideal $\mathfrak{p}$ of $W(L)$ such that $\mathfrak{p} \cap B = \mathfrak{q}$. For the existence of $\mathfrak{p}$ we apply exercise §4.12 from vol. I to the multiplicative subset $S = B \smallsetminus \mathfrak{q}$ of the ring $R = W(L)$; a maximal ideal of $S^{-1}R$ then yields a prime ideal $\mathfrak{p}$ of R satisfying $\mathfrak{p} \cap B \subseteq \mathfrak{q}$. It follows that $\mathfrak{p} \cap B = \mathfrak{q}$, because $\mathfrak{q}$ is minimal. We have $\mathbb{Z} \simeq B/\mathfrak{q} \to W(L)/\mathfrak{p}$. Therefore, by Theorem 2, $B/\mathfrak{q} \to W(L)/\mathfrak{p}$ must be an isomorphism. But then it is clear that there is a map $t : W(L) \to \mathbb{Z}$ that maintains the commutativity of (26). $\qquad\square$

We now wish to draw two more consequences from Theorem 2:

F3. *The Witt ring $W(K)$ is a Jacobson ring; that is, every prime ideal $\mathfrak{p}$ of $W(K)$ is the intersection of the maximal ideals containing $\mathfrak{p}$. Therefore the Jacobson radical and the nilradical of $W(K)$ coincide.*

Proof. Let R be a commutative ring with unity. The *nilradical* of R is by definition the set of all nilpotent elements of R. It is also the intersection of all prime ideal of R (see exercise §4.14 in vol. I). The *Jacobson radical* of R is the intersection of all maximal ideals of R. Thus the second statement in F3 follows from the first. To prove the first statement, we just have to show that $W(K)/\mathfrak{p}$ has Jacobson radical 0. But this is clear from (19). $\qquad\square$

The following remarkable facts about quadratic forms were discovered by A. Pfister in 1965. We expound them here as examples of the applicability of our Theorem 2.

Theorem 4. *If K is not real, $I(K)$ is the set of all nilpotent elements in $W(K)$ and there exists $n \in \mathbb{N}$ such that*

$$2^n \times f = 0 \quad \text{for all } f \in W(K).$$

Thus for K not real the additive group of $W(K)$ is a 2-torsion group of finite exponent.

Proof. Suppose K is not real. By Theorem 2, $I(K)$ is the only prime ideal of $W(K)$, and so it coincides with the nilradical of $W(K)$. Since $1 + 1 = \langle 1, 1 \rangle \in I(K)$, there is some $n \in \mathbb{N}$ such that $(1+1)^n = 0$ in $W(K)$. It follows that $2^n \times f = (1+1)^n f = 0$ for all $f \in W(K)$. $\qquad\square$

Lemma 1. *Let L/K be a quadratic field extension, so $L = K(\sqrt{d})$, $d \in K^\times \smallsetminus K^{\times 2}$. Then the kernel of $r_{L/K} : W(K) \to W(L)$ is the principal ideal of $W(K)$ generated by $\langle 1, -d \rangle$.*

Proof. Since $r_{L/K}\langle 1, -d \rangle = \langle 1, -d \rangle_L = \langle 1, -1 \rangle_L = 0$, the ideal generated by $\langle 1, -d \rangle$ is contained in the kernel of $r_{L/K}$. Now let q be an *anisotropic* quadratic space over K. We assume that q_L is *isotropic*. Thus there exist vectors x, y over K satisfying $q_L(x + y\sqrt{d}) = 0$, where x, y do not both vanish. It follows that

$$(27) \qquad\qquad q(x) + dq(y) = 0 \quad \text{and} \quad q(x, y) = 0.$$

Thus x and y are orthogonal with respect to q. We set $a = q(x)$ and $b = q(y)$. Since $a = -db$ and at least one side is nonzero because q is anisotropic, both are nonzero. Therefore q contains the subspace

$$[a, b] \simeq [a, -ad] = [a] \otimes [1, -d].$$

The assertion now follows easily by induction on $\dim q$. $\qquad\square$

Theorem 5. *If K is real, the following statements are equivalent for $f \in W(K)$:*

(i) $r_{L/K}(f) = 0$ *for every real closure L of K.*

(ii) $\mathrm{sgn}_P(f) = 0$ *for every order P of K.*

(iii) f *is nilpotent.*

(iv) f *is a torsion element of (the additive group of) $W(K)$.*

(v) *There exists $n \in \mathbb{N}$ such that $2^n \times f = 0$.*

Proof. If L is a real closed field, $\mathrm{Sign}(L)$ consists of a single element, which we denote by sgn^L; obviously $\mathrm{sgn}^L : W(L) \to \mathbb{Z}$ is an isomorphism.

Now, the orders P of K are in correspondence with the real closures L of K, where

(28)
$$
\begin{array}{ccc}
W(K) & \xrightarrow{\;r_{L/K}\;} & W(L) \\
 & \underset{\mathrm{sgn}_P}{\searrow}\quad\underset{\mathrm{sgn}^L}{\swarrow} & \\
 & \mathbb{Z} &
\end{array}
$$

is commutative; hence the equivalence of (i) and (ii). An element of a commutative ring with unity is nilpotent if and only if it belongs to every prime ideal; together with Theorem 2, this implies the equivalence of (ii) and (iii). Implications (v) $\Rightarrow$ (iv) and (iv) $\Rightarrow$ (ii) are trivial. Thus what is left is to show that (iii) implies (v).

Suppose to the contrary that there is a nilpotent element $f \in W(K)$ such that

(29)
$$2^i \times f \neq 0 \quad \text{for } i = 0, 1, 2, 3, \ldots$$

In an algebraic closure of K there is, by Zorn's Lemma, a *maximal* extension E over which (29) still holds (with $r_{E/K} f$ instead of f). By Theorem 4, E is real. Thus, after renaming, we can assume that K is *maximal*, in the sense that (29) no longer holds in any finite extension of K of degree at least 2.

Now, K must contain more than two square classes, because if a real field K only has the square classes $[1]$ and $[-1]$, it satisfies $W(K) \simeq \mathbb{Z}$, so $W(K)$ has no (nonzero) nilpotent. Therefore there exist $a, b \in K^\times$ such that $1, a, b, ab$ represent four distinct square classes. In general, if $d \in K^\times$ is not a square in K, there is by the maximality of K some integer i such that

(30)
$$2^i \times f = 0 \quad \text{over } K(\sqrt{d}).$$

We can choose i in such a way that (30) is simultaneously satisfied for $d = a, b, ab$. Now take the element $g = 2^i \times f$ of $W(K)$. If (30) holds, Lemma 1 implies that $g = (1 - \langle d \rangle)h$ for some $h \in W(K)$. Although h depends on d, it follows in any case that $\langle d \rangle g = -g$. Thus

$$-\langle a \rangle g = g, \quad -\langle b \rangle g = g, \quad -\langle ab \rangle g = g.$$

Multiplying the first of these equations by $\langle b \rangle$ and applying the others, we get $g = -g$. Therefore $2 \times g = 0$ and hence $2^{i+1} \times f = 0$ in $W(K)$. Contradiction! $\square$

Theorem 6. *If $f \in W(K)$ is a torsion element, its order is a power of 2.*

Proof. This follows from Theorem 5 when K is real and from Theorem 4 when K is not.   □

We now describe the set of *zero divisors* in $W(K)$—a less important question perhaps, but one whose investigation led us originally to our Theorem 2. First:

Lemma 2. *Let R be a nonzero commutative ring with unity, and let N be the set of zero divisors of R, including 0. Then N is a union of prime ideals in R:*

$$(31) \qquad N = \bigcup \mathfrak{p}_i.$$

Proof. Let $S = R \smallsetminus N$ be the multiplicative subset of elements of R that are not zero divisors. Each $x \in N$ lies in a maximal ideal $\mathfrak{P}$ of $S^{-1}R$. Then $\mathfrak{p} = \mathfrak{P} \cap R$ is a prime ideal of R, with $x \in \mathfrak{p}$ and $\mathfrak{p} \cap S = \varnothing$ (see again Exercise §4.12 in vol. I). Since $\mathfrak{p}$ and S are disjoint, $\mathfrak{p}$ consists only of zero divisors of R.   □

We apply this result to $R = W(K)$. First we show that every *minimal* prime ideal $\mathfrak{p}$ contains only zero divisors. If $f \in \mathfrak{p}$, we have $\mathrm{sgn}_P(f) = 0$ for the corresponding order P, and hence $f = \sum_{i=1}^n (\langle a_i \rangle + \langle -b_i \rangle)$ with $a_i, b_i \in P$. If we now set $g = \prod_{i=1}^n (\langle a_i \rangle + \langle b_i \rangle)$, we have $fg = 0$, but $g \neq 0$ because $\mathrm{sgn}_P(g) > 0$.

Thus we can assume that all minimal prime ideals occur in the union (31). On the other hand, any maximal ideal $\mathfrak{p}_p$ with $p \neq 2$ cannot occur in (31): indeed, the form $p \times 1 \in \mathfrak{p}_p$ is not a zero divisor, because by Theorem 6 there are no elements of odd order p in $W(K)$. Finally, the ideal $I(K)$ occurs in (31) if and only if the element $2 \times 1 = 1 + 1$ of $W(K)$ is a zero divisor.

We claim that a necessary and sufficient condition for 2×1 *not* to be a zero divisor of $W(K)$ is that K be *real* and *pythagorean*. (A field is called *pythagorean* if every sum of squares is already a square.) Indeed, if K is real and pythagorean and q is an anisotropic form over K, clearly $[1, 1] \otimes q \simeq q \perp q$ is anisotropic. Conversely, if 2×1 is not a zero divisor in $W(K)$, then K is *real* by Theorem 4; to show that it is pythagorean, suppose $d = a^2 + b^2$ with $a, b \in K^\times$. Using *Witt's relation* (22), we obtain

$$[1, 1] \simeq [a^2, b^2] \simeq [a^2 + b^2, (a^2 + b^2)a^2b^2] \simeq [a^2 + b^2, a^2 + b^2] \simeq [d, d] = [d] \otimes [1, 1],$$

so

$$\langle d \rangle (2 \times 1) = 2 \times 1 \quad \text{in } W(K).$$

Because 2×1 is not a zero divisor, d is a square in K, as claimed.

In sum:

Theorem 7. *If K is real and pythagorean, the set N of zero divisors in $W(K)$ is the union of all minimal prime ideals of $W(K)$, and hence an element $f \in W(K)$ is a zero divisor if and only if there is an order P such that $\mathrm{sgn}_P(f) = 0$.*

For other fields K we have $N = I(K)$; that is, the zero divisors of $W(K)$ are precisely the Witt classes of even-dimensional forms.

To conclude this chapter we will broach one more question that has a certain bearing on Theorem 5. It is suggested by Theorem 9 in Chapter 20 (see also Exercise §20.10); according to that theorem, the trace form $q = s_{L/K}$ of any finite field extension L/K satisfies

(32) $\text{sgn}_P(q) \geq 0$ for any order P of K.

Now consider, in the opposite direction, an arbitrary (nondegenerate) quadratic form q over K that satisfies (32). Let $f \in W(K)$ be the Witt class of q. It is natural to ask whether f is represented by a *trace form*; that is, whether there is a finite extension L/K such that q is Witt-equivalent to $s_{L/K}$. This question was first answered in the affirmative by P. E. Conner and R. Perlis for $K = \mathbb{Q}$, and then by W. Scharlau and W. Krüskemper for arbitrary *algebraic number fields* K. But later a beautiful and much sharper result was proved by M. Epkenhans (*Arch. Math.* **60**, 1993, 527–529):

Theorem 8. *Let K be an algebraic number field. Every (nondegenerate) quadratic form q over K that satisfies condition (32) and has dimension $n \geq 4$ is **equivalent** to a trace form $s_{L/K}$.*

As to forms q of dimension $n \leq 3$ over K and satisfying (32), this theorem implies that they are at least Witt-equivalent to trace forms over K (because one can take the sum of q with $[1, -1]$). As a matter of fact, it is easy to work through the possible cases: the trace forms of dimension at most 3 are represented by $[1]$, $[2, 2d]$ with $d \in K^\times \setminus K^{\times 2}$, and $[1, 2, 2d]$ with $d \in K^\times$.

We will have to forgo a proof of Theorem 8, even in outline, because it lies beyond our scope. But we mention that certain methods that play a role in the proof are connected with Inverse Galois Theory, especially Hilbert's Irreducibility Theorem (see p. 180 in vol. I).

On the arithmetic side, moreover, everything hinges on *Meyer's Theorem*:

Theorem (Meyer). *Let q be a quadratic form of dimension at least 5 over an algebraic number field K. If q is indefinite with respect to any order P of K, then q is isotropic over K.*

And this, in turn, is a consequence of a local-global principle that holds for every algebraic number field K, namely the famous *Hasse–Minkowski Theorem* (see for instance W. Scharlau, *Quadratic and hermitian forms*, p. 223):

Theorem (Hasse–Minkowski). *A quadratic form q over an algebraic number field K is isotropic over K if and only if q is isotropic over every completion of K.*

Absolute Values on Fields

1. In the field $\mathbb{R}$ of real numbers, the ordinary absolute value function $|\ |$ is defined via the *order* $\leq$; but for many purposes only the absolute value function actually comes into the picture, not the underlying order. Unlike the order on $\mathbb{R}$, the absolute value function can be extended to $\mathbb{C}$: a fact of fundamental importance. It is therefore natural to try to carry over the notion of absolute values to arbitrary fields, and this turns out to be a fruitful approach. We will restrict ourselves to the case where the image of the absolute value function is contained in the field $\mathbb{R}$ (so for instance the absolute value function for arbitrary ordered fields in the sense of §20.7 is not encompassed by the definition below).

Definition 1. An *absolute value* on a field K is a map

$$|\ | : K \to \mathbb{R}_{\geq 0}$$

from K into the set $\mathbb{R}_{\geq 0}$ of positive real numbers, satisfying the following properties:

(i) $|a| = 0 \iff a = 0$.

(ii) $|ab| = |a||b|$.

(iii) $|a + b| \leq |a| + |b|$ (the *triangle inequality*).

Remarks. (a) On the field $\mathbb{C}$ of complex numbers we are familiar with the *ordinary absolute value*, for which we use the notation

$$|\ |_\infty$$

to distinguish it from others. It induces an absolute value on each subfield K of $\mathbb{C}$, which we also denote by $|\ |_\infty$ in the case of $K = \mathbb{Q}$ and $K = \mathbb{R}$.

(b) Every absolute value $|\ |$ of a field K gives rise to a homomorphism from the multiplicative group $K^\times$ of K into the group $\mathbb{R}_{>0}$ of strictly positive real numbers; in particular, $|1| = 1$. The image $|K^\times|$ of this homomorphism is called the *valuation group* of $|\ |$.

(c) On every field K there is a *trivial absolute value* $|\ |$, defined by $|0| = 0$ and $|a| = 1$ for all $a \in K^\times$. It follows from (b) that a *finite field* can have no other absolute value.

(d) If $|\ |$ is an absolute value of K and ρ is a real number such that $0 < \rho \leq 1$, every map $|\ |^\rho : a \mapsto |a|^\rho$ is an absolute value on K (see exercise §23.1).

(e) It can easily be seen from (i), (ii), (iii) that any absolute value $|\ |$ also satisfies $|-1| = 1$, $|-a| = |a|$, $|a| - |b| \leq |a - b|$; and this last inequality implies

(1)
$$\big||a| - |b|\big|_\infty \leq |a - b|.$$

(f) On the subfield $K = \mathbb{Q}(\sqrt{2})$ of $\mathbb{R}$, setting $|a + b\sqrt{2}|_\sigma = |a - b\sqrt{2}|_\infty$ defines an absolute value $|\ |_\sigma$ that is essentially distinct from the ordinary absolute value on $\mathbb{Q}(\sqrt{2})$.

Already the example in this last remark suggests the versatility of the notion introduced in Definition 1. Here is another example, this one of great importance:

Example. Set $K = \mathbb{Q}$ and let p be a *prime number*. For every integer $a \neq 0$, denote by $w_p(a)$ the exponent of p in the prime factorization of a. Also define $w_p(0) = \infty$. Then

(2)
$$w_p(ab) = w_p(a) + w_p(b),$$

(3)
$$w_p(a + b) \geq \min(w_p(a), w_p(b)).$$

The function $w_p : \mathbb{Z} \to \mathbb{Z} \cup \{\infty\}$ can be extended to a well defined function $w_p : \mathbb{Q} \to \mathbb{Z} \cup \{\infty\}$ through the condition $w_p(a/b) = w_p(a) - w_p(b)$, and then relations (2) and (3) hold for every $a, b \in \mathbb{Q}$ (see Section 4.4 in vol. I, p. 40). We now define a function $|\ |_p : \mathbb{Q} \to \mathbb{R}_{\geq 0}$ by setting

(4)
$$|a|_p = \left(\frac{1}{p}\right)^{w_p(a)},$$

and in addition $|0|_p = 0$. One can show directly that $|\ |_p$ satisfies properties (i), (ii) and (iii) of an absolute value: (i) is obvious, (ii) follows easily from (2), and as for (iii), we actually have the following much sharper inequality as a consequence of (3):

(5)
$$|a + b|_p \leq \max(|a|_p, |b|_p).$$

We call $|\ |_p$ the *p-adic absolute value* of $\mathbb{Q}$.

Thus the field $\mathbb{Q}$ of rational numbers admits a whole series of absolute values: besides $|\ |_\infty$, there is, for every prime p, a corresponding p-adic absolute value $|\ |_p$. We will see in Theorem 1 that this list in fact exhausts essentially all possibilities for nontrivial absolute values on $\mathbb{Q}$. What is meant by "essentially" will soon be made precise.

To begin with, note that one key implication of an absolute value $|\ |$ of a field K resides in that it allows one to talk about convergence in K. For if we set

$$d(x, y) := |x - y| \quad \text{for all } x, y \in K,$$

the properties of $|\ |$ immediately imply that $d : K \times K \to \mathbb{R}_{\geq 0}$ is a *metric* on K, inasmuch as the metric axioms are satisfied:

$$d(x, y) = 0 \iff x = y,$$
$$d(x, y) \leq d(x, z) + d(y, z).$$

Thus (K, d) is a *metric space*, and as such lends itself to the use of basic topological notions such as *open sets, closed sets, convergent sequences* and so forth. Thus a sequence $(x_n)_n$ of elements of K is said to converge to an element $a \in K$ if $d(x_n, a) = |x_n - a|$ converges to 0 in $\mathbb{R}$. In this case we write $a = \lim x_n$, or, if we wish to make the dependence on the underlying absolute value $|\ |$ explicit,

(6) $$|\ |\text{-}\lim x_n = a.$$

When this equation holds with $a = 0$, we call $(x_n)_n$ a *null sequence*, or $|\ |$-null sequence. By definition, (6) is satisfied if and only if $(x_n - a)_n$ is a $|\ |$-null sequence. Also, $(x_n)_n$ is a $|\ |$-null sequence if and only if $(|x_n|)_n$ is a null sequence in $\mathbb{R}$ (where the absolute value on $\mathbb{R}$ is the ordinary one, as always). For practice you should persuade yourself that $|\ | : K \to \mathbb{R}$ is a *continuous* map; just consider equation (1).

Exactly as in the case of the ordinary absolute value, one can prove that the *addition, subtraction, multiplication* and *division* maps $K \times K \to K$ (or $K \times K^\times \to K$) are continuous. In symbols, if $(a_n)_n$ and $(b_n)_n$ are $|\ |$-convergent sequences in K and a, b are their $|\ |$-limits in K, then $\lim(a_n \pm b_n) = a \pm b$ and $\lim(a_n b_n) = ab$; and if, moreover, b is nonzero (implying that b_n is nonzero for almost all n), then $\lim(a_n/b_n) = a/b$.

We now declare that two absolute values on a field K are *not* essentially distinct if they lead to the same notion of convergence in K:

Definition 2. Two absolute values $|\ |_1$ and $|\ |_2$ on a field K are called *equivalent*, and the notation

$$|\ |_1 \sim |\ |_2$$

is used, if every $|\ |_1$-null sequence is also a $|\ |_2$-null sequence and vice versa.

Remark. For $K = \mathbb{Q}$ and p any prime number, consider the sequence $(p^n)_n$ in $\mathbb{Q}$. In view of (4) we have

$$|p^n|_p = \left(\frac{1}{p}\right)^n.$$

Thus $(p^n)_n$ is a $|\ |_p$-*null sequence*. In contrast, $(p^n)_n$ is obviously not convergent with respect to $|\ |_\infty$, so $|\ |_p$ and $|\ |_\infty$ are not equivalent. Given another prime $q \neq p$ we have $|p^n|_q = 1$ for all n, so $|\ |_p$ is not equivalent to $|\ |_q$ either.

In general we have the following striking criterion for the equivalence between absolute values:

F1. *Given absolute values* $|\ |_1$ *and* $|\ |_2$ *on a field* K, *the following statements are equivalent*:

(i) $|\ |_1 \sim |\ |_2$.

(ii) *There exists* $\rho > 0$ *in* $\mathbb{R}$ *such that* $|\ |_2 = |\ |_1^\rho$.

If $|\ |_1$ *is nontrivial, condition* (i) *is also equivalent to*:

(iii) *For every* $a \in K$, *the condition* $|a|_1 < 1$ *implies* $|a|_2 < 1$.

Proof. Suppose (i) holds and $|\ |_1$ is trivial. We claim that $|\ |_2$ is also trivial (so (ii) holds). Otherwise there exists $a \in K^\times$ such that $|a|_2 \neq 1$; by replacing a with $1/a$ if necessary, we can assume that $|a|_2 < 1$. Then $(a^n)_n$ is a null sequence with respect to $|\ |_2$, but not with respect to $|\ |_1$.

The implication (ii) $\Rightarrow$ (i) is obvious. It is also clear that (i) implies (iii): If $|a|_1 < 1$, then $(a^n)_n$ is an $|\ |_1$-null sequence. By assumption (i) the same sequence is also $|\ |_2$-null. This precludes the possibility that $|a|_2 \geq 1$; that is, $|a|_2 < 1$.

We are now entitled to assume that $|\ |_1$ is *nontrivial* and just have to show that (iii) implies (ii). So assume (iii). Then

$$(7) \qquad\qquad |b|_1 > 1 \implies |b|_2 > 1$$

for every $b \in K$: just apply (i) to $a = b^{-1}$. Next, since $|\ |_1$ is assumed nontrivial, there exists $c \in K$ such that $|c|_1 > 1$. Because of (7), we have $|c|_2 > 1$, so $|c|_2 = |c|_1^\rho$ for some $\rho > 0$ in $\mathbb{R}$. To show that the same relation holds for all elements of K, we take an arbitrary $a \in K^\times$ and relate its absolute values to those of c. First we have

$$(8) \qquad\qquad |a|_1 = |c|_1^\alpha \quad \text{for some } \alpha \in \mathbb{R}.$$

Suppose $m \in \mathbb{Z}$ and $n \in \mathbb{N}$ satisfy $m/n < \alpha$. Then $|a|_1 = |c|_1^\alpha > |c|_1^{m/n}$; that is, $|a|_1^n > |c|_1^m$, or again $|a^n/c^m|_1 > 1$. Using (7) we then get $|a^n/c^m|_2 > 1$, which means that

$$|a|_2 > |c|_2^{m/n}.$$

But by continuity we have

$$|a|_2 \geq |c|_2^\alpha.$$

We see likewise that $m/n > \alpha$ implies $|a|_2 < |c|_2^{m/n}$, whence, again by continuity, we get

$$|a|_2 \leq |c|_2^\alpha.$$

Combining both inequalities we conclude that (8) implies

$$(9) \qquad\qquad |a|_2 = |c|_2^\alpha.$$

Since $|c|_2 = |c|_1^\rho$ by definition, (9) and (8) lead, for all $a \in K^\times$, to the equation

$$|a|_2 = |a|_1^\rho,$$

which proves (ii). $\qquad\qquad\qquad\qquad\qquad\qquad\qquad\qquad\qquad\qquad$ $\square$

In studying an absolute value $|\ |$ on a field, it is natural to investigate first the behavior of $|\ |$ on the positive integer multiples $n1_K$ of the unity 1_K of K. As usual we denote these elements simply by n (and in the case of nonzero characteristic, one must watch out in each case for whether such a symbol denotes a natural number or an element of K).

Definition 3. An absolute value $|\ |$ on a field K is called *archimedean* if the set $\{|n| : n \in \mathbb{N}\}$ is unbounded; otherwise it is called *nonarchimedean*.

F2. *If K is a field and $|\ |$ an absolute value on K, there is equivalence between*:

(i) $|n| \leq 1$ *for every $n \in \mathbb{N}$.*

(ii) $|\ |$ *is nonarchimedean.*

(iii) $|\ |$ *satisfies the **strong triangle inequality**, which says that*

$$|a+b| \leq \max(|a|, |b|).$$

(iv) *For every real number $\rho > 0$, the map $|\ |^\rho$ is an absolute value on K.*

Proof. (i) $\Rightarrow$ (ii) is clear by definition. Assume (ii), and take $C > 0$ in $\mathbb{R}$ such that

(10) $$|n| \leq C \quad \text{for all } n \in \mathbb{N}.$$

Let a, b be elements of K with (say) $|a| \geq |b|$. For every $m \in \mathbb{N}$ the expansion $(a+b)^m = \sum \binom{m}{i} a^i b^{m-i}$, together with the triangle inequality and the bound (10), show that

$$|a+b|^m \leq C(m+1)|a|^m.$$

Now taking the m-th root and the limit as $m \to \infty$, we get

$$|a+b| \leq |a| = \max(|a|, |b|),$$

which is (iii).

Next assume (iii) and write

$$|a+b|^\rho \leq \max(|a|, |b|)^\rho = \max(|a|^\rho, |b|^\rho)$$

to show that $|\ |^\rho$ satisfies the strong triangle inequality. That it satisfies the first two defining properties of an absolute value is obvious regardless of (iii).

Finally, assume (iv) and take $n \in \mathbb{N}$. We know that $|\ |^m$ is an absolute value for any $m \in \mathbb{N}$; hence $|n|^m \leq n$ by the triangle inequality. Taking the m-th root and the limit as $m \to \infty$ yields $|n| \leq 1$, proving (i). $\square$

Remark. Let p be a prime number. By (5), the p-adic absolute value $|\ |_p$ on $\mathbb{Q}$ is nonarchimedean. For any real number $0 < c < 1$, we can define an absolute value equivalent to $|\ |_p$ by setting

(11) $$|a| = c^{w_p(a)};$$

this is because we can find for each such c a number $\rho > 0$ such that $c = p^{-\rho}$, and then $|\ | = |\ |_p^\rho$. In many situations, especially regarding convergence, it is immaterial what constant $0 < c < 1$ is chosen in (11). Still, there are good reasons to formulate the definition of $|\ |_p$ with the value $c = 1/p$ rather than some other; see, for example, formula (15) on page 45.

Before we go into the particular properties of nonarchimedean absolute values, we address the already mentioned classification result for absolute values on the rationals:

Theorem 1. *Every nontrivial absolute value $|\ |$ of the field $\mathbb{Q}$ is equivalent either to $|\ |_\infty$ or to $|\ |_p$ for some prime p.*

Proof. (a) First assume that the given absolute value $|\ |$ on $\mathbb{Q}$ is *nonarchimedean*. Then $|m| \leq 1$ for every $m \in \mathbb{N}$. Since $|\ |$ is assumed nontrivial, there must be some $n \in \mathbb{N}$ such that $|n| < 1$. Let p be the smallest natural number such that $|p| < 1$; clearly p is prime. We claim that

$$(12) \qquad\qquad |a| < 1,\ a \in \mathbb{Z} \implies p \,|\, a.$$

Indeed, division with rest yields $a = mp + r$, with $0 \leq r < p$. But since $r = a - mp$, the strong triangle inequality then implies that $|r| \leq \max(|a|, |m||p|) < 1$, which, in view of the minimality of p, can only be true if $r = 0$.

Now let a nonzero integer a be given. We have

$$(13) \qquad\qquad a = p^{w_p(a)} a_1, \quad \text{with } p \nmid a_1.$$

Since a_1 is an integer, $|a_1|$ is at most 1; but it cannot be *less* than 1, otherwise a_1 would be divisible by p — see (12). Thus $|a_1| = 1$, and (13) implies

$$(14) \qquad\qquad |a| = |p|^{w_p(a)}.$$

This holds for every $a \in \mathbb{Z}$; but since both sides behave multiplicatively, the same equation holds for all $a \in \mathbb{Q}$. Thus $|\ | = |\ |_p^\rho$, where ρ denotes the (strictly positive) real number such that $|p| = p^{-\rho}$.

(b) We now assume that $|\ |$ is an *archimedean* absolute value in $\mathbb{Q}$, and we have to show that it is equivalent to the ordinary absolute value $|\ |_\infty$ on $\mathbb{Q}$. Suppose this is not the case. By F1, there exists $q \in \mathbb{Q}$ such that $|q| < 1$ and $|q|_\infty \geq 1$. Clearly we can assume that $q > 0$, and thus that

$$|q| < 1 \quad \text{and} \quad q > 1.$$

Suppose $q = c/d$, where c and d are natural numbers with $c > d$. It is easy to show by induction that any natural number n has a *q-adic* representation of the form

$$n = a_0 + a_1 q + \cdots + a_r q^r, \quad \text{with } a_i \in \{0, 1, \ldots, c-1\}$$

(see §23.16). An application of the triangle inequality then yields

$$|n| \leq c\big(1 + |q| + \cdots + |q|^r\big) \leq \frac{c}{1 - |q|}.$$

Hence | | is bounded on $\mathbb{N}$ — impossible, because | | is *archimedean*.   □

Remark. Up to equivalence, $\mathbb{Q}$ admits a unique archimedean absolute value, namely | |$_\infty$. Next to it we now have — on an equal footing with | |$_\infty$ from the viewpoint of Definition 1 — the nonarchimedean p-adic absolute values | |$_p$, for all primes p. As we have seen (see F1, for example), all these absolute values are *pairwise inequivalent*. Nonetheless, there is among them a certain dependence relation, called the *product formula for* $\mathbb{Q}$. It says that, for every $a \in \mathbb{Q}^\times$,

$$\text{(15)} \qquad \prod_v |a|_v = 1,$$

where v runs over all primes p plus ∞. To prove it we merely observe that every integer (and thus every rational number) $a \neq 0$ has the representation

$$\text{(16)} \qquad a = \operatorname{sgn}(a) \cdot \prod_p p^{w_p(a)},$$

by the prime factorization of a. But since $|a|_p = p^{-w_p(a)}$ and $a = \operatorname{sgn}(a)|a|_\infty$, equation (16) says exactly the same as (15).

We now investigate in more depth the properties of *nonarchimedean* absolute values.

F3 (Scholium on the strong triangle inequality). *For any nonarchimedean absolute value* | | *on a field* K,

$$|a| \neq |b| \implies |a + b| = \max(|a|, |b|).$$

Proof. Suppose that $|b| < |a|$ and assume the conclusion is false. Then $|a + b| < |a|$. Again from the strong triangle inequality, we obtain $|a| = |a + b - b| \leq \max(|a+b|, |b|)$, hence the contradiction $|a| < |a|$.   □

How far-reaching the consequences of the strong triangle inequality are can be glimpsed from the next result.

F4 and Definition 4. *Let* | | *be a nonarchimedean absolute value on the field* K.

(i) $R := \{a \in K : |a| \leq 1\}$ *is a subring of* K, *called the **valuation ring** of* K (*with respect to* | |).

(ii) K *is the fraction field of* R.

(iii) $\mathfrak{p} := \{a \in K : |a| < 1\}$ *is an ideal of* R, *called the **valuation ideal** of* K (*with respect to* | |).

(iv) *An element $a \in K$ is a unit of R if and only if $|a| = 1$; thus $R^\times = R \smallsetminus \mathfrak{p}$. It follows that $\mathfrak{p}$ is a maximal ideal of R and every ideal of R distinct from R is contained in $\mathfrak{p}$.*

(v) *$R/\mathfrak{p}$ is a field, called the **residue field** of K (with respect to $|\ |$).*

(vi) *For $a, b \in R$, a divides b if and only if $|b| \le |a|$.*

The proof is straightforward and is left to the reader. It is worth adding that according to (iv) the valuation ring R is a *local ring*; its maximal ideal is the valuation ideal $\mathfrak{p}$ of $|\ |$ (see Exercise §4.13 in vol. I).

Let's explore the preceding situation in the special case of the nonarchimedean absolute value $|\ |_p$ of $\mathbb{Q}$:

F5. *The valuation ring R of $\mathbb{Q}$ with respect to $|\ |_p$ consists of the elements a/s, with $a, s \in \mathbb{Z}$ such that $s \not\equiv 0 \bmod p$. The corresponding valuation ideal is the principal ideal pR of R. The residue field is canonically isomorphic to the field $\mathbb{F}_p = \mathbb{Z}/p\mathbb{Z}$.*

Proof. By the definition of $|\ |_p$, every $x \in \mathbb{Q}^\times$ has a unique representation

$$(17) \qquad\qquad x = p^{w_p(x)} u, \quad \text{with } |u|_p = 1.$$

Because $w_p(u) = 0$, we can write $u = r/s$, with integers r, s not divisible by p. This yields the first two assertions. As for the residue field R/pR, the inclusion $\mathbb{Z} \subseteq R$ passes to a canonical homomorphism

$$(18) \qquad\qquad \mathbb{Z}/p\mathbb{Z} \to R/pR$$

of quotient rings, and all that remains to show is that this map is surjective. Take any element $a/s \in R$, with $a, s \in \mathbb{Z}$ and $s \not\equiv 0 \bmod p$. Because s is a unit in $\mathbb{Z}/p\mathbb{Z}$, we can find $b \in \mathbb{Z}$ such that $sb \equiv a \bmod p\mathbb{Z}$, hence also $\bmod\, pR$. But now we can divide by the unit $s \in R$ to obtain $b \equiv a/s \bmod pR$.   □

Remark. Let $|\ |$ be a nonarchimedean absolute value on K. It is often convenient to work with an additive function rather than the multiplicative function $|\ |$. For this one chooses a real constant $0 < c < 1$ and considers the function $w : K \to \mathbb{R} \cup \{\infty\}$ defined by

$$(19) \qquad\qquad |x| = c^{w(x)} \quad \text{for every } x \in K,$$

with, logically enough, $w(0) = \infty$. The function w has these properties:

(i) $w(a) = \infty \iff a = 0$.

(ii) $w(ab) = w(a) + w(b)$.

(iii) $w(a + b) \ge \min(w(a), w(b))$.

Conversely, if we have a function $w : K \to \mathbb{R} \cup \{\infty\}$ satisfying properties (i)–(iii) and we define, for any choice of $0 < c < 1$, a function $|\ | : K \to \mathbb{R}$ using (19), this function is a nonarchimedean absolute value, and the use of a different c leads to an equivalent absolute value.

Definition 5. A function $w : K \to \mathbb{R} \cup \{\infty\}$ with properties (i), (ii), (iii) above is called a *valuation* of the field K. The subgroup $w(K^\times)$ of the additive group of R is called the *valuation group* of w.

If a valuation w of K corresponds to an absolute value $|\ |$ of K in the way just described, any statement regarding $|\ |$ can be translated into a statement about w and vice versa. For instance, in the notation of F4, we have

$$R = \{x \in K \mid w(x) \geq 0\}, \quad \mathfrak{p} = \{x \in K \mid w(x) > 0\}.$$

Also, a sequence $(x_n)_n$ in K is $|\ |$-null if and only if $w(x_n) \to \infty$ as $n \to \infty$. The reader is encouraged to work out and keep in mind such restatements in the case of p-adic absolute values $|\ |_p$ and the corresponding valuations w_p. Incidentally, the valuation group of w_p is $\mathbb{Z}$.

2. We now fix a field K and an absolute value $|\ |$ on K.

Definition 6. A sequence $(a_n)_n$ in K is called a *Cauchy sequence* (with respect to $|\ |$, or a $|\ |$-Cauchy sequence) if, for every real number $\varepsilon > 0$, there exists $N \in \mathbb{N}$ such that

$$|a_n - a_m| < \varepsilon \quad \text{for all } m, n > N.$$

If every $|\ |$-Cauchy sequence in K converges with respect to $|\ |$ to an element of K, we say that K is *complete* (with respect to $|\ |$). In this case we also say that $|\ |$ is a *complete absolute value*.

Remark. As in real analysis, one shows:

   (i) Every $|\ |$-convergent sequence in K is a $|\ |$-Cauchy sequence.

   (ii) If $(a_n)_n$ is a $|\ |$-Cauchy sequence in K, then $(|a_n|)_n$ is a Cauchy sequence in $\mathbb{R}$.

   (iii) Every $|\ |$-Cauchy sequence in K is bounded.

   (iv) If a $|\ |$-Cauchy sequence $(a_n)_n$ has a subsequence $(a_{n_k})_k$ that is $|\ |$-null, then $(a_n)_n$ is itself a $|\ |$-null sequence.

   (v) The set $\mathfrak{C}$ of all $|\ |$-Cauchy sequences in K is a commutative ring with unity.

   (vi) The set $\mathfrak{N}$ of all $|\ |$-null sequences in K is an ideal of the ring $\mathfrak{C}$.

As the reader knows, the field $\mathbb{Q}$ is not complete with respect to the ordinary absolute value, and it is exactly this that prompts the extension of $\mathbb{Q}$ to the field $\mathbb{R}$ of real numbers, which is complete. We now show that such an extension is also possible for any absolute value $|\ |$ on a field K. (We will make use as needed of known properties of $\mathbb{R}$, so the process will be easier than the construction of $\mathbb{R}$; see also Exercise §20.8.)

Definition 7. A *completion* of K with respect to the absolute value $|\ |$ is a pair $(\hat{K}, |\ \hat{}\ |)$ consisting of an extension $\hat{K}$ of K and an absolute value $|\ \hat{}\ |$ on $\hat{K}$ satisfying the following properties:

(i) $|\ \hat{|}$ is an extension of $|\ |$.

(ii) K is *dense* in $\hat{K}$ with respect to $|\ \hat{|}$.

(iii) $\hat{K}$ is *complete* with respect to $|\ \hat{|}$.

Theorem 2.   (I) *There exists a completion* $(\hat{K}, |\ \hat{|})$ *of K with respect to* $|\ |$.

(II) *If* $(K_1, |\ |_1)$ *and* $(K_2, |\ |_2)$ *are both completions of K with respect to* $|\ |$, *there exists a unique K-isomorphism* $\sigma : K_1 \to K_2$ *such that* $|\sigma x|_2 = |x|_1$ *for all* $x \in K_1$.

Proof. (a) Consider the quotient ring

$$\hat{K} := \mathfrak{C}/\mathfrak{N}$$

of the ring $\mathfrak{C}$ of all $|\ |$-Cauchy sequences in K modulo the ideal $\mathfrak{N}$ of all $|\ |$-null sequences in K. This quotient is a commutative ring with unity, and is clearly not the zero ring. We will show that $\hat{K}$ is a *field*.

Let α be a nonzero element of $\hat{K}$, and let $(a_n)_n \in \mathfrak{C}$ be a representative of α. Since $(a_n)_n$ is not a null sequence, there exists $\varepsilon > 0$ and $N \in \mathbb{N}$ such that

(20) $$|a_n| \geq \varepsilon \quad \text{for every } n > N;$$

otherwise there would be a subsequence of $(a_n)_n$ converging to zero, which is impossible by statement (iv) after Definition 6. Now (20) implies, in particular, that $a_n \neq 0$ for every $n > N$. Hence there is a sequence $(b_n)_n$ in K such that

(21) $$b_n = \frac{1}{a_n} \quad \text{for every } n > N.$$

We claim that $(b_n)_n$ is a Cauchy sequence. For $m, n > N$, we have

$$|b_m - b_n| = \left| \frac{a_n - a_m}{a_n a_m} \right| = \frac{|a_n - a_m|}{|a_n||a_m|} \leq \frac{|a_n - a_m|}{\varepsilon^2},$$

where the last inequality uses (20). Let β be the class of $(b_n)_n$ in $\hat{K}$. Then $\alpha\beta = 1$, because $(a_n b_n - 1)_n$ is indeed a null sequence, by (21).

(b) Define the map $\iota : K \to \hat{K}$ that assigns to each $a \in K$ the class of the constant sequence $(a, a, \dots)$. Clearly ι is a field homomorphism, hence injective.

(c) We now define an absolute value $|\ \hat{|}$ on $\hat{K}$. Given $\alpha \in \hat{K}$, take a representative $(a_n)_n$ of α in $\mathfrak{C}$. By statement (ii) after Definition 6, $(|a_n|)_n$ is a Cauchy sequence in $\mathbb{R}$ and thus converges to some real number

(22) $$|\alpha|\hat{|} = \lim |a_n|.$$

If $(b_n)_n$ is another representative of α, the difference sequence $(a_n - b_n)_n$ is $|\ |$-null; hence $(|a_n - b_n|)_n$ is a null sequence in $\mathbb{R}$ and so also $(|a_n| - |b_n|)_n$, by inequality (1). Thus $\lim |a_n| = \lim |b_n|$, which shows that $|\alpha|\hat{|}$ is well defined by (22). One easily checks that the map $|\ |^{\widehat{}} : \widehat{K} \to \mathbb{R}$ has all the properties of an absolute value. Also it is clear that

(23) $$|\iota a|^{\widehat{}} = |a| \quad \text{for every } a \in K,$$

where ι is the field embedding defined in part (b). If we identify K with a subfield of $\widehat{K}$ via ι, what (23) says is that the absolute value $|\ |^{\widehat{}}$ is an extension of $|\ |$, so condition (i) in the definition of a completion is met.

(d) Next we show that K is $|\ |^{\widehat{}}$-*dense* in $\widehat{K}$. Take $\alpha \in \widehat{K}$ and a representative $(a_n)_n$ of α; we claim that

(24) $$\alpha = \lim_{m \to \infty} a_m.$$

Indeed, for fixed m we have, by (22),

(25) $$|\alpha - a_m|^{\widehat{}} = \lim_{n \to \infty} |a_n - a_m|;$$

but since $(a_n)_n$ is a Cauchy sequence, the right-hand side of (25) converges to 0 as m goes to ∞.

(e) To complete our existence proof, all that remains to be shown is that the absolute value $|\ |^{\widehat{}}$ on $\widehat{K}$ is *complete*. Let $(\alpha_n)_n$ be a $|\ |^{\widehat{}}$-Cauchy sequence in $\widehat{K}$; we must show that it is $|\ |^{\widehat{}}$-convergent to some limit in $\widehat{K}$. By (d) we know that for every n there exists $a_n \in K$ such that $|\alpha_n - a_n|^{\widehat{}} < 1/n$; but then $(\alpha_n - a_n)_n$ is a $|\ |^{\widehat{}}$-null sequence in $\widehat{K}$. Since $a_n = (a_n - \alpha_n) + \alpha_n$, it follows that $(a_n)_n$ is a Cauchy sequence in $(\widehat{K}, |\ |^{\widehat{}})$, hence also in $(K, |\ |)$. Let α be the class of $(a_n)_n$ in $\widehat{K}$. Since

$$\alpha - \alpha_n = (\alpha - a_n) + (a_n - \alpha_n),$$

we obtain using (24) that $(\alpha_n)_n$ converges with respect to $|\ |^{\widehat{}}$ to $\alpha \in \widehat{K}$.

Note that if we apply the preceding construction of a completion to an absolute value equivalent to $|\ |$, we evidently get the same $\widehat{K}$, with an absolute value equivalent to $|\ |^{\widehat{}}$.)

Regarding the unique part of the theorem, we derive it immediately from the next statement. $\square$

F6. *Let $(K_1, |\ |_1)$ be a $|\ |$-completion of K. Let K_2 be another field and $|\ |_2$ a* **complete** *absolute value on K_2. Every homomorphism*

$$\rho : K \to K_2 \quad \text{such that } |\rho x|_2 = |x| \text{ for every } x \in K$$

can be extended uniquely to a homomorphism

$$\sigma : K_1 \to K_2 \quad \text{such that } |\sigma x|_2 = |x|_1 \text{ for every } x \in K_1.$$

Proof. Given $\alpha \in K_1$, there is a sequence $(a_n)_n$ in K such that $\alpha = |\ |_1$-$\lim a_n$. Now, $(a_n)_n$ is in any case a $|\ |_1$-Cauchy sequence. By assumption, then $(\rho a_n)_n$ is a $|\ |_2$-Cauchy sequence in K_2. But K_2 is $|\ |_2$-complete, so $(\rho a_n)_n$ has a (uniquely defined) limit in K_2. Set

$$\sigma\alpha = \lim \rho a_n;$$

then $\sigma\alpha$ is well defined, as a routine argument shows. It is also easy to check that the map $\sigma : K_1 \to K_2$ thus obtained has the required properties. Uniqueness is likewise straightforward. $\qquad\square$

Remark. Theorem 2 says that a given field K with absolute value $|\ |$ always has a completion $(\hat{K}, |\ \hat{\ }\ |)$, and this completion is unique up to an isometric isomorphism. Hence we will speak from now on of *the* completion of $(K, |\ |)$. And since the absolute value $|\ \hat{\ }\ |$ on $\hat{K}$ is uniquely determined by $|\ |$, we will simply say that $\hat{K}$ is the *completion of K* with respect to $|\ |$, denoting the unique extension of $|\ |$ to an absolute value on $\hat{K}$ also by $|\ |$.

Definition 8. For p a prime or the symbol ∞, we denote by

$$\mathbb{Q}_p$$

the completion of $\mathbb{Q}$ with respect to the absolute value $|\ |_p$. We still denote by $|\ |_p$ the corresponding absolute value on $\mathbb{Q}_p$. When p is not ∞ we call $\mathbb{Q}_p$ the *field of p-adic numbers.*

With these conventions,

$$\mathbb{R} = \mathbb{Q}_\infty$$

is the field of real numbers and $|\ |_\infty$ is the usual absolute value on $\mathbb{R}$. We have seen that next to this completion of $\mathbb{Q}$ there stands a whole series of p-adic completions $\mathbb{Q}_p$, for p prime, with their nonarchimedean absolute values $|\ |_p$ — an insight we owe to K. Hensel. Our plan is now to wade a little way into this "nonarchimedean sea", where algebra rather than analysis is the focus (at least for us). Our goals will be modest, for although many aspects of this study are easier than in the real case, the flip side is that we can no longer rely on familiar intuitions.

Remark. The construction used in the proof of Theorem 2 will not play any further role, and we will simply use the fact that $\mathbb{Q}_p$ is the $|\ |_p$-completion of $\mathbb{Q}$ in the sense of Definition 7 (see also the remark before Definition 8). As will soon become clear, $\mathbb{Q}_p$ is in fact distinct from $\mathbb{Q}$.

F7. *Let $\hat{K}$ be the $|\ |$-completion of K. If $|\ |$ is nonarchimedean, K and $\hat{K}$ have the same valuation group and canonically isomorphic residue fields with respect to $|\ |$.*

Proof. Take $\alpha \in \hat{K}$. Since K is dense in $\hat{K}$, there is a sequence $(a_n)_n$ in K converging to α in $\hat{K}$. Thus, if $\alpha \neq 0$, we have $|a_n - \alpha| < |\alpha|$ for n large enough. For such n we can then write $|a_n| = |\alpha|$, by applying the scholium F3 on the strong triangle inequality to $a = a_n$, $b = a_n - \alpha$. This shows that $|\hat{K}^\times| = |K^\times|$.

Let R and $\hat{R}$ be the valuation rings of K and $\hat{K}$ with respect to $|\ |$, and let $\mathfrak{p}$ and $\hat{\mathfrak{p}}$ be the corresponding valuation ideals. The inclusion $R \subseteq \hat{R}$ gives rise to a canonical homomorphism

$$R/\mathfrak{p} \to \hat{R}/\hat{\mathfrak{p}}$$

of residue fields, which we must show is surjective. Given $\alpha \in \hat{R} \smallsetminus \{0\}$ there exists, as we saw earlier, some $a \in K$ such that $|a - \alpha| < |\alpha| \le 1$ and $|a| = |\alpha| \le 1$. It follows that $a \in R$ and $\alpha \equiv a \bmod \hat{\mathfrak{p}}$. $\qquad\square$

Remark. Clearly, the function $w_p : \mathbb{Q}^\times \to \mathbb{Z}$ can be uniquely extended to $\mathbb{Q}_p$ in such a way that

(26)
$$|\alpha|_p = \left(\frac{1}{p}\right)^{w_p(\alpha)}$$

for every $\alpha \in \mathbb{Q}_p^\times$; the extension will still be denoted by w_p. By F7, the valuation group $\mathrm{im}\, w_p$ does not get any bigger. The situation for an arbitrary valuation w on a field K is wholly analogous: w can be extended to the completion $\hat{K}$ of K with respect to w, and $w(\hat{K}) = w(K)$ (see after Definition 5).

The next result shows that the convergence of *infinite series* in nonarchimedean analysis is often a less troublesome affair than in real analysis; in particular, the subtle phenomenon of *conditional convergence* does not occur in nonarchimedean fields at all.

F8. *If a field K is endowed with a **complete**, **nonarchimedean** absolute value $|\ |$, an infinite series*

(27)
$$\sum_{n=1}^{\infty} a_n$$

($a_n \in K$) converges if and only if $(a_n)_n$ is a null sequence. In other words, a sequence $(b_n)_n$ in K converges if and only if $(b_{n+1} - b_n)_n$ is a null sequence.

Proof. Let b_n be the n-th partial sum of (27). For every $n > m$ we have

$$b_n - b_m = a_{m+1} + a_{m+2} + \cdots + a_n.$$

Because of the *strong triangle inequality*, $(b_n)_n$ is a Cauchy sequence if and only if $(a_n)_n$ is a null sequence. Since K is complete, this proves the assertion. $\qquad\square$

Corollary. *Under the assumptions of F8, any power series $\sum_{n=0}^{\infty} a_n x^n$ whose coefficients a_n lie in the valuation ring R of $(K, |\ |)$ converges for every $x \in K$ such that $|x| < 1$, and the limit is an element of R.*

Proof. We have $|a_n x^n| \le |a_n| |x|^n \le |x|^n$, so the series converges for $|x| < 1$ by F8. The ring $R = \{y : |y| \le 1\}$ is *closed* with respect to $|\ |$, so the limit lies in R. $\qquad\square$

Definition 9. For the field $\mathbb{Q}_p$ of p-adic numbers, let

$$\mathbb{Z}_p$$

be the valuation ring of $\mathbb{Q}_p$, that is, the set of $a \in \mathbb{Q}_p$ such that $|a|_p \leq 1$. The elements of $\mathbb{Z}_p$ are called *p-adic integers*.

Remark. The inclusion $\mathbb{Z} \subset \mathbb{Z}_p$ gives rise to a homomorphism $\mathbb{Z}/p^n\mathbb{Z} \to \mathbb{Z}_p/p^n\mathbb{Z}_p$ for each $n \in \mathbb{N}$, and this map is obviously injective. It is easy to show that it is also surjective; see F5. It follows that $\mathbb{Z}$ is dense in $\mathbb{Z}_p$; that is, every p-adic number a of absolute value $|a|_p \leq 1$ can be approximated arbitrarily closely by ordinary integers. All this can be read off from the next statement:

F9. *Every $a \in \mathbb{Z}_p$ has a unique representation*

$$(28) \qquad a = \sum_{i=0}^{\infty} a_i p^i, \quad \text{with } a_i \in \{0, 1, \ldots, p-1\}.$$

Every $a \in \mathbb{Q}_p$ has a unique representation

$$(29) \qquad a = \sum_{-\infty \ll i} a_i p^i, \quad \text{with } a_i \in \{0, 1, \ldots, p-1\},$$

where the notation $-\infty \ll i$ shall be used to indicate that the sum runs over $i \in \mathbb{Z}$, but a_i vanishes for almost every $i < 0$. We then have $w_p(a) = \min\{i \mid a_i \neq 0\}$.

Proof. (1) The set $S = \{0, 1, \ldots, p-1\}$ contains one representative of each residue class in $\mathbb{Z}_p/p\mathbb{Z}_p = \mathbb{Z}/p\mathbb{Z}$ (see F7 and F5). In other words, for every $a \in \mathbb{Z}_p$, there is a unique $a_0 \in S$ such that

$$(30) \qquad\qquad\qquad a \equiv a_0 \bmod p.$$

We now define recursively a sequence $a_0, a_1, a_2, \ldots$ of elements of S such that, for each $n = 0, 1, 2, \ldots$, the congruence

$$(31) \qquad\qquad a \equiv a_0 + a_1 p + \cdots + a_n p^n \bmod p^{n+1}$$

is satisfied; the representation (28) of a as an infinite sum is simply a restatement of the validity of (31) for every n. The start of the recursion comes from (30). Now assume (31) holds as written; then

$$(32) \qquad\qquad a = a_0 + a_1 p + \cdots + a_n p^n + a' p^{n+1},$$

with a' in $\mathbb{Z}_p$. Write $a' \equiv a_{n+1} \bmod p$ for some $a_{n+1} \in S$. Then

$$a \equiv a_0 + a_1 p + \cdots + a_n p^n + a_{n+1} p^{n+1} \bmod p^{n+2},$$

completing the recursion step for existence. The construction also shows that the sequence $(a_n)_n$ is unique, since a_{n+1} is fully determined by $a_0, a_1, \ldots, a_n$ via (32) and the congruence $a' \equiv a_{n+1} \bmod p$.

(2) Every $a \in \mathbb{Q}_p^\times$ has a unique representation

$$a = p^n a', \quad \text{with } n \in \mathbb{Z} \text{ and } |a'|_p = 1;$$

note that $w_p(a) = n$. Using part (1) of the proof, write a' as

$$a' = \sum_{i=0}^{\infty} a_i' p^i, \quad \text{with } a_i' \in S,$$

thus obtaining

$$a = p^n a' = \sum_{i=0}^{\infty} a_i' p^{n+i} = \sum_{i=n}^{\infty} a_{i-n}' p^i = \sum_{i=n}^{\infty} a_i p^i,$$

with $a_i := a_{i-n}'$. Hence every $a \in \mathbb{Q}_p$ can be represented in the form (29). Conversely, if such a representation is given and we set $n := \min\{i \mid a_i \neq 0\}$, we have (if $n \neq \infty$)

$$a = p^n \sum_{i=n}^{\infty} a_i p^{i-n} = p^n \sum_{i=0}^{\infty} a_{n+i} p^i = p^n \sum_{i=0}^{\infty} a_i' p^i = p^n a',$$

with $a_0' \neq 0$. It follows that $|a'|_p = 1$, so $w_p(a) = n$. The uniqueness of the coefficients a_i', and hence of the a_i, follows from part (1) of the proof. □

3. We continue to work with a field K and an absolute value $|\ \ |$ on K. We now inquire whether $|\ \ |$ can be extended to algebraic extensions of K.

Definition 10. Let V be a vector space over K. A *norm* of V over $|\ \ |$ is a map $\|\ \ \| : V \to \mathbb{R}_{\geq 0}$ with the following properties:

(i) $\|x\| = 0 \iff x = 0$.

(ii) $\|\alpha x\| = |\alpha| \|x\|$ for $\alpha \in K$ and $x \in V$.

(iii) $\|x + y\| \leq \|x\| + \|y\|$.

We call V, or more explicitly $(V, \|\ \ \|)$, a *normed space* over $(K, |\ \ |)$.

Remarks. (1) If $(V, \|\ \ \|)$ is a normed space, the map $(x, y) \mapsto \|x - y\|$ is a *metric* on V, and all accompanying topological notions are then available for use. It is clear, for instance, what is meant by a Cauchy sequence in $(V, \|\ \ \|)$.

(2) Let V be an n-dimensional vector space over K, and take a basis $b_1, \ldots, b_n$ of V. For $x = \sum x_i b_i$, set

$$\|x\| = \max(|x_1|, \ldots, |x_n|).$$

This yields a norm on V over $|\ \ |$, called the *maximum norm* (with respect to the chosen basis). A sequence $(x^{(k)})_k$ in V is null with respect to this norm if and only if each of the sequences of coordinates $(x_i^{(k)})_k$, for $i = 1, \ldots, n$, is $|\ \ |$-null. Thus convergence in $(V, \|\ \ \|)$ is expressible in terms of convergence in $(K, |\ \ |)$. The same is true regarding Cauchy sequences. Hence V is complete with respect to the norm $\|\ \ \|$ if and only if K is complete with respect to $|\ \ |$.

(3) Let $K[X]$ be a polynomial ring in one variable over K. For every polynomial $f = a_0 + a_1 X + \cdots + a_n X^n$ in $K[X]$, set

$$(33) \qquad \qquad \|f\| = \max(|a_0|, \ldots, |a_n|);$$

this gives a *norm* on $K[X]$ over $|\ |$. The subspace V consisting of polynomials of degree less than m is an m-dimensional K-vector space; when $|\ |$ is complete, then, V is also complete with respect to $\|\ \|$. If $|\ |$ is *nonarchimedean*, the norm $\|\ \|$ is *ultrametric*, meaning that $\|f + g\| \leq \max(\|f\|, \|g\|)$ for all f, g. Incidentally, we have $\|fg\| = \|f\| \cdot \|g\|$, so by extending $\|\ \|$ to the fraction field $K(X)$ of $K[X]$ we obtain a (nonarchimedean) *absolute value* on $K(X)$. Since $\|\ \|$ is a natural extension of $|\ |$, one also writes $|\ |$ instead of $\|\ \|$.

(4) If E/K is a field extension and $\|\ \|$ is an absolute value on E that extends $|\ |$, $\|\ \|$ is a norm over $|\ |$ on the K-vector space E.

Here is a fundamental result, also in real analysis:

F10. *Let $(V, \|\ \|_0)$ be a normed space over $(K, |\ |)$, where K is $|\ |$-complete. Suppose that V is finite-dimensional over K. If $\|\ \|$ is any norm on V over $|\ |$, there exist real constants $a > 0$ and $b > 0$ such that*

$$(34) \qquad \qquad a\|x\|_0 \leq \|x\| \leq b\|x\|_0 \quad \text{for every } x \in V.$$

Furthermore, $(V, \|\ \|)$ is complete.

Remark. When (34) holds, the norms $\|\ \|$ and $\|\ \|_0$ are called *equivalent*, but if $\|\ \|$ and $\|\ \|_0$ are absolute values over an extension field, their equivalence as norms is a stronger condition than the equivalence of absolute values defined earlier.

Proof of F10. The last assertion follows from (34), since V is complete with respect to any maximum norm, say. To prove the rest we can assume by transitivity that $\|\ \|_0$ in (34) is the maximum norm in some basis $e_1, \ldots, e_n$ of V. We apply induction on n. The case $n = 0$ is trivial, so suppose $n > 0$. For any

$$x = x_1 e_1 + \cdots + x_n e_n$$

we have $\|x\| \leq |x_1| \|e_1\| + \cdots + |x_n| \|e_n\|$, so $\|x\| \leq b\|x\|_0$, where $b := \|e_1\| + \cdots + \|e_n\|$. Finding a is harder. We consider for each $1 \leq i \leq n$ the subspace U_i of V spanned by all the e_j with $j \neq i$. By the induction assumption, U_i is *complete* with respect to $\|\ \|$, hence *closed*. Thus $e_i + U_i$ is also a closed subset of $(V, \|\ \|)$, for each i. Therefore there exists an open $\|\ \|$-ball around 0, of radius $\varepsilon > 0$, that is *disjoint* from all the subsets $e_i + U_i$. Now take any $x \in V \smallsetminus \{0\}$, and let x_i be a coordinate of x with $\|x\|_0 = |x_i|$. The i-th coordinate of $x_i^{-1}x$ equals 1; that is, $x_i^{-1}x \in e_i + U_i$, so $\|x_i^{-1}x\| \geq \varepsilon$. It follows that $\|x\| \geq \varepsilon|x_i| = \varepsilon\|x\|_0$, proving (34) with $a = \varepsilon$. $\qquad \square$

F11. *Let E/K be a field extension and let $|\ |_1, |\ |_2$ be absolute values on E that restrict to the same absolute value $|\ |$ on K.*

(a) *If* | | *is nontrivial, the condition* | $|_1 \sim$ | $|_2$ *implies* | $|_1 =$ | $|_2$.

(b) *If* E/K *is finite and* K *is* | |-*complete,* | $|_1$ *and* | $|_2$ *necessarily coincide; moreover* E *is* | $|_1$-*complete.*

Proof. (a) Suppose | $|_1 \sim$ | $|_2$. By F1 we have | $|_2 =$ | $|_1^\rho$, with $\rho > 0$. If | | is nontrivial, there exists $a \in K^\times$ such that $|a|_2 = |a|_1 = |a| \neq 1$. But then $\rho = 1$, so | $|_2 =$ | $|_1$.

(b) By F10, we know that | $|_1 \sim$ | $|_2$. Then part (a) yields the equality of | $|_1$ and | $|_2$, unless | | is trivial. But in the latter case, F10 says that | $|_1$ and | $|_2$ are both equivalent to the trivial absolute value on E, and hence coincide with it. The completeness of | $|_1$ also follows from F10. $\square$

Remark. Let | | be an *archimedean* absolute value on K. Then K has characteristic 0, and we can regard $\mathbb{Q}$ as a subfield of K. Thus | | induces on $\mathbb{Q}$ an absolute value, which must be equivalent to | $|_\infty$ by Theorem 1; by F1, there exists $\rho > 0$ such that the function | $|^\rho$ restricts to | $|_\infty$ on $\mathbb{Q}$. Exercise §23.2 implies that this function is an absolute value on K; hence, by replacing | | with | $|^\rho$ if necessary, we can assume that | | coincides with | $|_\infty$ on $\mathbb{Q}$.

Next, if K is | |-complete, F6 allows us to view the | $|_\infty$-completion $\mathbb{R} = \mathbb{Q}_\infty$ of $\mathbb{Q}$ as a subfield of K. It can then be shown (§23.11) that the only possibilities in this case are $K = \mathbb{R}$ or $K = \mathbb{C}$. Hence:

Ostrowski's Theorem. *A field that is complete with respect to an* **archimedean** *absolute value* | | *coincides with either* $\mathbb{R}$ *or* $\mathbb{C}$, *and* | | *is equivalent to the ordinary absolute value thereon.*

If the field is not assumed complete, one can take its completion and apply the theorem. Hence *a field endowed with an* **archimedean** *absolute value* | | *can be regarded as a subfield of* $\mathbb{C}$ *in such a way that* | | *is equivalent to the restriction of the ordinary absolute value on* $\mathbb{C}$.

We now discuss a result of fundamental important for the arithmetic of fields that are *complete* with respect to a *nonarchimedean* absolute value. We start with the setup of F4, and introduce some additional notation. Denote by $\bar{R} := R/\mathfrak{p}$ the residue field of K with respect to | |, and by

(35) $$R \to \bar{R}, \quad a \mapsto \bar{a}$$

the corresponding quotient homomorphism. We extend this map in the usual way to a homomorphism

(36) $$R[X] \to \bar{R}[X], \quad f \mapsto \bar{f}$$

of polynomial rings. The kernel of (36) is the ideal $\mathfrak{p}R[X]$ of $R[X]$; but if $\bar{f} = \bar{g}$ we will also write simply

(37) $$f \equiv g \mod \mathfrak{p}.$$

Similarly, if $c \in R$, we express congruence modulo the ideal $cR[X]$ by

(38) $$f \equiv g \bmod c.$$

Thus, in terms of $| \ |$, the meaning of (38) is that all coefficients c_i of $f - g$ satisfy $|c_i| \leq |c|$. Equivalently, (38) says that $|f - g| \leq |c|$, where $| \ |$ is the polynomial norm introduced in Remark 3 of page 54.

Theorem 3 (Hensel's Lemma). *Let K be complete with respect to a nonarchimedean absolute value $| \ |$. If $f \in R[X]$ is a polynomial with coefficients in the valuation ring R of K, and if there is over $\bar{R} = R/\mathfrak{p}$ a factorization $\bar{f} = \varphi\psi$ with $\varphi, \psi \in \bar{R}[X]$ **relatively prime**, there exist polynomials g and h in $R[X]$ such that*

$$f = gh, \quad \bar{g} = \varphi, \quad \bar{h} = \psi, \quad \deg g = \deg \varphi.$$

Proof. For convenience, we set

(39) $$m = \deg f, \quad r = \deg \varphi.$$

Since $\bar{f} = \varphi\psi$, there exist polynomials g_0, h_0 in $R[X]$ such that

(40) $$\bar{g}_0 = \varphi, \quad \bar{h}_0 = \psi, \quad \deg g_0 = r, \quad \deg h_0 \leq m - r,$$

(41) $$f \equiv g_0 h_0 \bmod \mathfrak{p}.$$

Because φ and ψ are relatively prime, we can find polynomials $a, b \in R[X]$ with

(42) $$ah_0 + bg_0 \equiv 1 \bmod \mathfrak{p}.$$

In view of (41) and (42) there must exist $\pi \in \mathfrak{p}$ satisfying

(43) $$f \equiv g_0 h_0 \bmod \pi,$$

(44) $$ah_0 + bg_0 \equiv 1 \bmod \pi.$$

We now want to improve the approximation (43) step by step, so that eventually it will lead to an equality. We use the trial expansions

$$g = g_0 + a_1 \pi + a_2 \pi^2 + \cdots,$$
$$h = h_0 + b_1 \pi + b_2 \pi^2 + \cdots$$

where the polynomials $a_i, b_i \in R[X]$ satisfy

(45) $$\deg a_i < r, \quad \deg b_i \leq m - r.$$

These degree conditions ensure that g and h really are well defined limits in $R[X]$. Moreover we have $g \equiv g_0 \bmod \pi$ and $h \equiv h_0 \bmod \pi$, so that $\bar{g} = \varphi$ and $\bar{h} = \psi$; and the degrees satisfy $\deg g = r$, $\deg h \leq m - r$. Denote by g_n and h_n the n-th partial sums of g and h. Suppose that a_i, b_i have been determined for $1 \leq i \leq n-1$ so the approximation

$$(46) \qquad\qquad f \equiv g_{n-1} h_{n-1} \bmod \pi^n$$

is satisfied; for $n = 1$ this is indeed the case — see (43). We seek a_n, b_n such that

$$(47) \qquad\qquad f \equiv g_n h_n \bmod \pi^{n+1}.$$

If we can do this for all n, we will have, by passing to the limit in (47), the desired equality $f = gh$. Since $g_n = g_{n-1} + a_n \pi^n$ and $h_n = h_{n-1} + b_n \pi^n$, the congruence (47) is equivalent to

$$f \equiv g_{n-1} h_{n-1} + (a_n h_{n-1} + b_n g_{n-1}) \pi^n \bmod \pi^{n+1}.$$

After dividing by π^n this amounts to

$$(48) \qquad\qquad a_n h_{n-1} + b_n g_{n-1} \equiv d_n \bmod \pi,$$

where

$$d_n := (f - g_{n-1} h_{n-1}) / \pi^n$$

is a polynomial in $R[X]$ because of (46). Clearly, $\deg d_n \leq m$. Multiplying (44) by d_n, we do in fact obtain a solution a_n, b_n of the congruence (48), because $g_{n-1} \equiv g_0 \bmod \pi$ and $h_{n-1} \equiv h_0 \bmod \pi$. However, we still have to make sure the degree conditions (45) are satisfied. For this we can obviously replace a_n in (48) by its smallest remainder $\bmod\, g_{n-1}$; hence we can assume $\deg a_n < \deg g_{n-1} = r$. Then (48) says that $b_n g_{n-1}$ is congruent modulo π to a polynomial of degree at most m. By adding to b_n a multiple of π if needed, we can assume that b_n either vanishes or has leading coefficient not divisible by π. Since g_{n-1} has degree r and its leading coefficient is a unit in R, we conclude that $\deg(b_n) + r \leq m$ as needed. $\square$

We draw from Hensel's Lemma a series of important corollaries:

F12. *Let K be complete with respect to a nonarchimedean absolute value $|\ |$, and let*

$$f(X) = a_n X^n + \cdots + a_1 X + a_0$$

be a polynomial of degree n over K. If

$$|a_n| < |f| \quad and \quad |a_i| = |f| \ for\ some\ i > 0,$$

then f is reducible in $K[X]$.

Proof. After multiplication by a scalar, we can assume that $|f| = 1$. Hence $f \in R[X]$. Let r be the largest index i such that $|a_i| = |f| = 1$. By assumption we have $0 < r < n$, and

$$\bar{f}(X) = (\bar{a}_r X^r + \cdots + \bar{a}_0) \cdot \bar{1}$$

with $\bar{a}_r \neq 0$. The assertion follows from Theorem 3. $\square$

The importance of the criterion in F12 is perhaps not obvious at first sight, but it will become clear in the proof of a fundamental *extension theorem* that rests upon F12 (see Theorems 4 and 4' below).

F13. *Let K be complete with respect to a nonarchimedean absolute value $|\ |$, and let $f \in R[X]$ be a polynomial over the valuation ring R of K. If there exists $a \in R$ such that*

$$(49) \qquad\qquad f(a) \equiv 0 \bmod \mathfrak{p} \quad and \quad f'(a) \not\equiv 0 \bmod \mathfrak{p},$$

there is a unique $b \in R$ such that

$$(50) \qquad\qquad f(b) = 0 \quad and \quad b \equiv a \bmod \mathfrak{p}.$$

Proof. What (49) says is that the element $\bar{a}$ of $\bar{R} = R/\mathfrak{p}$ is a *simple* root of the polynomial $\bar{f} \in \bar{R}[X]$. Now apply Theorem 3 with $\varphi(X) := X - \bar{a}$.   □

F14. *Let K and f be as in F13. If there exists $a \in R$ such that*

$$(51) \qquad\qquad |f(a)| < |f'(a)|^2,$$

there is a unique $b \in R$ such that

$$(52) \qquad\qquad f(b) = 0 \quad and \quad |b - a| < |f'(a)|.$$

Proof. This assertion is a useful sharpening of F13, but it can actually be deduced from F13 as follows: Expand f around a in powers of $f'(a)X$, obtaining first

$$(53) \qquad f(a + f'(a)X) = f(a) + f'(a)f'(a)X + f'(a)^2 X^2 h(X)$$

for some $h(X) \in R[X]$. Set $c = f(a)/f'(a)^2$ and consider the polynomial

$$g(X) = c + X + X^2 h(X)$$

obtained by dividing the right-hand side of (53) by $f'(a)^2$. By assumption, $|c| < 1$. Thus g lies in $R[X]$, and we have $g(0) = c \equiv 0 \bmod \mathfrak{p}$ and $g'(0) = 1 \not\equiv 0 \bmod \mathfrak{p}$. By F13, therefore, g has a unique root x such that $|x| < 1$. Hence $b := a + f'(a)x$ is a root of f; it satisfies $|b - a| < |f'(a)|$ and is uniquely determined by this condition.   □

Remark. The content of F14 is reminiscent of *Newton's method* from calculus. Indeed, Newton's method *can* be used in nonarchimedean analysis to obtain a sequence of approximate roots of f converging to a true root. And in contrast to the real case, the process can never go astray, regardless of the choice of an initial value a, so long as it satisfies (51).

F15. *Let $a = p^{w_p(a)} a_1$ be an element of $\mathbb{Q}_p^\times$, and choose $c \in \mathbb{Z}$ representing a_1 modulo p. Then a has a square root in $\mathbb{Q}_p$ if and only if $w_p(a)$ is even and the following additional condition is met:*

• *If $p \neq 2$, the congruence*

(54) $$X^2 \equiv c \mod p$$

has a solution in $\mathbb{Z}$.
• *If $p = 2$,*

(55) $$a_1 \equiv 1 \mod 8.$$

Proof. Apply F13 (or F14 if $p = 2$) to the polynomial $f(X) = X^2 - a_1$. □

Thus for $p \neq 2$ the question of whether a is a square in $\mathbb{Q}_p$ boils down to the existence of a solution in $\mathbb{Z}$ to the congruence (54). This question in turn — of whether c is a quadratic residue modulo p — is computationally easy to solve, thanks to the *quadratic reciprocity law* (vol. I, pp. 111–113).

We now come to the real goal of this section:

Theorem 4. *Let K be complete with respect to an absolute value $|\ |$. If E/K is a finite extension of degree n, we can extend $|\ |$ to a unique absolute value on E, again denoted by $|\ |$. For every α in E, we have*

(56) $$|\alpha| = \left| N_{E/K}(\alpha) \right|^{1/n},$$

and E is $|\ |$-complete.

Proof. (1) The uniqueness and the completeness of the extension follow from F11.

(2) If $|\ |$ is *archimedean*, there is nothing more to show in view of *Ostrowski's Theorem* (page 55).

(3) Let $|\ |$ be *nonarchimedean*. Equation (56) defines a function $|\ | : E \to \mathbb{R}_{\geq 0}$. For $\alpha \in K$ we have $N_{E/K}(\alpha) = \alpha^n$, so this function agrees with the original absolute value on K. We must show that $|\ | : E \to \mathbb{R}$ satisfies properties (i), (ii), (iii) of an absolute value (Definition 1). For (i) and (ii) this is clear from well known properties of $N_{E/K}$. There remains to show that $|\ |$ satisfies the (strong) triangle inequality on E. Obviously it suffices to prove that

(57) $$|\alpha| \leq 1 \quad \Longrightarrow \quad |\alpha + 1| \leq 1$$

for every $\alpha \in E$. Since $N_{E/K}(x) = N_{K(\alpha)/K}(x)^{E:K(\alpha)}$ for every $x \in K(\alpha)$, we can just as well assume that $E = K(\alpha)$. Now let

$$f(X) = a_n X^n + a_{n-1} X^{n-1} + \cdots + a_0, \quad \text{with } a_n = 1,$$

be the *minimal polynomial* of α over K. Since $a_0 = \pm N_{E/K}(\alpha)$, there is equivalence between $|\alpha| \leq 1$ and $|a_0| \leq 1$. Next, since f is *irreducible* and has leading term 1, we can apply Corollary F12 to Hensel's Lemma and conclude that

$$|a_i| \leq 1 \quad \text{for all } i.$$

Hence all coefficients of f lie in the valuation ring R of K. The same is true of the minimal polynomial $f(X-1)$ of $\alpha+1$ over K. In particular, the degree-0 coefficient b_0 of the latter polynomial satisfies $|b_0| \leq 1$. By definition, we have $|\alpha+1| = |b_0|^{1/n}$, so we obtain $|\alpha+1| \leq 1$ as desired. $\square$

Theorem 4'. *Let K be complete with respect to a nonarchimedean absolute value $|\ |$, and let C be an algebraic closure of K. Then $|\ |$ has a unique extension to an absolute value $|\ |$ on C, given by*

$$(58) \qquad |\alpha| = \left| N_{K(\alpha)/K}(\alpha) \right|^{1/K(\alpha):K} \quad \text{for all } \alpha \in C.$$

Let R be the valuation ring of K. For $\alpha \in C$ there is equivalence between:

(i) $|\alpha| \leq 1$.

(ii) *All coefficients of the minimal polynomial of α over K lie in R.*

(iii) α *is integral over R.*

Proof. The first paragraph follows easily from Theorem 4. That (i) implies (ii) has already been seen in the proof of Theorem 4; or, using only the *statement* of Theorem 4, we can reason as follows: Let $\alpha_1 = \alpha, \alpha_2, \ldots, \alpha_n$ be the K-conjugates of α in C; then $|\alpha| = |\alpha_i|$ by (58). Hence $|\alpha| \leq 1$ implies $|\alpha_i| \leq 1$. Since the coefficients of f can be expressed polynomially in terms of $\alpha_1, \ldots, \alpha_n$, they too have absolute value at most 1.

The implication (ii) $\Rightarrow$ (iii) is trivial. There remains to derive (i) from (iii). Suppose therefore that $\alpha^m + a_{m-1}\alpha^{m-1} + \cdots + a_0$ vanishes, where the a_i lie in R. If $|\alpha| > 1$, we have $|a_i \alpha^i| \leq |\alpha|^i < |\alpha|^m$ for $i < m$; but then F3 leads to the contradiction $|\alpha|^m = |\alpha^m + a_{m-1}\alpha^{m-1} + \cdots + a_0| = 0$. Therefore $\alpha \leq 1$. $\square$

Using Theorems 4 and 4' it is now possible to achieve an overview of the possible extensions of an absolute value to a finite field extension, even in the noncomplete case:

Theorem 5. *Let E/K be a finite extension of degree n and let $|\ |$ be an absolute value on K.*

(a) $|\ |$ *can be extended to an absolute value on E.*

(b) *There are at most n extensions of $|\ |$ to distinct absolute values on E.*

(c) *Let $|\ |_1, \ldots, |\ |_r$ be the list of all the distinct extensions of $|\ |$ to E. For each $1 \leq i \leq r$, let $\hat{E}_i$ be a completion of E with respect to $|\ |_i$; further, let $(\hat{K}, |\ |)$ be a fixed completion of K with respect to $|\ |$. Then the canonical homomorphism*

$$(59) \qquad E \otimes_K \hat{K} \to \prod_{i=1}^{r} \hat{E}_i$$

*of $\hat{K}$-algebras is surjective, and in particular $\sum_{i=1}^{r} \hat{E}_i : \hat{K} \leq E : K$. If E/K is **separable**, the map (59) is an isomorphism and hence*

$$(60) \qquad E:K = \sum_{i=1}^{r} \hat{E}_i : \hat{K}.$$

Proof. (1) Let C be an algebraic closure of $\hat{K}$. By Theorems 4 and 4′, | | has a unique extension to an absolute value on C, which we again denote by | |. Let $\sigma_1, \ldots, \sigma_m$ be all the distinct K-homomorphisms from E into C. (Recall that $m \le n$, and $m = n$ if E/K is separable.) Every σ_i defines an absolute value | $|_i$ on E via the rule

(61) $$|x|_i = |\sigma_i x| \quad \text{for all } x \in E.$$

(It can happen that | $|_i = |$ $|_j$ for $i \ne j$.) Clearly each | $|_i$ agrees with | | on K. Now let | |′ be any absolute value on E agreeing with | | on K, and let $(E', |$ $|')$ be a | |′-completion of E. Denote by K' the completion of K with respect to | | *in E'.* The extension EK'/K' is finite, so EK' is complete with respect to | |′, and hence

(62) $$E' = EK'.$$

Since $\hat{K}$ and K' are both | |-completions of K, we have by Theorem 2 a well defined K-isomorphism $\rho : K' \to \hat{K}$ such that $|\rho x| = |x|'$. This can be extended to a K-homomorphism

$$\sigma : E' \to C.$$

Because extensions of absolute values are unique for complete fields, we have

$$|x|' = |\sigma x| \quad \text{for every } x \in E'.$$

This is true, in particular, for every $x \in E$; but since σ restricts on E to one of the maps σ_i, we see that | |′ = | $|_i$. This proves assertions (a) and (b) of the theorem.

Moreover we have seen that the absolute value extensions in question are all of the form (61) (in particular, there is only one if E/K is *purely inseparable*). If we wish, we can reorder $\sigma_1, \sigma_2, \ldots, \sigma_m$ so that (61) accounts for all the distinct extensions of | | to E as i ranges from 1 to r.

(2) Now let $(\hat{E}_1, |$ $|_1)$, ..., $(\hat{E}_r, |$ $|_r)$ be as in statement (c) of the theorem. Since $\hat{E}_i$ is | $|_i$-complete, we know from F6 that there is a unique homomorphism $\hat{K} \to \hat{E}_i$ preserving absolute values. Via these homomorphisms each $\hat{E}_i$ acquires a canonical $\hat{K}$-algebra structure (so that, with due attention, $\hat{E}_i$ can be regarded as an extension of $\hat{K}$). Thus the diagonal map $E \to \prod_{i=1}^r \hat{E}_i$ provides the canonical homomorphism of $\hat{K}$-algebras (59). The image of $\hat{K} \to \hat{E}_i$ is the completion of K with respect to | | *in $\hat{E}_i$.* When E/K is *purely inseparable*, therefore, assertion (c) is proven, in view of (62).

(3) If F is an intermediate field of E/K, we have

$$E \otimes_K \hat{K} = E \otimes_F (F \otimes \hat{K}).$$

Using this fact and the preceding paragraph, one easily reduces assertion (c) to the case of a *separable extension* E/K. Then $E = K(\alpha)$, and the prime factorization

(63) $$f = f_1 f_2 \cdots f_s$$

of $f = \mathrm{MiPo}_K(\alpha)$ *over* $\hat{K}$ has no multiple factors. Choose for each f_i a root α_i of f_i in C. Reorder the homomorphisms $\sigma_1, \ldots, \sigma_m$ so that

$$\sigma_i \alpha = \alpha_i \quad \text{for } 1 \leq i \leq s.$$

For $j > s$ the image $\sigma_j \alpha$ is conjugate over $\hat{K}$ to exactly one α_i with $1 \leq i \leq s$, and so there exists $\tau \in G(C/\hat{K})$ such that $\sigma_j = \tau \sigma_i$. But for every $x \in E$ we then have

$$|\sigma_j x| = |\tau \sigma_i x| = |\sigma_i x| \quad \text{(why?)};$$

that is, $|\ |_j = |\ |_i$. The number r of distinct extensions of $|\ |$ to E is thus at most s. Now suppose that $1 \leq i \leq s$. Since $|\sigma_i x| = |x|_i$ for every $x \in E$, we can extend the K-homomorphism $\sigma_i : E = K(\alpha) \to \hat{K}(\alpha_i)$ in a natural way to a K-homomorphism $\hat{\sigma}_i : \hat{E}_i \to \hat{K}(\alpha_i)$ on the $|\ |_i$-completion $\hat{E}_i$ of E (since $\hat{K}(\alpha_i)$ is *complete*, being a finite extension of $\hat{K}$).

Next we show that $r = s$; in other words, the absolute values $|\ |_1, \ldots, |\ |_s$ are all distinct. If $|\ |_i = |\ |_j$ for $1 \leq i, j \leq s$, consider the K-homomorphism

$$\hat{\sigma}_j \hat{\sigma}_i^{-1} : \hat{K}(\alpha_i) \to \hat{E}_i = \hat{E}_j \to \hat{K}(\alpha_j).$$

This map preserves $|\ |$ and is therefore a $\hat{K}$-homomorphism. Hence α_i and α_j are conjugate over $\hat{K}$, which only leaves the possibility $i = j$.

(4) We are now in the home stretch: the preceding observations and an application of the *Chinese remainder theorem* yield the commutative diagram

(64)

$$
\begin{array}{ccc}
\hat{K}[X]/f & \longrightarrow & \prod_{i=1}^{r} \hat{K}[X]/f_i \\
\downarrow & & \downarrow \\
E \otimes_K \hat{K} & \longrightarrow & \prod_{i=1}^{r} \hat{E}_i
\end{array}
$$

of canonical maps, where the upper horizontal arrow and both vertical ones are known to be isomorphisms.   □

Remark. In the setup of Theorem 5, the bijectivity of the map $E \otimes_K \hat{K} \to \prod_{i=1}^{r} \hat{E}_i$ of (59) — or the validity of equation (60), to which it is equivalent by Theorem 5 — often holds even when E/K is not separable. That it does not always hold can be seen from the counterexample in §24.2. As a matter of fact, the kernel of (59) can be seen to coincide with the *nilradical* of $E \otimes_K K$; see §23.15.

F16. *In the situation of Theorem 5, assume the canonical homomorphism of $\hat{K}$-algebras* (59) *is an* **isomorphism***, as for instance when E/K is separable. Then for every $\alpha \in E$ the characteristic polynomial $P_{E/K}(\alpha; X)$ equals the product of the characteristic polynomials $P_{\hat{E}_i/\hat{K}}(\alpha; X)$. In particular,*

$$(65) \qquad S_{E/K}(\alpha) = \sum_{i=1}^{r} S_{\hat{E}_i/\hat{K}}(\alpha) \quad \text{and} \quad N_{E/K}(\alpha) = \prod_{i=1}^{r} N_{\hat{E}_i/\hat{K}}(\alpha).$$

If we set $n_i := \hat{E}_i : \hat{K}$ for $1 \le i \le r$, then

(66)
$$|N_{E/K}(\alpha)| = \prod_{i=1}^{r} |\alpha|_i^{n_i}.$$

Proof. Since
$$P_{E/K}(\alpha; X) = P_{E \otimes_K \hat{K}/\hat{K}}(\alpha; X),$$

our first assertion follows immediately from the isomorphism

$$E \otimes_K \hat{K} \simeq \prod_{i=1}^{r} \hat{E}_i$$

of $\hat{K}$-algebras (see vol. I, p. 135). Let $\hat{K}_i$ be the canonical image of $\hat{K}$ in $\hat{E}_i$. Then

$$\left|N_{\hat{E}_i/\hat{K}}(\alpha)\right| = \left|N_{\hat{E}_i/\hat{K}_i}(\alpha)\right|_i = |\alpha|_i^{n_i},$$

where we used Theorem 4 for the last equality. Now (66) follows from the second equality in (65). $\qquad\square$

Residue Class Degree and Ramification Index

We consider a field extension E/K and a *nonarchimedean* absolute value $|\ |$ on E. The *valuation rings* of E and K are denoted by A and R, respectively, and the corresponding *valuation ideals* by $\mathfrak{P}$ and $\mathfrak{p}$. Abusing notation (in contrast to the more precise practice in Theorem 3 of Chapter 23, for instance) the residue fields will often be denoted by

$$\overline{K} = R/\mathfrak{p}, \qquad \overline{E} = A/\mathfrak{P}.$$

The natural homomorphism $R/\mathfrak{p} \to A/\mathfrak{P}$ is injective, because $R \cap \mathfrak{P} = \mathfrak{p}$; thus we can and will always regard $\overline{K}$ as a subfield of $\overline{E}$.

Definition 1. In the situation just outlined, the degree

$$f = \overline{E} : \overline{K}$$

of the field extension $\overline{E}/\overline{K}$ is called the *residue class degree* of E/K (with respect to $|\ |$). The index

$$e = |E^\times| : |K^\times|$$

of one valuation group in the other is called the *ramification index* of E/K (with respect to $|\ |$). We also write $e(E/K)$ for e and $f(E/K)$ for f, where of course $|\ |$ is to be regarded as fixed.

Remarks. (a) If F is an intermediate field of E/K, we obviously have

$$e(E/K) = e(E/F) \cdot e(F/K), \quad f(E/K) = f(E/F) \cdot f(F/K).$$

(b) If $\widehat{E}$ is a $|\ |$-completion of E and $\widehat{K}$ is the $|\ |$-completion of K inside $\widehat{E}$, it immediately follows from F7 in the previous chapter that

$$e(\widehat{E}/\widehat{K}) = e(E/K), \quad f(\widehat{E}/\widehat{K}) = f(E/K).$$

Thus the residue class degree and the ramification index are preserved by passing to completions.

(c) Let E/K be finite. Then, with the notation above, $\widehat{E}/\widehat{K}$ is also finite, and $\widehat{E}:\widehat{K} \le E:K$ by Theorem 5 in Chapter 23. We call $\widehat{E}:\widehat{K}$ the *local degree* of E/K with respect to $|\ |$.

F1. *Let the assumptions and notation be as in Definition 1. If E/K is finite of degree n, we have*

$$(1) \qquad\qquad ef \le n.$$

Thus e and f are also finite in this case.

Proof. Let $\alpha_1, \ldots, \alpha_r$ be elements of A whose images $\overline{\alpha}_1, \ldots, \overline{\alpha}_r$ in $\overline{E}$ are linearly independent over $\overline{K}$; also let $\pi_1, \ldots, \pi_s$ be elements of $E^\times$ whose absolute values modulo $|K^\times|$ are pairwise distinct. We will show that the elements $\alpha_i \pi_j$ are all linearly independent over K. From this we get $rs \le n$, and hence also (1).

First we establish that for any $a_1, \ldots, a_r \in K$ we have

$$(2) \qquad\qquad \left| \sum_{i=1}^{r} a_i \alpha_i \right| = \max(|a_1|, \ldots, |a_r|).$$

Leaving aside the trivial case where all the a_i vanish, we can, multiplying by an appropriate factor, assume that $\max(|a_1|, \ldots, |a_r|) = 1$. Now, if (2) does not hold, the left-hand side is < 1, so $\sum \overline{a}_i \overline{\alpha}_i = 0$. But at least one a_i has absolute value 1, so this contradicts the linear independence of the $\overline{\alpha}_i$.

Now suppose

$$(3) \qquad\qquad \sum_{i,j} a_{ij} \alpha_i \pi_j = 0, \quad \text{with } a_{ij} \in K.$$

We rewrite this as

$$(4) \qquad\qquad \sum_{j} c_j \pi_j = 0, \quad \text{with } c_j = \sum_{i} a_{ij} \alpha_i.$$

Because of (2), each nonzero c_j has absolute value in $|K^\times|$. By our choice of the π_j, we have $|c_i \pi_i| \ne |c_j \pi_j|$ for $i \ne j$ and $c_j \ne 0$. From the scholium on the strong triangle inequality (F3 in Chapter 23) we therefore get

$$\left| \sum c_j \pi_j \right| = \max(|c_1 \pi_1|, \ldots, |c_s \pi_s|).$$

But this is compatible with (4) only if all the c_j vanish. By (2) this implies that all the a_{ij} vanish, as we wished to prove. □

Remark. Let E/K have finite degree n. If an absolute value $|\ |$ on K has r distinct extensions $|\ |_1, |\ |_2, \ldots, |\ |_r$ to absolute values on E, we have

$$(5) \qquad\qquad \sum_{i=1}^{r} e_i f_i \le n,$$

where the e_i and f_i are the ramification indexes and residue class degrees of E/K with respect to each $|\ |_i$. This follows from F1 and the inequality $\sum n_i \le n$ (see Theorem 5 in Chapter 23), taking into account remark (b) after Definition 1.

If K is $|\ |$-*complete*, there is only *one* extension of $|\ |$ to E. But equality does not necessarily follow in (1); for a counterexample, see Problem §24.3. There are mild sufficient conditions, involving the notion of *discreteness*, that do guarantee the equality $ef = n$; our next goal is to work our way up to a proof of this fact (see Theorem 1 on page 69).

Definition 2. An absolute value $|\ |$ of K is called *discrete* if $|K^\times|$ is a discrete nontrivial subgroup of $\mathbb{R}_{>0}$.

Remarks. (a) Any discrete absolute value $|\ |$ is nonarchimedean. For if $|\ |$ were archimedean, $\mathbb{Q}$ would be a subfield of K. But an archimedean absolute value on $\mathbb{Q}$ already has a dense subgroup of $\mathbb{R}_{>0}$ as its valuation group; see Theorem 1 in Chapter 23.

(b) Let $|\ |$ be a *discrete absolute value* on K. Being a discrete subgroup of $\mathbb{R}_{>0}$, the group $|K^\times|$ is *cyclic*, and hence of the form $|K^\times| = \{c^n \mid n \in \mathbb{Z}\}$, with $c > 0$ (see §24.1). Since $|K^\times|$ is nontrivial, $c \neq 1$. We can in fact assume that $c < 1$. Now consider the corresponding *valuation* $w : K \to \mathbb{R} \cup \{\infty\}$, defined by

(6) $$|x| = c^{w(x)};$$

then $w(K^\times) = \mathbb{Z}$. Such a valuation (coming from a discrete absolute value) is called *normalized*. Now choose $\pi \in K^\times$ such that $|\pi| = c$, hence $w(\pi) = 1$; we call any such π a *prime element* of K with respect to $|\ |$ (or w). Indeed, for every $x \in K^\times$ we have a representation

(7) $$x = \pi^{w(x)} u, \quad \text{with } w(u) = 0,$$

and this is the only representation of the form $x = \pi^i u$ with $w(u) = 0$, since an application of w to both sides yields $w(x) = i$. We thus see that the valuation ring R is a *unique factorization domain* and that π is the only prime in the ring R, up to multiplication by units of R. In fact one can check that R *is a principal ideal domain*, and every nonzero ideal of R is of the form $\pi^i R = (\pi^i)$, with $i \geq 0$ uniquely determined.

(c) Given an extension E/K, an absolute value on E with $e = e(E/K)$ *finite* is discrete if and only if its restriction to K is. Indeed, if w is a corresponding valuation, we have $ew(E^\times) \subseteq w(K^\times) \subseteq w(E^\times)$; thus $w(K^\times)$ is isomorphic to $\mathbb{Z}$ if and only if $w(E^\times)$ is.

F2. *Suppose K is complete with respect to a discrete absolute value $|\ |$, and let w be the corresponding normalized valuation. Let $S \subset K$ contain one representative of each element of $\overline{K}$, and suppose $0 \in S$. For every $i \geq 0$ in $\mathbb{Z}$, choose π_i such that $w(\pi_i) = i$. (If π is a prime for w, we can set $\pi_i = \pi^i$, for instance.) Then every $a \in K$ has a unique representation*

(8) $$a = \sum_{-\infty \ll i} a_i \pi_i \quad \text{with } a_i \in S,$$

and the smallest i in (8) such that $a_i \neq 0$ is $i = w(a)$.

Proof. The existence and uniqueness of the representation (8) for an element a in the valuation ring R of K are proved in a way wholly analogous to the case $K = \mathbb{Q}_p$ and $\pi_i = p^i$, treated in F9 of Chapter 23. Now suppose $a = \pi^m a'$ is any element of K such that $m = w(a)$. Then $\pi_i' = \pi_{i+m}\pi^{-m}$ likewise satisfies $w(\pi_i') = i$. Hence there is a unique representation

$$a' = \sum_{i=0}^{\infty} a_i' \pi_i' \quad \text{with } a_i' \in S.$$

Because

$$a = \pi^m a' = \sum_{i=0}^{\infty} a_i' \pi_{i+m}' = \sum_{i=m}^{\infty} a_{i-m}' \pi_i,$$

we then obtain the desired representation of a. What's left to show can again be handled in analogy with Chapter 23, F9. $\square$

In the case of $\pi_i = \pi^i$ for every i, we thus see that any $a \in K$ has a unique representation as a *Laurent series* in π with coefficients in S and finite principal part. Addition and multiplication of elements of K obviously corresponds to the same operations on Laurent series; but since we cannot always arrange for S to be additively and multiplicatively closed, a sum or product Laurent series need not have coefficients in S, and hence need not itself be a π-adic development of the form (8). However we *can* arrange for it to be such in an important special case:

Example. Let $F(X)$ be the field of rational functions in a variable X over a field F. For a nonzero *polynomial* $f \in F[X]$, let $w(f)$ be the degree of f in X, and let $w(0) = \infty$. The map w thus defined can be extended in an obvious way to a (normalized) *valuation* w of $F(X)$. It is easy to check that the subfield F of $F(X)$ is isomorphic, via the residue class homomorphism, to the residue field of $F(X)$ with respect to w. Now let $F((X))$ be the w-*completion* of $F(X)$; then, by F2, every $f \in F((X))$ has a unique representation

$$(9) \qquad\qquad f = \sum_{-\infty \ll i} a_i X^i, \quad \text{with } a_i \in F.$$

The valuation ring $F[[X]]$ of $F((X))$ with respect to w is the *ring of formal power series in X over F*, consisting of those f such that $a_i = 0$ for $i < 0$. For this reason $F((X))$ is called the *field of formal Laurent series* (*with finite principal part*) *in X over F*. In this case, therefore, the subfield F provides a multiplicatively and additively closed set of representatives.

By contrast, the field $\mathbb{Q}_p$ of p-adic numbers has no addition-preserving set of representatives, since $\mathbb{Q}_p$ and $\overline{\mathbb{Q}}_p$ have distinct characteristics. (But by Theorem 4 below, $\mathbb{Q}_p$ does have a multiplication-preserving set of representatives.)

We now turn to the proof, already alluded to, of the formula $n = ef$ for *complete, discrete* valuations. First:

Lemma. *Let the setup be as in Definition* 1, *and let* π *be an element of* K *such that* $|\pi| < 1$. *Assume* K *is* $|\ |$-*complete. If* $\beta_1, \ldots, \beta_r$ *are representives in* A *for the elements of a generating set of* $A/\pi A$ *over* $R/\pi R$, *then* $\beta_1, \ldots, \beta_r$ *generate* A *over* R.

Proof. Set $M = R\beta_1 + \cdots + R\beta_r$. By assumption,

$$(10) \qquad\qquad A = M + \pi A.$$

Take $x \in A$. From (10) we obtain by recursion sequences $(x_i^{(n)})_n$ of elements of R such that

$$(11) \qquad x \equiv x_1^{(n)}\beta_1 + \cdots + x_r^{(n)}\beta_r \mod \pi^n A,$$

$$(12) \qquad\qquad x_i^{(n+1)} \equiv x_i^{(n)} \mod \pi^n R.$$

Because of (12), each sequence $(x_i^{(n)})_n$ converges to an element x_i in R (see F8 in Chapter 23), while (11) gives, upon passing to the limit, $x = x_1\beta_1 + \cdots + x_r\beta_r$. $\square$

Theorem 1. *Let* E/K *be a field extension and* $|\ |$ *a **discrete** absolute value on* E. *Assume that* E/K *has finite ramification index* e *and finite residue class degree* f *with respect to* $|\ |$. *If* K *is* $|\ |$-***complete**, *the extension* E/K *is finite, and its degree* n *is given by*

$$(13) \qquad\qquad n = ef.$$

Proof. Let π and Π be $|\ |$-prime elements of K and E, respectively. The ideal πA of A has the form $\pi A = \Pi^m A$, where $m \in \mathbb{N}$; see Remark (b) after Definition 2. Because $|E^\times| : |K^\times| = \langle|\Pi|\rangle : \langle|\Pi^m|\rangle = m$, we then get $m = e$, and hence

$$(14) \qquad\qquad \pi A = \Pi^e A.$$

Now let $\alpha_1, \ldots, \alpha_f$ be representatives of a basis of $\overline{E} = A/\Pi A$ over $\overline{K} = R/\pi R$. Consider the elements

$$(15) \qquad \alpha_i \Pi^j, \quad \text{for } 1 \leq i \leq f \text{ and } 0 \leq j < e;$$

we claim they generate A over R (and hence also E over K). But this follows directly from the preceding lemma, because if we set $M := R\alpha_1 + \cdots + R\alpha_f$ we have $A = M + \Pi A = M + \Pi(M + \Pi A) = \cdots = M + \Pi M + \cdots + \Pi^{e-1}M + \Pi^e A$, so the elements (15) do generate A modulo $\Pi^e A = \pi A$.

So far we have shown that the extension E/K is finite and its degree n satisfies $n \leq ef$. But F1 says that $ef \leq n$, so equality (13) is proved. $\square$

Furthermore, we see that the elements in (15) form a *basis* of E/K.

Theorem 2. *Let* E/K *be a **separable**, **finite** field extension. Let* $|\ |_1, \ldots, |\ |_r$ *be all the distinct extensions of a **discrete** absolute value* $|\ |$ *on* K *to absolute values* on E. If we denote by e_i and f_i the residue class degree and ramification index of E/K with respect to $|\ |_i$, for $i = 1, \ldots, r$, the degree n of E/K satisfies

$$(16) \qquad\qquad n = \sum_{i=1}^{r} e_i f_i.$$

Proof. By Equation (60) in Chapter 23 we have $n = \sum_{i=1}^{r} n_i$, and Theorem 1 gives $n_i = e_i f_i$. Taking into account Remark (b) after Definition 1, we obtain the desired equality (see also the remark preceding Definition 2). $\square$

We now introduce a notion of great significance:

Definition 3. Let the situation be as in Definition 1. Assume E/K is finite. Denote by $\hat{E}$ the $|\ |$-completion of E and by $\hat{K}$ the $|\ |$-completion of K inside $\hat{E}$. We say that E/K is *unramified* (with respect to $|\ |$) if

$$\overline{E} : \overline{K} = \hat{E} : \hat{K}$$

and $\overline{E}/\overline{K}$ is *separable*.

Remarks. Keep the assumptions and notation from the definition. Since $ef \leq \hat{E} : \hat{K}$, the condition that E/K is unramified with respect to $|\ |$ implies that $e = 1$. By Theorem 1, then, *if $|\ |$ is discrete, E/K is unramified (with respect to $|\ |$) if and only if $e = 1$ and $\overline{E}/\overline{K}$ is separable.* If L is an intermediate field of E/K, it is clear that *E/K is unramified with respect to $|\ |$ if and only if E/L and L/K are.*

In the case of a *complete*, unramified extension E/K, the degree of the extension coincides with the residue class degree f, that is, the degree of the extension $\overline{E}/\overline{K}$ of residue fields. The next theorem shows that in this case the extension E/K is effectively determined by the extension $\overline{E}/\overline{K}$.

Theorem 3. *Let K be $|\ |$-complete, where $|\ |$ is nonarchimedean. On the algebraic closure C of K, consider the unique extension of $|\ |$ (also denoted by $|\ |$).*

(i) *The residue field $\overline{C}$ of C is an algebraic closure of $\overline{K}$.*

(ii) *For every finite extension $F/\overline{K}$ in $\overline{C}/\overline{K}$ there is an extension L/K in C/K such that $\overline{L} = F$ and $L : K = \overline{L} : \overline{K}$.*

(iii) *The map taking each intermediate field of C/K to its residue field affords a bijection between the set of all finite **unramified** extensions L/K in C/K and that of all finite separable extensions $F/\overline{K}$ in $\overline{C}/\overline{K}$. This bijection preserves inclusion (in both directions). Every unramified L/K is separable.*

(iv) *Every finite extension E/K in C/K has a largest unramified subextension L/K. If $\overline{E}/\overline{K}$ is separable, this maximal extension satisfies $\overline{E} = \overline{L}$ and $L : K = \overline{E} : \overline{K}$.*

(v) *If L/K is unramified, L/K is Galois if and only if $\overline{L}/\overline{K}$ is, and then there is a natural isomorphism $G(L/K) \to G(\overline{L}/\overline{K})$ of Galois groups.*

Proof. (a) Let $\bar{\alpha}$, with $\alpha \in C$ and $|\alpha| \leq 1$, be an arbitrary element of $\bar{C}$. The minimal polynomial f of α over K has all its coefficients in the valuation ring R of K (Chapter 23, Theorem 4′). Hence we can consider $\bar{f} \in \bar{K}[X]$; we have $\bar{f}(\bar{\alpha}) = 0$, and $\bar{f}$ is normalized. Thus $\bar{\alpha}$ is algebraic over $\bar{K}$.

(b) An arbitrary normalized polynomial in $\bar{K}[X]$ is of the form $\bar{f}$, where $f \in R[X]$ is normalized. Since C is algebraically closed, we have $f(X) = \prod_i (X - \alpha_i)$, with the α_i in C. By Theorem 4′ of Chapter 23, all the α_i lie in the valuation ring of C. It follows that $\bar{f}(X) = \prod_i (X - \bar{\alpha}_i)$; that is, $\bar{f}$ is a product of linear factors over $\bar{C}$, which shows that $\bar{C}$ is algebraically closed.

(c) Take $\bar{\alpha} \in \bar{C}$ and let $\bar{f}$ be the minimal polynomial of $\bar{\alpha}$ over $\bar{K}$; we can assume f is normalized. Since $f \in R[X]$ splits into linear factors over C, an element $\alpha \in C$ representing $\bar{\alpha}$ can be taken such that $f(\alpha) = 0$. Set $L = K(\alpha)$. Then $R[\alpha]$ is contained in the valuation ring of L, so $\bar{K}[\bar{\alpha}] \subseteq \bar{L}$. By F1 we then have

$$\bar{L} : \bar{K} \leq L : K \leq \deg f = \deg \bar{f} = \bar{K}(\bar{\alpha}) : \bar{K} \leq \bar{L} : \bar{K},$$

showing that equality holds throughout the chain; on particular, $\bar{L} = \bar{K}(\bar{\alpha})$ and $L : K = \bar{L} : \bar{K}$. Moreover, $K(\alpha) : K$ equals $\deg f$, so f is *irreducible* over K. This proves (ii) in the case of a *simple* extension $F = \bar{K}(\bar{\alpha})$. The general case follows easily by induction.

(d) Now let $F/\bar{K}$ be a finite separable extension in $\bar{C}/\bar{K}$. By the primitive element theorem, $F = \bar{K}(\bar{\alpha})$. Hence (c) gives us an *unramified* L/K such that $\bar{L} = F$. Moreover, if we write $f(X) = \prod(X - \alpha_i)$ with $\alpha_i \in C$ and $\alpha_1 = \alpha$, it follows that $\bar{f}(X) = \prod(X - \bar{\alpha}_i)$. Hence f is separable if $\bar{f}$ is.

(e) Let E/K be any finite extension in C/K, and let F be an intermediate field of the maximal separable extension $(\bar{E})_s/\bar{K}$ in $\bar{E}/\bar{K}$. The preceding construction provides an unramified extension $L = K(\alpha)$ of K with $\bar{L} = F$. We claim that $L \subseteq E$. This will complete the proof of (iii): indeed, if E/K is itself *unramified* and $F = \bar{E}$, the inclusion $L \subseteq E$ implies $E = L$ (compare degrees). It is now clear that the correspondence preserves inclusions. Finally, this also proves (iv); to see this, consider the unramified extension L/K, where $\bar{L} = (\bar{E})_s$.

To prove our claim that $L \subseteq E$, we must show that $\alpha \in E$. Consider f as a polynomial over E. The polynomial $\bar{f}$ has $\bar{\alpha}$ as a *simple* root in $F \subseteq \bar{E}$. Since E is complete, it follows from F13 in Chapter 23 that f has a root β in E with $\bar{\beta} = \bar{\alpha}$. But the polynomial f was so arranged that $\bar{\alpha}_i \neq \bar{\alpha}_j$ if α_i, α_j are distinct roots of f in C: see part (d). It follows that $\beta = \alpha$; that is, $\alpha \in E$.

(f) Now to part (v). By assumption, $\bar{L}/\bar{K}$ is separable, so $\bar{L}$ is of the form $\bar{L} = \bar{K}(\bar{\alpha})$. If f and α are as above, we have $L = K(\alpha)$ and $f = \mathrm{MiPo}_K(\alpha)$. In addition, L/K is separable. Now suppose L/K is normal; this amounts to saying that f splits into linear factors over L. Then $\bar{f}$ also splits fully over $\bar{L}$, and $\bar{L}/\bar{K}$ is likewise normal.

Conversely, suppose $\bar{L}/\bar{K}$ is normal. We know that f can in any case be written as $f = \prod(X - \alpha_i)$ over C. Set $L_i = K(\alpha_i)$. By (c) we have $\bar{L}_i = \bar{K}(\bar{\alpha}_i)$. In view of the assumption, we have $\bar{L} = \bar{K}(\bar{\alpha}) = \bar{K}(\bar{\alpha}_i)$, hence $\bar{L} = \bar{L}_i$. It follows that $L_i = L$, by (iii), showing that L/K is normal.

(g) Take $\sigma \in G(L/K)$. Then $|\sigma x| = |x|$ for every $x \in L$ (see again Theorem 4' in Chapter 23). It follows that σ gives rise to an automorphism $\bar{\sigma}$ of the residue field $\bar{L}$; for any x in the valuation ring of L we then have

$$\bar{\sigma}(\bar{x}) = \overline{\sigma(x)}.$$

Clearly, $\bar{\sigma} \in G(\bar{L}/\bar{K})$. The map $\sigma \mapsto \bar{\sigma}$ is then a natural homomorphism $G(L/K) \to G(\bar{L}/\bar{K})$. To prove that it is an isomorphism, it is enough, in view of the equality $L:K = \bar{L}:\bar{K}$, to show that $\bar{\sigma} = 1$ only when $\sigma = 1$. We can take α and f as above, with $L = K(\alpha)$. For $\sigma \in G(L/K)$, the image $\sigma\alpha$ equals one of the roots $\alpha_1 = \alpha, \alpha_2, \ldots, \alpha_n$ of f. Now let $\sigma\alpha = \alpha_i$. If $\bar{\sigma} = 1$, we have $\bar{\alpha} = \bar{\sigma}\bar{\alpha} = \overline{\sigma\alpha} = \bar{\alpha}_i$. It follows that $i = 1$, so $\sigma\alpha = \alpha$ and hence $\sigma = 1$. □

Theorem 4. *Suppose that K is complete with respect to a nonarchimedean absolute value $|\ \ |$ and that the corresponding residue field $\bar{K}$ is a **finite field** with q elements. Let the notation be as in Theorem 3.*

(i) *For every $n \in \mathbb{N}$ there exists **exactly one** unramified extension of degree n in C/K, namely the extension $K(\zeta)/K$ obtained by adjoining to K a primitive (q^n-1)-st root of unity ζ.*

(ii) *Suppose $m \in \mathbb{N}$ and $p = \text{char } \bar{K}$ are relatively prime, and let ζ_m be a primitive m-th root of unity in C. The extension $K(\zeta_m)/K$ is unramified, and its degree is the order n of q modulo m. Thus the group $W(K)_{p'}$ of all roots of unity of K whose order is relatively prime to p has order $q - 1$; this group is a full set of representatives for the multiplicative group of $\bar{K}$.*

(iii) *Every finite unramified extension L/K is Galois, and the Galois group $G(L/K)$ possesses a unique element φ that gives rise to the map $x \mapsto x^q$ on $\bar{L}$. This element, also written $\varphi_{L/K}$, is called the **Frobenius automorphism** (or just the **Frobenius**) of L/K. It has order $L:K$. Thus $G(L/K)$ is **cyclic** and $\varphi_{L/K}$ is a canonical generator for it.*

Proof. (a) $\bar{K}$ has a unique extension F of degree n in $\bar{C}$, namely the field $F = \mathbb{F}_{q^n}$ with q^n elements (Chapter 9, Theorem 1 in vol. I). Since every extension of a finite field is separable, Theorem 3(iii) yields the existence of a unique unramified extension L/K of degree n in C/K. Consider over L the polynomial $g(X) = X^m - 1$, with $m = q^n - 1$. The multiplicative group of $\bar{L} = \mathbb{F}_{q^n}$ is cyclic (Chapter 9, Theorem 2); let ω be a generator for it. Over $\bar{L}$, the polynomial $\bar{g}(X)$ is the product of distinct linear factors $X - \omega^i$, for $0 \leq i < q^n-1$. By F13 in Chapter 23, therefore, we can associate with ω a root $\zeta \in L$ of $g(X)$ such that $\bar{\zeta} = \omega$. Then ζ is a *primitive* (q^n-1)-st root of unity, because $\zeta^i = 1$ implies $\omega^i = \bar{\zeta}^i = 1$. Thus the powers ζ^i, for $0 \leq i < q^n-1$, make up a full set of representatives for $\bar{L}^\times$ in L.

(b) For a given $m \in \mathbb{N}$ relatively prime to q, let n be the smallest natural number such that $q^n \equiv 1 \mod m$. Because ζ_m is at all events a (q^n-1)-st root of unity, $K(\zeta_m)$ is contained in $K(\zeta)$, where ζ is a primitive (q^n-1)-st root of unity as in (a). Since ζ_m is a power of ζ, part (a) implies that $\text{ord}(\bar{\zeta}_m) = m$; by the choice of n, therefore, $\mathbb{F}_{q^n}$ is the smallest extension of $\bar{K} = \mathbb{F}_q$ containing $\bar{\zeta}_m$. Hence $\bar{K}(\bar{\zeta}_m) = \mathbb{F}_{q^n} = \bar{K}(\zeta)$. Since $K(\zeta_m) \subseteq K(\zeta)$ it follows that $K(\zeta_m) = K(\zeta)$. This proves (ii) as well.

(c) We know from Theorem 4 in Chapter 9 that any extension of finite fields is Galois, with cyclic Galois group generated by the automorphism $x \mapsto x^q$, where q is the number of elements of the ground field. Thus (iii) follows from part (v) of Theorem 3.   □

Incidentally, if $n = L:K$ and ζ is as in (i), we have

(17) $$\varphi_{L/K}(\zeta) = \zeta^q;$$

indeed, $\overline{\varphi(\zeta)}$ is a (q^n-1)-st root of unity (because ζ is one), and we see using part (a) that $\overline{\varphi(\zeta)} = \overline{\zeta^q}$ implies (17).

As mentioned earlier, the notion of *unramified extensions* is of great importance, and Theorem 4 plays a key role in arithmetic. To conclude this chapter we will briefly look at the opposite extreme, *purely ramified extensions*:

F3. *Let $|\ \ |$ be a discrete absolute value on K, with normalized valuation w.*

(a) *Let $f(X) = X^n + a_{n-1}X^{n-1} + \cdots + a_0 \in K[X]$ be an Eisenstein polynomial (meaning that $w(a_i) \geq 1 = w(a_0)$), and let Π be a root of f. Then $|\ \ |$ has exactly one extension to an absolute value on $E = K(\Pi)$; the extension E/K is **purely ramified**, meaning that $e(E/K) = E:K$; the polynomial f is irreducible; and Π is a prime element of E.*

(b) *Conversely, let E/K be purely ramified with respect to an extension to E of the absolute value $|\ \ |$ on K, and let Π be a prime element of E. Then $E = K(\Pi)$, and the minimal polynomial f of Π over K is an Eisenstein polynomial.*

Proof. (a) From $\Pi^n + a_{n-1}\Pi^{n-1} + \cdots + a_0 = 0$ and the Eisenstein property, we see that $|\Pi| < 1$ (where we denote an extension of $|\ \ |$ to E again by $|\ \ |$; see Theorem 5 in Chapter 23). Because $|a_0| > |a_i||\Pi^i|$ for $i > 0$, we then get

$$|a_0| = |\Pi^n| = |\Pi|^n.$$

Consequently the ramification degree $e(E/K)$ is at least n, because a_0 is a prime element of K (recall that $w(a_0) = 1$). Therefore $E:K \geq e(E/K) \geq n \geq E:K$, proving equality; and this concludes the proof of (a), taking (5) into consideration. The irreducibility of f is also assured, of course, by the *Eisenstein criterion* (vol. I, Chapter 5, F10).

(b) From the assumptions we get $K(\Pi):K \geq e(K(\Pi)/K) \geq e(E/K) = E:K$, which shows first of all that $E = K(\Pi)$. Let $\widehat{E}/\widehat{K}$ be the corresponding extension of completions. Since $\widehat{E}:\widehat{K} \geq e(\widehat{E}/\widehat{K}) = e(E/K) = E:K \geq \widehat{E}:\widehat{K}$ and $\widehat{E} = \widehat{K}(\Pi)$, the polynomial f is irreducible over $\widehat{K}$ as well. Thus, over an algebraic closure C of $\widehat{E}$, we have

$$f(X) = \prod_i (X - \Pi_i), \quad \text{with } |\Pi_i| = |\Pi|,$$

and so the coefficients a_i of f (apart from the leading coefficient) all have absolute value less than 1; moreover $|a_0| = |\Pi|^n$, so a_0 is a prime element of K.   □

Example. Let p^k be a prime power (with $k > 0$) and let ζ be a primitive p^k-th root of unity. The extension $\mathbb{Q}_p(\zeta)/\mathbb{Q}_p$ is purely ramified of degree $n = (p-1)p^{k-1}$, and $1 - \zeta$ is a prime element of $\mathbb{Q}_p(\zeta)$.

Proof. Consider the p^k-th cyclotomic polynomial

$$g(X) = \frac{X^{p^k} - 1}{X^{p^{k-1}} - 1} = 1 + X^{p^{k-1}} + \cdots + X^{(p-1)p^{k-1}},$$

whose roots are precisely the primitive p^k-th roots of unity:

$$g(X) = \prod_i{}'(X - \zeta^i),$$

where the product runs over all integers i relatively prime to p with $1 \le i \le p^k$. Now $\zeta - 1$ is a root of the polynomial $f(X) := g(X + 1)$; but it is easy to see that f is an Eisenstein polynomial, which proves the claim by F3.

Here is an alternative way to cap the argument without using F3: We have $g(1) = p = \prod_i'(1 - \zeta^i)$; but since for each i in question the expression $1 - \zeta^i$ differs from $1 - \zeta$ only by an invertible factor, we see that $(1 - \zeta)^n$ is conjugate to p, so the ramification index is at least n.   □

In the case $k = 1$, when $\zeta = \zeta_p$ is a primitive p-th root of unity, we can say more. Since $\mathbb{Q}_p$ contains the $(p-1)$-st roots of unity, the cyclic extension $\mathbb{Q}_p(\zeta_p)/\mathbb{Q}_p$ must be generated by a single $(p-1)$-st root of an element of $\mathbb{Q}_p$. Indeed, we claim that

$$(18) \qquad \mathbb{Q}_p(\zeta_p) = \mathbb{Q}_p\left(\sqrt[p-1]{-p}\right).$$

To prove this, we start as above from

$$(19) \qquad p = \prod_{i=1}^{p-1}(1 - \zeta^i) = (1 - \zeta)^{p-1}\varepsilon$$

and notice that

$$(20) \qquad \varepsilon = \prod_{i=1}^{p-1}(1 + \zeta + \cdots + \zeta^{i-1}) \equiv \prod_{i=1}^{p-1} i = (p - 1)! \bmod \mathfrak{p},$$

with $\mathfrak{p} = (1 - \zeta_p)$. Since $(p - 1)! \equiv -1 \bmod p$, there follows

$$(21) \qquad -\varepsilon \equiv 1 \bmod \mathfrak{p}.$$

But then, by Hensel's Lemma (see F13 in Chapter 23, or else §23.13), $-\varepsilon$ is a $(p-1)$-st power in $\mathbb{Q}_p(\zeta_p)$. The same is true about $-p$, because of (19). Now the equality (18) is clear, since both fields there have degree $p - 1$ over $\mathbb{Q}_p$ (see F3).

25

Local Fields

1. We start with a short and pithy definition, one deserving of later elaboration:

Definition 1. By a *local field* we understand a field K that is *locally compact* with respect to some nontrivial absolute value $|\ |$ on K. In this situation we also say that K is $|\ |$-local.

Let K be a field with a nontrivial absolute value $|\ |$, and let K be locally compact with respect to $|\ |$. The condition amounts to saying that, for some — equivalently, for any — constant $c > 0$, the subset

$$(1) \qquad \{x \in K \mid |x| \le c\}$$

is *compact*. Then K is also *complete*. Indeed, a Cauchy sequence $(x_n)_n$ in K is necessarily bounded, and thus, by the compactness of (1), it has a convergent subsequence. But a Cauchy sequence having a convergent subsequence is itself convergent.

We now have two cases. If $|\ |$ is *archimedean*, then $K = \mathbb{R}$ or $K = \mathbb{C}$, and $|\ |$ is equivalent to the ordinary absolute value (see Remark after F11 in Chapter 23).

Otherwise, $|\ |$ is *nonarchimedean*. The valuation ring $R = \{x \in K \mid |x| \le 1\}$ is then the disjoint union of the cosets (residue classes) modulo the valuation ideal $\mathfrak{p} = \{x \in K \mid |x| < 1\}$. Since $\mathfrak{p}$ is open, so is each coset $a + \mathfrak{p}$. Since R is compact as well, $R/\mathfrak{p}$ must be finite. Hence K has a *finite residue field* $\overline{K}$.

Moreover, a nonarchimedean $|\ |$ is *discrete*. For suppose the valuation group $|K^\times|$ has an accumulation point $c > 0$ in $\mathbb{R}$. There is a sequence $(x_n)_n$ in K, with all the $|x_n|$ distinct, such that $(|x_n|)_n$ converges toward c in $\mathbb{R}$. Being bounded, the sequence $(x_n)_n$ possesses a subsequence that converges to some $a \in K$. We have $|a| = c$, but also $|x_n| = |a|$ for infinitely many n, because $x_n = a + (x_n - a)$ and $|\ |$ is nonarchimedean. Contradiction!

F1. (a) *Let K be a local field with respect to a nonarchimedean absolute value $|\ |$.*
  (i) *K is $|\ |$-complete.* (ii) *$|\ |$ is discrete.*
  (iii) *The residue field $\overline{K}$ of K with respect to $|\ |$ is finite.*

(b) *Conversely: If | | is an absolute value on a field K with properties* (i)–(iii) *of part* (a), *then K is | |-local.*

Proof. Only part (b) remains to be proved. Let π be a *prime element* of K with respect to | |, and let q be the cardinality of $\overline{K}$. We first show that, for every $n \in \mathbb{N}$, the quotient ring $R/\pi^n R$ is finite, and in fact

$$(2) \qquad\qquad \mathrm{Card}(R/\pi^n R) = q^n.$$

To this end consider the chain $R \supseteq \pi R \supseteq \pi^2 R \supseteq \cdots \supseteq \pi^n R$ of additive subgroups, whose factors $\pi^i R/\pi^{i+1} R$ are each isomorphic to $R/\pi = \overline{K}$.

Proving that K is *locally compact* amounts to showing that $R = \{x \mid |x| \le 1\}$ is *compact*. Thus, let $\mathfrak{U}$ be an open cover of R, and assume, contrary to the claim, that R cannot be covered by finitely many sets $U \in \mathfrak{U}$. Since R/π is finite, there exists a coset $a_1 + \pi R$ that cannot be covered by a finite subset of $\mathfrak{U}$. Because R/π^2 is also finite, there is likewise a coset $a_2 + \pi^2 R$, with $a_2 \equiv a_1 \bmod \pi R$, that also does not admit a finite subcover. By recursion, we obtain a sequence $(a_n)_n$ in R with

$$(3) \qquad\qquad a_{n+1} \equiv a_n \bmod \pi^n R,$$

where each $a_n + \pi^n R$ cannot be covered by a finite subset of $\mathfrak{U}$. In view of (3), the sequence $(a_n)_n$ converges toward some $a \in R$, satisfying $a \equiv a_n \bmod \pi^n R$ for all n. But a lies in some element $U \in \mathfrak{U}$; since U is open, there exists n such that $a + \pi^n R \subseteq U$. Hence $a_n + \pi^n R = a + \pi^n R$ in fact admits the one-element cover $\{U\}$, contrary to the construction. $\qquad\qquad\qquad\qquad\qquad\qquad\qquad\qquad\square$

Theorem 1. *Let K be a field with an absolute value | | having properties* (i), (ii) *and* (iii) *from* F1.

(I) *If K has characteristic zero, it is a finite extension of $\mathbb{Q}_p$, where $p = \mathrm{char}\,\overline{K}$, and | | is equivalent to the unique extension of | $|_p$ to an absolute value on K.*

(II) *If K has nonzero characteristic, it equals the field $\mathbb{F}_q((X))$ of formal Laurent series with finite principal part in one variable X over some finite field $\mathbb{F}_q = \overline{K}$, and | | belongs to the canonical valuation of $\mathbb{F}_q((X))$.*

Proof. (I) When char $K = 0$ we can regard $\mathbb{Q}$ as a subfield of K. Set $p = \mathrm{char}\,\overline{K}$. Then $|p| < 1$; by substituting an equivalent absolute value we can assume that $|p| = 1/p$. But now the restriction of | | to $\mathbb{Q}$ can only be the p-adic absolute value on $\mathbb{Q}$: see Theorem 1 and F1 in Chapter 23. Since K is | |-*complete* — see condition (i) — one can view $\mathbb{Q}_p$ as a subfield of K. By (iii), the residue class degree f of $K/\mathbb{Q}_p$ is necessarily finite; the ramification index e of $K/\mathbb{Q}_p$ is also finite, because the valuation group $|K^\times|$ is cyclic, by (ii), and hence any nontrivial subgroup of it has finite index. From Theorem 1 in Chapter 24 we conclude that $K/\mathbb{Q}_p$ is finite.

(II) Now suppose char $K = p > 0$. Then char $\overline{K} = \mathrm{char}\,K = p$. The prime field of K is $\mathbb{F}_p$. Let q be the cardinality of $\overline{K}$. By Theorem 4(ii) in Chapter 24, K contains a primitive $(q-1)$-st root of unity, and its image $\overline{\zeta}$ in $\overline{K}$ generates the multiplicative group of $\bar{K}$. Take the subfield $\mathbb{F}_p(\zeta)$ of K; the residue class map restricts to a field homomorphism $\mathbb{F}_p(\zeta) \to \bar{K}$. By our choice of ζ, this homomorphism is surjective, hence an isomorphism. Thus we can identify $\bar{K}$ with the subfield $\mathbb{F}_p(\zeta)$ of K; that is, $\mathbb{F}_p(\zeta) = \bar{K} = \mathbb{F}_q$ is the field with q elements.

Let X be a prime element of K with respect to $|\ |$. By F2 in Chapter 24, every $a \in K$ admits a unique representation of the form

$$a = \sum_{-\infty \ll i} a_i X^i, \quad \text{with } a_i \in \mathbb{F}_q,$$

and conversely, any such infinite series determines an element of K. Since $\mathbb{F}_q$ is a subfield of K, we have $K = \mathbb{F}_q((X))$: see Chapter 24, Example following F2. $\square$

To summarize:

F2. *The local fields are*

(1) *the fields* $\mathbb{R}$ *and* $\mathbb{C}$,

(2) *the finite extensions of the fields* $\mathbb{Q}_p$, *and*

(3) *the rational function fields* $\mathbb{F}_q((X))$ *over finite fields* $\mathbb{F}_q$.

We will now choose a privileged absolute value $|\ |_K$ on each local field K. In case (1) we take $|\ |_K = |\ |_\infty$. In cases (2) and (3), the field K has a canonical *normalized valuation* w_K; we define $|\ |_K$ by setting

(4) $$|\alpha|_K = \left(\frac{1}{q}\right)^{w_K(\alpha)}, \quad \text{with } q = \text{Card } \bar{K}.$$

In case (2), $|\ |_K$ does *not* coincide with the unique extension $|\ |_p$ to K of the p-adic absolute value on $\mathbb{Q}_p$, unless $K : \mathbb{Q}_p = 1$. More precisely,

(5) $$|\alpha|_K = |\alpha|_p^n = \left| N_{K/\mathbb{Q}_p}(\alpha) \right|_p,$$

where $n := K : \mathbb{Q}_p$. To see this, take $e = e(K/\mathbb{Q}_p)$: we have

(6) $$w_K = e w_p,$$

which, together with the equalities $q = p^f$ and $ef = n$ (Theorem 4 in Chapter 23), immediately yields (5).

Remark. We mention here, without proof, that the definition of a local field can be cast in even more fundamental terms. Let K be a *topological field* (that is, a field endowed with a nondiscrete topology with respect to which addition, multiplication and inversion are continuous), and suppose K is *locally compact* in that topology. Then K admits a canonical absolute value $|\ |_K$ that determines the given topology.

To elaborate: Being a locally compact topological group, the additive group of K has a translation-invariant measure μ_K, which is uniquely determined up to a constant factor. Hence, any $a \in K$ can be assigned a well defined expansion factor $\|a\|_K$ describing the effect of multiplication by a:

(7) $\mu(aM) = \|a\|_K \, \mu(M)$ for all μ-measurable sets M.

Then $\| \ \|_K$ is an absolute value on K, except when $K = \mathbb{C}$, in which case $\| \ \|_K = | \ |^2_\infty$ instead. It is easy to check that in the remaining cases $\| \ \|_K$ coincides with the previously defined absolute value $| \ |_K$.

In this connection, the reader is encouraged to (re)read problems §24.12–13.

2. Next we obtain another and very satisfying characterization of local fields. To prepare the ground:

Definition 2. The following fields are called *global fields*:

  (a) Any *algebraic number field*; that is, any finite extension of $\mathbb{Q}$.

  (b) Any *function field in one variable over a finite field*; that is, any finite extension of the rational function field $\mathbb{F}_p(X)$.

Apart from the archimedean exceptions, all nontrivial absolute values of a global field are *discrete, with finite residue field*; indeed, this is true for $\mathbb{Q}$ and for $\mathbb{F}_p(X)$ (see Chapter 23, Theorem 1 and Problem §23.5), and both properties are inherited by finite extensions (see Chapter 24, F1 and Remark (c) after Definition 2). Therefore, in view of F1, every completion of a global field with respect to a nontrivial absolute value is a *local field*. Even more is true:

Theorem 2. *Local fields are precisely the completions of global fields with respect to nontrivial absolute values.*

Proof. We go down the list of local fields: For $\mathbb{R}$ and $\mathbb{C}$ the assertion is justified by considering the global fields $\mathbb{Q}$ and $\mathbb{Q}(i)$, respectively, with respect to $| \ |_\infty$. Also $\mathbb{F}_q((X))$ is a completion of $\mathbb{F}_q(X)$. For finite extensions of $\mathbb{Q}_p$ the claim will be proved with F3 below.   □

To prepare the ground we need to delve into certain peculiarities of nonarchimedean valuated fields, which are also of interest in themselves. We begin with:

Krasner's Lemma. *Let K be complete with respect to a nonarchimedean absolute value, and let C be an algebraic closure of K with absolute value $| \ |$ extending that of K. Suppose $\alpha \in C$ has a separable minimal polynomial*

$$f(X) = \prod_j (X - \alpha_j), \quad with \ \alpha_1 = \alpha.$$

Then, whenever $\beta \in C$ satisfies

$$|\beta - \alpha| < |\alpha_j - \alpha| \quad for \ every \ j \neq 1,$$

we have $K(\alpha) \subseteq K(\beta)$.

Proof. Otherwise there would be $\sigma \in G(C/K)$ with $\sigma(\alpha) \neq \alpha$ but $\sigma(\beta) = \beta$. Since σ is an isometry, we would get the contradiction

$$|\beta - \alpha| = |\sigma(\beta - \alpha)| = |\beta - \sigma\alpha| = |\beta - \alpha_j| > |\beta - \alpha|. \qquad \square$$

Theorem 3. *Let K be complete with respect to a nonarchimedean absolute value $|\ |$ and let $f \in K[X]$ be a normalized, irreducible, separable polynomial of degree n. There exists a constant $\delta > 0$ with the following property: If $g \in K[X]$ is normalized of degree n and if $|g - f| < \delta$, then g is irreducible, and for any root α of f there is a root β of g satisfying $K(\alpha) = K(\beta)$.*

(Problem §24.22 contains a significant generalization of this result.)

Proof. (a) First let a root β of g be given; we claim that $|\beta| \leq |g|$. For, letting b_i denote the coefficients of g, we'd get if $|\beta| > |g|$ the inequality $|\beta^n| > |b_i||\beta^i|$ for $i < n$, hence the contradiction $0 = |\beta^n + b_{n-1}\beta^{n-1} + \cdots + b_0| = |\beta^n|$ (keep in mind that $|g| \geq 1$).

(b) Clearly, $|g - f| < 1$ implies $|g| = |f|$.

(c) Now suppose $|g - f| < \delta \leq 1$ for some yet to be chosen $\delta > 0$, and let a_i denote the coefficients of f. Using (a) and (b), we see that any root β of g satisfies

$$|f(\beta)| = |f(\beta) - g(\beta)| = \left|\sum_i (a_i - b_i)\beta^i\right| < \delta|g|^n = \delta|f|^n.$$

Writing $f(X) = \prod_i (X - \alpha_i)$ over C we get $\left|\prod_i (\beta - \alpha_i)\right| < \delta|f|^n$; hence we can find i such that

$$|\beta - \alpha_i| < \sqrt[n]{\delta}|f|.$$

For small enough $\delta > 0$, then,

$$|\beta - \alpha_i| < |\alpha_j - \alpha_i| \quad \text{for every } j \neq i,$$

which by Krasner's Lemma implies that $K(\alpha_i) \subseteq K(\beta)$. But then, by a degree argument, we actually have

(8) $$K(\alpha_i) = K(\beta),$$

so g, like f, is irreducible (and separable). If α is a fixed root of f, there exists a K-isomorphism $\sigma: K(\alpha_i) \to K(\alpha)$ such that $\sigma\alpha_i = \alpha$, and (8) yields $K(\alpha) = K(\sigma\beta)$: this is what we need, since $\sigma\beta$ is a root of g. $\qquad \square$

F3. *For every finite extension $K/\mathbb{Q}_p$ there is a subfield F of K such that*

  (i) *$F/\mathbb{Q}$ is finite of degree $F : \mathbb{Q} = K : \mathbb{Q}_p$, and*

  (ii) *F is $|\ |_p$-dense in K; in other words, K is the $|\ |_p$-completion of the algebraic number field F.*

Proof. By the primitive element theorem we can write $K = \mathbb{Q}_p(\alpha)$. Set $f = \text{MiPo}_{\mathbb{Q}_p}(\alpha)$. Since $\mathbb{Q}$ is dense in $\mathbb{Q}_p$, Theorem 3 provides an irreducible polynomial $g \in \mathbb{Q}[X]$ of degree $n = \deg f = K : \mathbb{Q}_p$, and also a root β of g such that $\mathbb{Q}_p(\beta) = \mathbb{Q}_p(\alpha) = K$. Then the subfield $F = \mathbb{Q}(\beta)$ of K satisfies $F : \mathbb{Q} = K : \mathbb{Q}_p$; moreover $F = \mathbb{Q}(\beta) = \mathbb{Q} + \mathbb{Q}\beta + \cdots + \mathbb{Q}\beta^{n-1}$ is dense in $\mathbb{Q}_p + \mathbb{Q}_p\beta + \cdots + \mathbb{Q}_p\beta^{n-1} = K$. $\square$

We now investigate yet another important fact in whose proof Krasner's Lemma plays a role.

F4. *The algebraic closure C of $\mathbb{Q}_p$ is* **not** *complete with respect to* $|\ |_p$.

Proof. Given a natural number n, let a_n be a root of unity of order $p^{n!} - 1$. Then $\mathbb{Q}_p(a_n)/\mathbb{Q}_p$ is *unramified* of degree $n!$, and $\mathbb{Q}_p(a_n) \subseteq \mathbb{Q}_p(a_{n+1})$ (see Chapter 24, Theorem 4). Consider the sequence $(s_n)_n$ given by

$$s_n = \sum_{k=0}^{n} a_k p^k \in \mathbb{Q}_p(a_n).$$

Since $s_n - s_{n-1} = a_n p^n$, this is a Cauchy sequence in C with respect to $|\ |_p$. Assume that $(s_n)_n$ is $|\ |_p$-convergent in C, say toward $\alpha \in C$. Write

$$f = \text{MiPo}_{\mathbb{Q}_p}(\alpha) = \prod_i (X - \alpha_i) \quad \text{over } C,$$

with $\alpha_1 = \alpha$. Since $|s_n - \alpha| \to 0$, *Krasner's Lemma* implies, for n large enough,

$$\alpha \in \mathbb{Q}_p(s_n) \subseteq \mathbb{Q}(a_n).$$

For a fixed such large n, set $K = \mathbb{Q}_p(a_n)$ and consider the p-adic expansion of α in K:

$$\alpha = \sum_{k=0}^{\infty} c_k p^k,$$

where $c_k \in \{0\} \cup \langle a_n \rangle$ is either a power of a_n or 0 (see Chapter 24, F2, together with Theorem 4). For $m > n$, we have in $\mathbb{Q}(a_m)$ the congruence

$$\alpha \equiv \sum_{k=0}^{m} a_k p^k \equiv \sum_{k=0}^{m} c_k p^k \mod p^{m+1}.$$

By the uniqueness of p-adic expansions in $\mathbb{Q}(a_m)$, we obtain $a_k = c_k$ for all $k \leq m$. In particular, a_m lies in $\mathbb{Q}(a_n)$. But this contradicts the inequality $\mathbb{Q}(a_m) : \mathbb{Q} > \mathbb{Q}(a_n) : \mathbb{Q}$. $\square$

Theorem 4. *The completion $\mathbb{C}_p$ of the algebraic closure C of $\mathbb{Q}_p$ with respect to $|\ |_p$ is algebraically closed.*

Note that this is a special case of §24.15.

Proof. Let $f \in \mathbb{C}_p[X]$ be a normalized and irreducible polynomial over $\mathbb{C}_p$, and let α be a root of f (in the algebraic closure of $\mathbb{C}_p$). The field C is dense in $\mathbb{C}_p$. By Theorem 3, applied to the complete field $K = \mathbb{C}_p$, there is then a polynomial $g \in C[X]$ and a root β of g such that $\mathbb{C}_p(\alpha) = \mathbb{C}_p(\beta)$. Since C is algebraically closed (and the coefficients of g lie in C), we see that $\beta \in C$, and hence $\alpha \in \mathbb{C}_p$. $\square$

The field $\mathbb{C}_p$, whose algebraic closedness we have just proved, plays a role in p-adic analysis similar to that of the complex numbers in archimedean analysis.

3. Galois extensions of local fields, as we shall now show, only admit *solvable* Galois groups. The proof requires a new refinement of the notion of ramified extensions and a study of how they arise. An extension E/K of local fields is called *tamely ramified* if its ramification index is relatively prime to the characteristic of the residue field $\overline{K}$; otherwise it is *wildly ramified*. (In this context we will disregard the archimedean local fields $\mathbb{R}$ and $\mathbb{C}$.)

F5. *Let E/K be a **purely ramified** extension of local fields of degree*

$$(9) \qquad\qquad e \not\equiv 0 \bmod p, \quad \text{where } p = \text{char } \overline{K}.$$

Then there exists a prime element π_0 of K such that

$$(10) \qquad\qquad E = K\left(\sqrt[e]{\pi_0}\right).$$

*More precisely: If π is a **predetermined** prime element of K, then*

$$(11) \qquad\qquad E = K\left(\sqrt[e]{\pi\zeta}\right)$$

*for an appropriate $(q-1)$-th root of unity ζ in K, where q is the cardinality of $\overline{K}$. If E/K is assumed to be **Galois**, K contains a primitive e-th root of unity; that is,*

$$(12) \qquad\qquad e \mid q-1,$$

*and $G(E/K)$ is **cyclic** of order e.*

Proof. Let Π be a prime of E. We first note that

$$\Pi^e = u\pi, \quad \text{where } u \in E \text{ is a unit.}$$

Because E/K is purely ramified, $\overline{E}$ and $\overline{K}$ coincide, and by Chapter 24, Theorem 4, the $(q-1)$-th roots of unity form a family — *contained in K* — of coset representatives of the multiplicative group of $\overline{E} = \overline{K}$. Hence there exists a $(q-1)$-st root of unity $\zeta \in K$ satisfying

$$u = \zeta u_1,$$

where u_1 is a 1-*unit* of E, that is, a unit satisfying $u_1 \equiv 1 \bmod \Pi$. But from (9) we obtain that $u_1 = c^e$ is an e-th power in E (see §23.13). Thus we have a prime element $\Pi_0 := c^{-1}\Pi$ of E satisfying

$$\Pi_0^e = \zeta\pi,$$

that is, an e-th root of the element $\pi_0 := \zeta\pi$ of K. This proves (11), since $E = K(\Pi_0)$ by F3 in Chapter 24.

Assume further that E/K is Galois. The polynomial $X^e - \pi_0$ is irreducible over K and splits into e distinct linear factors over E. Thus E contains a primitive e-th root of unity, which must already lie in K, because of (9) and the equality $\overline{E} = \overline{K}$ (see again Theorem 4 in Chapter 24). Hence, in view of (10), $G(E/K)$ is cyclic (see F1 in Chapter 14, vol. I).   □

Theorem 5. *Every Galois extension E/K of local fields has a solvable Galois group.*

Proof. Since $G(\mathbb{C}/\mathbb{R}) = \mathbb{Z}/2$, we only have to worry about the nonarchimedean case. Consider the maximal intermediate field L of E/K that is unramified over K (Chapter 24, Theorem 3). The extension E/L is purely ramified and $G(L/K)$ is cyclic (Chapter 24, Theorem 4). Thus we can restrict ourself for the rest of the proof to the case that E/K is *purely ramified*.

Let H be a *Sylow p-group* of $G = G(E/K)$, where $p = \operatorname{char}\overline{K}$. Denote by F the corresponding intermediate field. Since $F:K = G:H \not\equiv 0 \bmod p$, we can apply F5 to deduce that $F = K\left(\sqrt[e]{\pi}\right)$, where π is a prime of K and $e = F:K$. The polynomial $X^e - \pi$ is irreducible over K and splits over E into e distinct linear factors (since E/K is normal); thus E contains a primitive e-th root of unity. Again, because E/K is purely ramified and p does not divide e, this root must already lie in K. Hence F/K is *Galois*, and by F5 its Galois group is cyclic. Thus, to prove that G is solvable, we need only show the solvability of H; but this just comes from H being a p-group.   □

Remarks. (1) In the proof of Theorem 5 we obtained something beyond the mere solvability of $G = G(E/K)$: we have established *canonical* initial elements of a *principal series*

(13) $$G \supseteq T \supseteq V \supseteq \cdots \supseteq 1$$

for G, namely T, corresponding to the largest intermediate field E/K unramified over K, and V, corresponding to the largest tamely ramified one. (Note that V is normal in G since it is the unique Sylow p-group of T.) One can define the remaining terms of (13) canonically as well, but we won't get into this matter or into the interesting properties of this principal series; however, see at least §24.10 on page 314.

(2) It also deserves to be mentioned that the structure of the solvable group $G(E/K)$ must in any case obey a sharply restrictive condition: The group T in (13) is the *semidirect* product of its normal Sylow p-group V with a *cyclic* subgroup C:

$$T = CV, \quad C \cap V = 1.$$

To derive the existence of such a subgroup C, observe that T/V is cyclic, with order relatively prime to p.

Whereas the field $\mathbb{Q}$, for example, has infinitely many extensions of degree 2, local fields *of characteristic zero* are more restrained:

Theorem 6. *A local field K of characteristic 0 admits (inside a fixed algebraic closure) only finitely many extensions of a given degree.*

Proof. We can assume that K is nonarchimedean and fix a prime element π of K. In view of Theorems 3 and 4 of Chapter 24, we need only concern ourselves with *purely ramified* extensions E/K of prescribed degree $n = E:K$. By Chapter 24, F3, any such extension arises by the adjunction of a root of an *Eisenstein polynomial*

$$(14) \qquad f(X) = X^n + \pi c_{n-1} X^{n-1} + \cdots + \pi c_0$$

of degree n over K, where the c_i all belong to the valuation ring R and c_0, in addition, lies in the group of units $U = R^\times$. Conversely, one can associate to each n-tuple $c = (c_0, c_1, \ldots, c_{n-1})$ in

$$(15) \qquad U \times R \times \cdots \times R = U \times R^{n-1}$$

the corresponding polynomial (14), and thus form finitely many extensions of degree n over K by adjoining (individually) the roots of that polynomial. Now, as can be seen easily from Theorem 3, each n-tuple c in (15) has a small enough neighborhood all of whose elements determine the same (finitely many) extensions as c. The set (15) being *compact*, there exist overall only finitely many purely ramified extensions E/K with $E:K = n$. $\qquad\square$

Remark. It is easy to see that the statement of Theorem 6 also holds in nonzero characteristic if only those extensions E/K such that $e(E/K)$ is not divisible by char K are allowed.

4. In this last section we investigate more closely the structure of the *multiplicative group* of a local field. For the reals, the logarithm function transforms $\mathbb{R}_{>0}$ isomorphically into the additive group $\mathbb{R}$. Thus

$$\mathbb{R}^\times = \{\pm 1\} \times \mathbb{R}_{>0} \simeq \mathbb{Z}/2 \times \mathbb{R}.$$

For $\mathbb{C}$, using the exponential function $e^{2\pi i x}$, one gets

$$(16) \qquad\qquad \mathbb{C}^\times = \mathbb{R}_{>0} \times S^1 \simeq \mathbb{R} \times \mathbb{R}/\mathbb{Z}.$$

Now let K be a *nonarchimedean local* field with normalized valuation w_K, valuation ring $R = R_K$, valuation ideal $\mathfrak{p} = \mathfrak{p}_K$ and residue field $\overline{K} = R/\mathfrak{p}$. Take

$$(17) \qquad\qquad q = q_K = \mathrm{Card}(\overline{K}) \quad \text{and} \quad p = \mathrm{char}\, \overline{K}.$$

Denote by $|\ \ |_K$ the canonical absolute value of K, given by

$$(18) \qquad\qquad |a|_K = \left(\frac{1}{q}\right)^{w_K(a)} \quad \text{for all } a \in K.$$

In the *group of units* $U = U_K = R_K^{\times} = \{a \in K \mid w_K(a) = 0\}$ of K, there is for every $n \in \mathbb{N}$ a *subgroup*

(19) $$U^{(n)} = U_K^{(n)} = 1 + \mathfrak{p}^n = \{x \in R \mid x \equiv 1 \bmod \mathfrak{p}^n\};$$

its elements are called the *n-units* of K. The decreasing sequence formed by the $U^{(n)}$ amounts to an *open neighborhood basis* of 1 in $K^{\times}$.

Set $U_K^{(0)} = U_K$. Also, for each natural number m, denote by

(20) $$W_m$$

the *group of m-th roots of unity* in the algebraic closure of K, and set $W_m(K) = W_m \cap K$.

F6. *Choose a prime element π of K. The group $K^{\times}$ has a direct product decomposition*

(21) $$K^{\times} = \langle \pi \rangle \times U_K = \langle \pi \rangle \times W_{q-1} \times U_K^{(1)}$$

(in the topological sense as well). The group $W'(K)$ of all roots of unity of K with order relatively prime to p satisfies $W'(K) = W_{q-1}$. There is an isomorphism

(22) $$U_K / U_K^{(1)} \simeq \overline{K}^{\times},$$

and for every $n \geq 1$ the factor group

(23) $$U_K^{(n)} / U_K^{(n+1)} \simeq \overline{K}$$

*is isomorphic to the **additive** group of the residue field $\overline{K}$, and hence an elementary abelian p-group of order q.*

Proof. Every $a \in K^{\times}$ has a decomposition

$$a = \pi^{w_K(a)} u, \quad \text{with } u \in U_K,$$

which is canonical (once π is chosen). We already know (Chapter 24, Theorem 4) that $W'(K) = W_{q-1}$, and W_{q-1} is isomorphically mapped to $\overline{K}^{\times}$ via the residue class homomorphism

(24) $$\begin{aligned} U_K &\to \overline{K}^{\times}, \\ x &\mapsto \bar{x}. \end{aligned}$$

The kernel of (24) is obviously $U_K^{(1)}$. It follows that

(25) $$U_K = W_{q-1} \times U_K^{(1)},$$

proving (22) . There remains to justify (23). The map $U_K^{(n)} \to \overline{K}$ defined by

$$1 + a\pi^n \mapsto \bar{a} := a \bmod \mathfrak{p}$$

is easily found to be a homomorphism from $U_K^{(n)}$ onto $\overline{K}$, and its kernel is $U_K^{(n+1)}$. This yields (23). $\qquad \square$

The composition of the multiplicative group $K^\times$ of a local field K is only roughly described by (21). From there on, the matter rests with the structure of the *group of* 1-*units* $U^{(1)}$. Regarding the latter, (23) gives nonetheless a certain amount of information.

We now investigate the obvious question of whether, for every natural number m, the index

(26) $$K^\times : K^{\times m}$$

of the subgroup $K^{\times m} = \{x^m \mid x \in K^\times\}$ of m-th powers in $K^\times$ can be calculated. The answer is easy for $K = \mathbb{R}$ and $K = \mathbb{C}$: For the reals it is 2 if m is even and 1 if odd; for $K = \mathbb{C}$ it is always 1. So assume again that K is nonarchimedean. Using F6 and considering the homomorphism $x \mapsto x^m$ on each of the subgroups involved, one easily checks that

(27) $$\frac{K^\times : K^{\times m}}{\operatorname{Card} W_m(K)} = m\,\frac{U : U^m}{\operatorname{Card} W_m(K)} = m\,\frac{U^{(1)} : U^{(1)m}}{\operatorname{Card} W_m(U^{(1)})},$$

where $W_m(U^{(1)}) = W_m \cap U^{(1)}$. So far, so good. But in a certain way things are not as simple as one might expect:

F7. *If K is a local field of characteristic $p > 0$, then*

(28) $$K^\times : K^{\times p} = \infty.$$

Proof. We have $K = \mathbb{F}_q((X))$; see F2. Assume that $K^\times : K^{\times p}$ is finite; then (27) implies that $U : U^p$ too is finite. Now, since U is compact, so is U^p; in particular, U^p is *closed*. Since $U : U^p$ is finite, this implies that U^p is also *open*. For large enough n, then, we must have

$$1 + X^n \in U^p \subseteq K^p = \mathbb{F}_q((X^p));$$

but this is impossible if p does not divide n.

   An alternative argument: Since char $K = p$, the set K^p is a *subfield* of K, and since $K = K^p(X)$ we obviously have

$$K : K^p = p < \infty.$$

But for a nontrivial finite field extension K/k with k infinite, the quotient $K^\times / k^\times$ can never be finite (see Problem §1.6 in vol. I). □

Theorem 7. *Let K be any local field. For every natural number m,*

(29) $$\frac{K^\times : K^{\times m}}{\operatorname{Card} W_m(K)} = \frac{m}{\|m\|_K}.$$

(Here $\| \ \|_K = | \ |_K$ except when $K = \mathbb{C}$; see the Remark following F2. Note also that in nonzero characteristic, say char $K = p = $ char $\overline{K}$, the denominator on the right vanishes if p divides m and the equality must be read as saying that $K^\times : K^{\times m}$ is infinite in this case.)

Proof. For $K = \mathbb{R}, \mathbb{C}$ the assertion is clear; therefore let K be nonarchimedean. First suppose that m is prime to $p = \operatorname{char} \bar{K}$; then $U^{(1)} = U^{(1)m}$, as can be seen by applying F13 in Chapter 23 (see also §23.13). Thus the conclusion can be read off from F6 and equality (27).

One checks equally easily that both sides of (29) behave multiplicatively for $m = m_1 m_2$, with m_1, m_2 relatively prime. Altogether, then, we need only verify the formula for $m = p^k$, where $p = \operatorname{char} \bar{K}$. If K, too, has characteristic p, this is taken care of by F7.

We are left with the case $\operatorname{char} K = 0$. Here we resort to the structure of the *group of 1-units* $U^{(1)}$, which can be exactly determined in this case: as proved in Theorem 9 below,

$$(30) \qquad U_K^{(1)} \simeq W_{p^\infty}(K) \times \mathbb{Z}_p^n,$$

where $W_{p^\infty}(K)$ is the group of roots of unity in K whose order is a power of p, and n is the degree of $K/\mathbb{Q}_p$. From (30) we get

$$(31) \qquad \frac{U_K^{(1)} : U_K^{(1)m}}{\operatorname{Card} W_m(U^{(1)})} = \mathbb{Z}_p^n : m\mathbb{Z}_p^n.$$

Since $\mathbb{Z}_p^n / m\mathbb{Z}_p^n \simeq (\mathbb{Z}_p/m\mathbb{Z}_p)^n = (\mathbb{Z}_p/p^k\mathbb{Z}_p)^n$ has order $(p^k)^n = p^{nk} = p^{nw_p(m)} = |m|_K^{-1}$ (see (5)), the desired equality (29) follows from (27). $\qquad \square$

F8. *Take $a \in U_K^{(1)}$ and $x \in \mathbb{Z}_p$. For every sequence $(x_n)_n$ of integers converging p-adically to x, the sequence $(a^{x_n})_n$ converges in K. The element*

$$(32) \qquad a^x = \lim_{n \to \infty} a^{x_n}$$

is independent of the choice of the sequence $(x_n)_n$, so the map $(x, a) \mapsto a^x$ is well defined; this makes $U_K^{(1)}$ into a $\mathbb{Z}_p$-module.

Proof. First note that $b \in U_K^{(n)}$ implies $b^p \in U_K^{(n+1)}$, in view of (23). More generally, therefore, we have for each $z \in \mathbb{Z}$:

$$(33) \qquad b \in U_K^{(n)} \implies b^z \in U_K^{(n+w_p(z))}.$$

Applying this to $a \in U_K^{(1)}$, we obtain for any sequence $(z_n)_n$ in $\mathbb{Z}$ the implication

$$(34) \qquad \lim z_n = 0 \implies \lim a^{z_n} = 1.$$

Now let $(x_n)_n$ be an arbitrary p-adic Cauchy sequence in $\mathbb{Z}$. By (34), the sequence with entries

$$a^{x_{n+1}} - a^{x_n} = a^{x_n}(a^{x_{n+1}-x_n} - 1)$$

converges to 0 in K. Hence $(a^{x_n})_n$ is a Cauchy sequence in K and so converges. Its limit lies in $U_K^{(1)}$, since this set is closed. Given $x \in \mathbb{Z}_p$, there is certainly a sequence $(x_n)_n$ in $\mathbb{Z}$ converging p-adically to x. If $(y_n)_n$ is another such sequence, the sequence with entries

$$a^{y_n} - a^{x_n} = a^{x_n}(a^{y_n - x_n} - 1)$$

is, again by (34), a null sequence in K. Thus a^x is well defined by (32).[1] Since the usual power laws $(a^x)^y = a^{xy}$, $a^0 = 1$, $(ab)^x = a^x b^x$, and $a^{x+y} = a^x a^y$ obviously hold, we have indeed made $U_K^{(1)}$ into a $\mathbb{Z}_p$-module. $\quad\square$

In $\mathbb{R}$, as we know, we have, for $a > 0$:

$$\log a = \lim_{h \to 0} \frac{a^h - 1}{h}.$$

Imitating the same pattern, we define, for every $a \in U_K^{(1)}$,

(35)
$$\log a = \lim_{n \to \infty} \frac{a^{p^n} - 1}{p^n},$$

where we must assume char $K = 0$ in view of the denominator.

Theorem 8. *Suppose* char $K = 0$. *For every* $a \in U_K^{(1)}$, *the value of* $\log a$ (*also denoted* $\log_K a$ *or* $\log_p a$) *is well defined by* (35). *The map* $\log : U_K^{(1)} \to K$ *is continuous and satisfies*

(36)
$$\log(ab) = \log a + \log b;$$

it is therefore a homomorphism. Its kernel is the group $W_{p^\infty}(K)$ *of all roots of unity in* K *whose order is a power of* p.

Let $e = w_K(p) > 0$ *be the ramification index of* $K/\mathbb{Q}_p$. *For every natural number*

(37)
$$r > \frac{e}{p-1},$$

the $\log$ *function induces an isomorphism from* $U_K^{(r)}$ *onto the additive group* $\mathfrak{p}_K^r$, *which is also a homeomorphism (that is, bicontinous). This isomorphism*

(38)
$$U_K^{(r)} \simeq \mathfrak{p}_K^r, \quad for \ r > \frac{e}{p-1},$$

is also an isomorphism of $\mathbb{Z}_p$-modules.

Proof. For $a \in U^{(1)}$ we already know that $a^{p^m} \in U^{(m)}$, by (33). Since

$$\frac{a^{p^n} - 1}{p^n} = \frac{1}{p^m} \frac{(a^{p^m})^{p^{n-m}} - 1}{p^{n-m}},$$

we can restrict our attention to some $U^{(m)}$, with $m \geq 1$ arbitrary.

---

[1] In fact, the map $x \mapsto a^x$ is *continuous*, because $w_p(x) \geq i$ implies $a^x \in U_K^{(i+1)}$.

Lemma. *With the preceding notations, we have, for $a \in U_K^{(1)}$:*

(39) $\qquad w_K(a-1) > e/p-1 \implies w_K(a^p - 1) = w_K(a-1) + e.$

Under the assumption $w_K(a-1) \geq e/p-1$, we have in any case $w_K(a^p - 1) \geq w_K(a-1) + e.$

Proof. Set $a = 1 + y$. The conclusions of the lemma can be read off from the equation

$$a^p = (1+y)^p = 1 + py + \sum_{k=2}^{p-1} \binom{p}{k} y^k + y^p,$$

since each $\binom{p}{k}$ is divisible by $p = \mathfrak{p}^e$ and the summand y^p satisfies $w_K(y^p) = p w_K(y) = (p-1) w_K(y) + w_K(y)$. $\qquad\square$

We resume the proof of Theorem 8, taking $a \in U^{(m)}$ with $m \geq e/p-1$. The lemma implies $a^{p^n} \equiv 1 \bmod \mathfrak{p}^{m+ne}$; hence

(40) $\qquad \dfrac{a^{p^n} - 1}{p^n} \equiv 0 \bmod \mathfrak{p}^m.$

To prove that the elements z_n on the left form a Cauchy sequence, consider that

$$z_{n+1} - z_n = z_n \left(\frac{1 + a^{p^n} + a^{2p^n} + \cdots + a^{(p-1)p^n}}{p} - 1 \right).$$

Since $a^{p^n} \equiv 1 \bmod \mathfrak{p}^{m+ne}$, the numerator of the fraction in parentheses is congruent to p modulo $\mathfrak{p}^{m+ne}$, so the fraction itself is equivalent to 1 modulo $\mathfrak{p}^{m+(n-1)e}$. Therefore the terms $z_{n+1} - z_n$ form a null sequence, proving the existence of the limit in (35). In view of the equality

$$\frac{(ab)^{p^n} - 1}{p^n} = b^{p^n} \frac{a^{p^n} - 1}{p^n} + \frac{b^{p^n} - 1}{p^n}$$

and the fact that b^{p^n} converges to 1, it is now clear that log is a homomorphism.

According to (40) we have

(41) $\qquad \log a \in \mathfrak{p}^m \quad$ for $a \in U^{(m)}$ with $m \geq e/p-1$.

In particular, log is *continuous* at $a = 1$, and in view of (36) also at every $a \in U^{(1)}$.

Now suppose $r > e/p-1$ and take $a \in U^{(r)}$. By the lemma, then, $w_K(a^{p^n} - 1) = w_K(a-1) + ne = w_K(a-1) + w_K(p^n)$. Then (35) implies

(42) $\qquad w_K(\log a) = w_K(a-1) \quad$ for $w_K(a-1) > e/p-1$.

To prove that $\log : U^{(r)} \to \mathfrak{p}^r$ is an isomorphism, it obviously suffices to show that, for each n, the induced map

(43) $\qquad\qquad U^{(r)}/U^{(r+n)} \to \mathfrak{p}^r/\mathfrak{p}^{r+n}$

arising from log is an isomorphism (since log is continuous and $U^{(r)}$ is compact). Because of (23), both groups in (43) have order q^n, so all that remains to show is that (43) is injective. But this is clear from (42).

For $a \in U^{(1)}$, given $a^n = 1$, we obtain $n \log a = \log a^n = 0$, so $\log a = 0$. Conversely, suppose $\log a = 0$. For $r > e/p-1$ as in (37) we have $a^{p^r} \in U^{(r)}$. From $\log a^{p^r} = p^r \log a = 0$ there now follows $a^{p^r} = 1$. We obtain

$$(44) \qquad \ker \log = W_{p^\infty}(K) = W_{p^r}(K),$$

where r is the smallest natural number satisfying (37). Notice in this connection that a root of unity of order a power of p necessarily lies in $U^{(1)}$ (see F6, for example).

We still have to prove the last sentence of the theorem. In fact we show, more generally, that $\log : U_K^{(1)} \to K$ is a homomorphism of $\mathbb{Z}_p$-modules. Indeed, applying the log function to (32) we get, by continuity,

$$(45) \qquad \log a^x = \lim \log a^{x_n} = \lim(x_n \log a) = x \log a. \qquad \square$$

Remarks. Let the situation be as in Theorem 8. We spell out some consequences:

(1) The only roots of unity contained in $U^{(1)}$ are those whose order is a p-power; more precisely,

$$(46) \qquad W(U_K^{(1)}) = W_{p^\infty}(K) = W_{p^r}(K),$$

where $r > e/p-1$. For such an r, $U^{(r)}$ contains no root of unity apart from 1.

(2) For $r > e/p-1$, the inverse of $\log : U^{(r)} \to p^r$ is a continuous map $\exp : p^r \to U^{(r)}$ satisfying $\exp(x+y) = \exp x \exp y$. We also write $\exp x = e^x$. For every $a \in U^{(r)}$ and $x \in \mathbb{Z}_p$, then, (45) yields

$$(47) \qquad\qquad a^x = e^{x \log a}.$$

(Note that this equality holds only for $w_K(a-1) > e/p-1$, so it does not suffice for defining a $\mathbb{Z}_p$-module structure on $U^{(1)}$.)

The log and exp functions satisfy the usual power series developments

$$(48) \qquad \log(1-x) = -\sum_{n=1}^{\infty} \frac{x^n}{n}, \qquad \exp(x) = \sum_{n=0}^{\infty} \frac{x^n}{n!}.$$

The former converges for $w_K(x) > 0$ and the latter for $w_K(x) > e/p-1$.

F9. *Suppose* char $K = 0$, *hence* $e := w_K(p) \in \mathbb{N}$. *For* $r > \dfrac{e}{p-1}$, *the map* $x \mapsto x^p$ *is an isomorphism form* $U^{(r)}$ *to* $U^{(r+e)}$. *In particular,*

$$(49) \qquad\qquad U^{(r)p} = U^{(r+e)}.$$

Proof. An application of log reduces the problem to showing that $y \mapsto py$ gives an isomorphism between p^r and $p^{r+e} = pp^r$. But this is obvious. $\qquad \square$

This result can also be justified by an appropriate use of Hensel's Lemma (F13 in Chapter 23). In turn, Theorem 7 can be proved using F9.

Theorem 9. *For* char $K = 0$, *the group* $U_K^{(1)}$ *of* 1*-units has, as a* $\mathbb{Z}_p$*-module (and a fortiori as an abelian group), the structure*

$$(50) \qquad U_K^{(1)} \simeq W_{p^\infty}(K) \times \mathbb{Z}_p^n, \quad \text{where } n = K : \mathbb{Q}_p.$$

Proof. Suppose $r > e/p-1$. The $\mathbb{Z}_p$-submodule $U^{(r)}$ of $U^{(1)}$ is isomorphic to $\mathfrak{p}^r = \pi^r R_K \simeq R_K$. At the same time, the $\mathbb{Z}_p$-module R_K has a *basis* with $n = K : \mathbb{Q}_p$ elements (see Chapter 24, proof of Theorem 1, and also Problem §24.6). Hence

$$(51) \qquad U_K^{(r)} \simeq \mathbb{Z}_p^n.$$

Because $U^{(1)}/U^{(r)}$ is finite, the $\mathbb{Z}_p$-module $U^{(1)}$ is in any case *finitely generated*. By the classification of finitely generated modules over a principal ideal domain (see vol. I, p. 164), we obtain

$$U_K^{(1)} = T \times F,$$

where T is the *torsion submodule* of $U_K^{(1)}$ and F is a *free* $\mathbb{Z}_p$-submodule, say $F \simeq \mathbb{Z}_p^m$. Being a finitely generated $\mathbb{Z}_p$-torsion module, T is a finite p-group, and so coincides with the group $W_{p^\infty}(K)$ of roots of unity of $U^{(1)}$. Hence

$$U_K^{(1)} \simeq W_{p^\infty}(K) \times \mathbb{Z}_p^m,$$

and all that's left to show is that $m = n$. To do this, set $B = U^{(1)}$, $A = U^{(r)}$. Because it is torsion-free, the submodule $A \simeq \mathbb{Z}_p^n$ of B is isomorphic to a submodule of $B/T \simeq F \simeq \mathbb{Z}_p^m$. This implies $n \leq m$, again by the arguments of Chapter 14 in vol. I (see in particular page 162).

By the finiteness of B/A and T, there exists $s \in \mathbb{N}$ such that $sB \subseteq A$ and $sT = 0$. Therefore, since

$$\mathbb{Z}_p^m \simeq F \simeq sF = sB \subseteq A \simeq \mathbb{Z}_p^n,$$

$\mathbb{Z}_p^m$ is isomorphic to a submodule of $\mathbb{Z}_p^n$. Consequently $m \leq n$ as needed.   □

Note that in spite of the isomorphisms (50) and (51), it need *not* be the case that $U_K^{(1)} = W_{p^\infty}(K) \times U_K^{(r)}$.

F10. *As above, let* $K/\mathbb{Q}_p$ *have degree* $n = ef$, *so* $U_K^{(r)}/U_K^{(r+e)}$ *is a* $\mathbb{Z}/p\mathbb{Z}$*-vector space of dimension* n *(see* (23) *in F6). Assume moreover that* $r > e/p-1$. *Any family* $b_1, \ldots, b_n$ *of elements of* $U_K^{(r)}$ *that projects to a basis of the* $\mathbb{Z}/p\mathbb{Z}$*-vector space* $U_K^{(r)}/U_K^{(r+e)}$ *is a basis of the* $\mathbb{Z}_p$*-module* $U_K^{(r)}$.

Proof. We work the proof so as to obtain another derivation of (51) at the same time. For economy we set $M = U^{(r)}$ and switch to additive notation for the $\mathbb{Z}_p$-module M. We must show that the map

$$(x_1, \ldots, x_n) \mapsto \sum_{i=1}^{n} x_i b_i$$

from $N = \mathbb{Z}_p^n$ to M is an isomorphism. For each $i \in \mathbb{N}$, this map gives rise to a commutative diagram

$$
\begin{array}{ccc}
p^i N / p^{i+1} N & \longrightarrow & p^i M / p^{i+1} M \\
\downarrow & & \downarrow \\
N / p^{i+1} N & \longrightarrow & M / p^{i+1} M \\
\downarrow & & \downarrow \\
N / p^i N & \longrightarrow & M / p^i M
\end{array}
$$

The assumption, together with F9, shows that the top map across is an isomorphism. From this it follows inductively that the bottom map is injective for every i and hence also an isomorphism. This obviously implies the claim. $\square$

Applying F10 to the special case $K = \mathbb{Q}_p$, we obtain:

F11. *Suppose $p \neq 2$. The multiplicative group of $\mathbb{Q}_p$ has the form*

$$
\mathbb{Q}_p^\times = \langle p \rangle \times W_{p-1} \times U^{(1)},
$$

with

(52) $$ U^{(1)} = (1 + p)^{\mathbb{Z}_p}; $$

every $a \in \mathbb{Q}_p^\times$ has a unique representation of the form

(53) $$ a = p^n \zeta (1 + p)^x, \quad \text{with } \zeta \in W_{p-1} \text{ and } x \in \mathbb{Z}_p. $$

For $p = 2$ we have $\mathbb{Q}_2^\times = \langle 2 \rangle \times U^{(1)} = \langle 2 \rangle \times \{\pm 1\} \times U^{(2)}$, with

(54) $$ U^{(2)} = (1 + 4)^{\mathbb{Z}_2} = 5^{\mathbb{Z}_2}; $$

every $a \in \mathbb{Q}_2^\times$ has a unique representation of the form

(55) $$ a = 2^n \, \varepsilon 5^x, \quad \text{with } \varepsilon = \pm 1 \text{ and } x \in \mathbb{Z}_p. $$

We see from (21) and (50) that the structural decomposition (16) for $\mathbb{C}$ has counterparts for p-adic fields K. In particular, for $K = \mathbb{Q}_p$ there is a nice analogy between the representations (53) and (55) for $a \in \mathbb{Q}_p^\times$ and the representation $z = r e^{2\pi i x}$ for $z \in \mathbb{C}$.

26

Witt Vectors

1. We now wish to discuss more systematically a question that was briefly broached in Chapter 24 (see F2 on page 67 and the observations following its proof.)

F1. *Let K be complete with respect to a discrete valuation w. Assume that the corresponding residue field $\overline{K}$ has characteristic zero: char $\overline{K} = $ char $K = 0$. There exists an additively and multiplicatively closed set S of representatives of $\overline{K}$. In other words, the valuation ring R of w contains a field $k = S$ that maps isomorphically onto $\overline{K}$ under the quotient map. Thus $K = k((X))$ is the field of formal Laurent series over k* (see F2 in Chapter 24 and the example following it).

Proof. In view of the assumption char $\overline{K} = $ char K, Zorn's Lemma guarantees the existence of a maximal subfield k of R. Let $\bar{k}$ be its image under the quotient map. The proposition will be proved if we can show that $\bar{k} = \overline{K}$.

Suppose there exists $t \in R$ such that $\bar{t}$ is transcendental over $\bar{k}$. Then, when restricted to the subring $k[t]$ of R, the residue class homomorphism is *injective*. It follows that $k(t) \subseteq R$, contradicting the maximality of k. Thus $\overline{K}/\bar{k}$ is certainly algebraic. Suppose there exists $a \in R$ such that $\bar{a} \notin \bar{k}$. Let $\bar{f}$ be the minimal polynomial of $\bar{a}$ over $\bar{k}$ (with $f \in R[X]$ normalized of degree $n > 1$). By Hensel's Lemma (F13 in Chapter 23), there exists some $\alpha \in R$ such that $f(\alpha) = 0$ and $\bar{\alpha} = \bar{a}$. Since $k(\alpha) = k[\alpha] \subseteq R$, we again obtain a contradiction with the maximality of k. $\square$

Remark. In F1, the field k is uniquely determined by its characterizing property only in the case that $\overline{K}$ is algebraic over its prime field.

The conclusion of F1 also holds when we merely assume char $\overline{K} = $ char K ("case of equal characteristic"). If $\overline{K}$ is *perfect*, this follows from Theorem 1 immediately below. If K is not perfect, we refer to the nice argument given in Zariski and Samuel, *Commutative algebra*, vol. II, p. 306.

Theorem 1. *Let K be complete with respect to a discrete valuation w. Assume that the associated residue field $\overline{K}$ is **perfect** of characteristic $p > 0$. Then there exists a* *unique* set S of representatives of $\overline{K}$ satisfying

(1) $$S^p = S.$$

This set is multiplicatively closed. An element α of the valuation ring R of w lies in S if and only if α is a p^n-th power in R for every $n \geq 0$. If char $K = p$, this set S is also additively closed.

Proof. Let π be a prime element of w. For $\alpha, \beta \in R$ and $i \geq 1$ we have

(2) $$\alpha \equiv \beta \bmod \pi^i \implies \alpha^p \equiv \beta^p \bmod \pi^{i+1},$$

as can be seen by writing $\alpha^p - \beta^p = (\alpha - \beta)(\alpha^{p-1} + \alpha^{p-2}\beta + \cdots + \beta^{p-1})$, for instance. Now fix $a \in \overline{K}$. For every n, let $\alpha_n \in R$ be a representative of $a^{p^{-n}} \in \overline{K}$. By definition we have $\alpha_{n+1}^p \equiv \alpha_n \bmod \pi$, and together with (2) this implies

$$\alpha_{n+1}^{p^{n+1}} \equiv \alpha_n^{p^n} \bmod \pi^{n+1}.$$

Thus there is a limit

(3) $$\langle a \rangle := \lim_{n \to \infty} \alpha_n^{p^n}$$

in K. Any other choice of the α_n leads to the same limit, because of (2). Since all terms in the sequence (3) lie in the residue class a, the same is true of the limit; thus $\langle a \rangle$ is a representative of a. It is easy to see from the definition that we also have

(4) $$\langle a^p \rangle = \langle a \rangle^p.$$

Altogether, we see that the image S_0 of the map $a \mapsto \langle a \rangle$ is a set of representatives for $\overline{K}$ having property (1). Let S be another such set and take $\alpha \in S$. If a is the residue class of α and we choose α_n in S, we must have

(5) $$\alpha_n^{p^n} = \alpha,$$

because of (1), and then (3) yields $\langle a \rangle = \alpha$. It follows that $S \subseteq S_0$ and hence $S = S_0$. One checks directly that $\langle ab \rangle = \langle a \rangle \langle b \rangle$, that is, S is multiplicatively closed. (In particular, 0 and 1 lie in S.) If $\alpha \in R$ is a p^n-th power in R for every n, one can choose the α_n so that (5) is satisfied, and we get $\langle a \rangle = \alpha$ as above, and hence $\alpha \in S$. From this we get also the last assertion of the theorem: If α, β are p^n-th powers, we have, when char $K = p$,

$$\alpha + \beta = \xi^{p^n} + \eta^{p^n} = (\xi + \eta)^{p^n}. \qquad \square$$

The preceding results go back three quarters of a century, to F. K. Schmidt, Helmut Hasse and Oswald Teichmüller. The elements of the uniquely determined set S described in Theorem 1 are called the *Teichmüller representatives* of K.

If we assume in Theorem 1 that char $K = $ char $\overline{K}$, then $k := S$ is a field, and $K = k((X))$ is the *field of formal Laurent series* over k. Since $k \simeq \overline{K}$, it follows that the isomorphism type of K is fully determined by the isomorphism type of its residue field.

2. In the Situation of Theorem 1, we now consider the case

$$(6) \qquad\qquad \text{char } K = 0$$

("case of unequal characteristics"). We assume w normalized and we call the natural number $e := w(p)$ the *(absolute) ramification index of* K. Now, in general, the structure of K is not wholly determined by that of the residue field $\overline{K}$ plus the ramification index e. We *would* expect it to be wholly determined if K is *(absolutely) unramified*, that is, if p is a prime of K; for then every $\alpha \in K$ has a unique representation of the form

$$(7) \qquad\qquad \alpha = \sum_{i=0}^{\infty} \langle a_i \rangle \, p^i,$$

where $a \mapsto \langle a \rangle$ gives the *Teichmüller representatives*. But how are the algebraic operations of K reflected in the coefficient sequences in $\overline{K}$? Take for instance addition:

$$(8) \qquad\qquad \sum \langle a_i \rangle \, p^i + \sum \langle b_i \rangle \, p^i = \sum \langle c_i \rangle \, p^i.$$

Each c_i is determined by all the a_j and b_k jointly. What kind of a dependence is it? If we take equation (8) modulo p, we obtain $\langle a_0 \rangle + \langle b_0 \rangle \equiv \langle c_0 \rangle \bmod p$, so

$$(9) \qquad\qquad c_0 = a_0 + b_0.$$

Now considering (8) modulo p^2, we obtain

$$\langle a_0 \rangle + \langle b_0 \rangle + (\langle a_1 \rangle + \langle b_1 \rangle) p \equiv \langle c_0 \rangle + \langle c_1 \rangle \, p \bmod p^2.$$

In light of (9) this yields

$$(10) \qquad \langle c_1 \rangle \equiv \langle a_1 \rangle + \langle b_1 \rangle + \frac{\langle a_0 \rangle + \langle b_0 \rangle - \langle a_0 + b_0 \rangle}{p} \bmod p.$$

Now, $\langle a_0^{p^{-1}} \rangle + \langle b_0^{p^{-1}} \rangle \equiv \langle a_0^{p^{-1}} + b_0^{p^{-1}} \rangle \bmod p$. Taking p-th powers leads to

$$\left(\langle a_0 \rangle^{p^{-1}} + \langle b_0 \rangle^{p^{-1}} \right)^p \equiv \langle a_0 + b_0 \rangle \bmod p^2,$$

since $\langle x^p \rangle = \langle x \rangle^p$; see also (2). Substituting this in (10), we obtain

$$\langle c_1 \rangle \equiv \langle a_1 \rangle + \langle b_1 \rangle + \frac{\langle a_0 \rangle + \langle b_0 \rangle - \left(\langle a_0 \rangle^{p^{-1}} + \langle b_0 \rangle^{p^{-1}} \right)^p}{p} \bmod p.$$

Thus we have found a polynomial $s_1(X_0, Y_0) \in \mathbb{Z}[X_0, Y_0]$ such that

$$\langle c_1 \rangle \equiv \langle a_1 \rangle + \langle b_1 \rangle + s_1\left(\langle a_0 \rangle^{p^{-1}}, \langle b_0 \rangle^{p^{-1}} \right) \bmod p.$$

After passing to residue classes, this becomes

$$c_1 = a_1 + b_1 + s_1\left(a_0^{p-1}, b_0^{p-1}\right),$$

and taking p-th powers yields

(11) $$c_1^p = a_1^p + b_1^p + s_1(a_0, b_0) =: S_1(a_0, b_0, a_1^p, b_1^p).$$

Up to raising to the p-th power, then, c_1 is the value of a well defined integer polynomial in a_0, b_0, a_1, b_1. One can make similar calculations modulo p^3, p^4 and so on, and this informally justifies our next result:

Theorem 2. *In the situation of Theorem 1, suppose K is **unramified**. There exist universal polynomials $S_n \in \mathbb{Z}[X_0, X_1, \ldots, X_n, Y_0, Y_1, \ldots, Y_n]$ (independent of K) such that (8) holds if and only if, for all $n = 0, 1, 2, \ldots$, the equations*

(12) $$c_n^{p^n} = S_n\left(a_0, a_1^p, \ldots, a_n^{p^n}, b_0, b_1^p, \ldots, b_n^{p^n}\right)$$

are satisfied. An analogous statement holds regarding (subtraction and) multiplication in K.

From Theorem 2 we derive that the structure of an *unramified* field K is fully determined by the associated residue field $\overline{K}$, if the latter is perfect. In particular, for the p-adic numbers, the theorem contains the remarkable fact that the field operations on $\mathbb{Q}_p$ are expressible in terms of nothing but rational operations in the field $\mathbb{F}_p = \mathbb{Z}/p\mathbb{Z}$ (note that $x^p = x$ in $\mathbb{F}_p$, so the exponent p^n in (12) can be omitted).

In proving Theorem 2 it is obviously desirable to bypass a morass of calculation by the use of the right formalism. This is what E. Witt (in *J. Reine Angew. Math.* **176** (1937), 126–140) managed to achieve. The calculus of *Witt vectors*, which he pioneered, is the appropriate tool for the task, and we turn our attention to it now. Conceptually, his technique is simple enough to allow a thorough understanding of its initial stages, yet so richly intricate that its manipulative potentialities extend far indeed.

3. In our discussion we will take a *fixed prime number p* as given. We work in the polynomial ring $\mathbb{Z}[X_0, Y_0, X_1, Y_1, \ldots, X_n, Y_n, \ldots]$ in countably many variables X_i, Y_i over $\mathbb{Z}$. If necessary we allow p-powers as denominators in the coefficients, that is, we adopt as our ring of coefficients the ring $\mathbb{Z}' = \mathbb{Z}[\frac{1}{p}]$ instead of $\mathbb{Z}$. We consider "vectors" $Z = (Z_0, Z_1, \ldots)$ with components Z_i in $\mathbb{Z}[X_0, Y_0, X_1, Y_1, \ldots]$. For $n = 0, 1, 2, \ldots$ we define

(13) $$Z^{(n)} := Z_0^{p^n} + p Z_1^{p^{n-1}} + \cdots + p^n Z_n.$$

The vector $(Z_0, Z_1, \ldots, Z_n, \ldots)$ is completely determined by its *ghost components* $Z^{(0)}, Z^{(1)}, \ldots, Z^{(n)}, \ldots$ thus defined (in German *Nebenkomponenten*, "side components"). Indeed, by recursion, each Z_n can be recovered as a well defined polynomial expression in $Z^{(0)}, Z^{(1)}, \ldots, Z^{(n)}$, with coefficients in $\mathbb{Z}'$. By applying the *Frobenius operator* F defined by

(14) $$F(Z_0, Z_1, \ldots, Z_n, \ldots) = (Z_0^p, Z_1^p, \ldots, Z_n^p, \ldots),$$

one obtains the recursions formulas

(15) $$Z^{(n)} = (FZ)^{(n-1)} + p^n Z_n, \quad Z^{(0)} = Z_0.$$

Thus, for example,

(16) $$Z_1 = \frac{1}{p} Z^{(1)} - \frac{1}{p} Z^{(0)p}.$$

Now if $*$ denotes one of the operations $+, \cdot, -$, let's define $X * Y$, where $X = (X_0, X_1, \ldots)$ and $Y = (Y_0, Y_1, \ldots)$, by means of the ghost components as follows:

(17) $$(X * Y)^{(n)} = X^{(n)} * Y^{(n)}.$$

The components $(X * Y)_n$ of $X * Y$ are likewise polynomials in $X_0, Y_0, \ldots, X_n, Y_n$ with coefficients in $\mathbb{Z}'$. We have

(18) $$(X \pm Y)_0 = X_0 \pm Y_0, \qquad (XY)_0 = X_0 Y_0,$$

and using (16) we obtain, for example,

$$(X + Y)_1 = \frac{1}{p}(X + Y)^{(1)} - \frac{1}{p}(X + Y)_0^p = \frac{1}{p}(X^{(1)} + Y^{(1)}) - \frac{1}{p}(X_0 + Y_0)^p$$
$$= \frac{1}{p}(X_0^p + pX_1 + Y_0^p + pY_1) - \frac{1}{p}(X_0 + Y_0)^p,$$

that is,

(19) $$(X + Y)_1 = X_1 + Y_1 + \frac{1}{p}\left(X_0^p + Y_0^p - (X_0 + Y_0)^p\right).$$

Compare this with (11). With a similar calculation we get

(20) $$(X \cdot Y)_1 = pX_1 Y_1 + X_0^p Y_1 + X_1 Y_0^p.$$

In these example calculations all the p-power denominators have vanished. It is a fact of capital importance that this phenomenon always takes place:

Lemma 1. *For each n, the polynomial $(X * Y)_n$ only has **integer** coefficients:*

$$(X * Y)_n \in \mathbb{Z}[X_0, Y_0, \ldots, X_n, Y_n].$$

We will write this polynomial also as

(21) $$(X * Y)_n = S_n^*(X; Y) = S_n^*(X_0, Y_0, \ldots, X_n, Y_n).$$

The proof of the lemma is deferred until a necessary stepping stone is in place:

Lemma 2. *Let integers $r \geq 1$ and $n \geq 0$ be given. If $A = (A_0, A_1, \ldots)$ and $B = (B_0, B_1, \ldots)$ are vectors with components in $\mathbb{Z}[X_0, Y_0, X_1, Y_1, \ldots]$, the congruences*

(22) $A_i \equiv B_i \bmod p^r$ $(0 \leq i \leq n)$

hold if and only if the corresponding ghost components satisfy

(23) $A^{(i)} \equiv B^{(i)} \bmod p^{r+i}$ $(0 \leq i \leq n)$

among ghost components are satisfied.

Proof. We apply induction on n. For $n = 0$ the statement is correct. Take $n > 0$. By the induction hypothesis we can assume that congruences (22) and (23) both hold for all $i \leq n-1$. By raising to the p-th power one then clearly gets

$$A_i^p \equiv B_i^p \bmod p^{r+1} \qquad \text{for } i \leq n-1.$$

Again by the induction assumption it follows that

$$(FA)^{(n-1)} \equiv (FB)^{(n-1)} \bmod p^{r+n}.$$

But by (15) we have

$$A^{(n)} = (FA)^{(n-1)} + p^n A_n, \quad B^{(n)} = (FB)^{(n-1)} + p^n B_n.$$

Hence congruences (22) and (23) are equivalent for $i = n$ as well.   □

We now carry out the proof of Lemma 1 by induction on n. For $n = 0$ the assertion is clear; see (18). Take $n > 0$. Set $Z := X * Y$. In view of (15) what we must show is that

(24) $Z^{(n)} \equiv (FZ)^{(n-1)} \bmod p^n.$

Since integer polynomials modulo p can be raised to the power p termwise, the induction assumption yields

$$(FZ)_i \equiv ((FX) * (FY))_i \bmod p \quad \text{for } i \leq n-1.$$

From Lemma 2 we then get

(25) $(FZ)^{(n-1)} \equiv ((FX) * (FY))^{(n-1)} \bmod p^n.$

We have the congruences $X^{(n)} \equiv (FX)^{(n-1)}$ and $Y^{(n)} \equiv (FY)^{(n-1)} \bmod p^n$, so

$$Z^{(n)} = X^{(n)} * Y^{(n)} \equiv (FX)^{(n-1)} * (FY)^{(n-1)} = ((FX) * (FY))^{(n-1)} \bmod p^n.$$

Comparing this with (25) we obtain (24).   □

By analogy with (17), the operations $*$ can be extended naturally to arbitrary vectors $A = (A_0, A_1, \dots)$ and $B = (B_0, B_1, \dots)$ with components in $\mathbb{Z}[X_0, Y_0, X_1, Y_1, \dots]$, as follows: Let $A * B$ the vector whose ghost components are

$$(26) \qquad\qquad (A * B)^{(n)} = A^{(n)} * B^{(n)}.$$

By substituting the variables $X_0, X_1, \dots$ and $Y_0, Y_1, \dots$ as appropriate in Lemma 1, one sees that

$$(27) \qquad\qquad (A * B)_n = S_n^*(A_0, B_0, \dots, A_n, B_n),$$

so the $(A * B)_n$ are integer polynomial expressions in $A_0, B_0, \dots, A_n, B_n$.

Remark. For vectors of the form $A = (A_0, 0, 0, \dots)$, the ghost components are given simply by

$$(28) \qquad\qquad A^{(n)} = A_0^{p^n}.$$

Thus, to multiply by vectors of this form, we can use the rule

$$(29) \quad (A_0, 0, 0, \dots) \cdot (B_0, B_1, \dots, B_n, \dots) = (A_0 B_0, A_0^p B_1, \dots, A_0^{p^n} B_n, \dots),$$

which is easily checked through a comparison of ghost components. It follows in the very same way that the vectors

$$(30) \qquad\qquad 0 = (0, 0, \dots) \quad \text{and} \quad 1 = (1, 0, 0, \dots)$$

satisfy (for any vector B) the equations

$$(31) \qquad\qquad B + 0 = B, \quad B \cdot 0 = 0, \quad B \cdot 1 = B.$$

Finally, one checks without difficulty the "decomposition rule"

$$(32) \qquad B = (B_0, \dots, B_n, 0, 0, \dots) + (0, \dots, 0, B_{n+1}, B_{n+2}, \dots).$$

For later applications it will be important to examine the p-fold sum

$$(33) \qquad\qquad pX = X + X + \dots + X$$

of $X = (X_0, X_1, \dots)$ with itself, under the addition defined by (26). By definition, pX is the vector whose ghost components are given by

$$(34) \qquad\qquad (pX)^{(n)} = pX^{(n)}.$$

Besides the Frobenius operator (14) it is convenient to use the *shift operator* V, defined by

$$(35) \qquad\qquad V(Z_0, Z_1, \dots) = (0, Z_0, Z_1, \dots).$$

We claim the congruence

(36) $$pX \equiv VFX \bmod p;$$

to see this, we must show for every n that $(pX)_n \equiv (VFX)_n \bmod p$. Taking into account Lemma 2 and (34), what's left to check is the congruence

$$pX^{(n)} \equiv (VFX)^{(n)} \bmod p^{n+1},$$

for every n, Since $FV = VF$, the recursive formula (15) does yield

$$(VFX)^{(n)} = (FVX)^{(n)} \equiv (VX)^{(n+1)} = pX^{(n)} \bmod p^{n+1},$$

as desired. Incidentally, the shift operator V behaves *additively*: $V(A + B) = VA + VB$. This is because $V(A + B)^{(n)} = VA^{(n)} + VB^{(n)}$ for all n (note that $(VZ)^{(n)} = pZ^{(n-1)}$).

4. To apply the tools developed in the previous section, let A be for the moment any *commutative ring with unity*, and denote by $\mathcal{W}(A)$ the set of all "vectors" $x = (x_0, x_1, x_2, \dots)$ with components x_i in A. In imitation of (13), let the ghost components of x be defined by

(37) $$x^{(n)} = x_0^{p^n} + px_1^{p^{n-1}} + \cdots + p^n x_n.$$

In this setting the ghost components no longer necessarily determine x; take for instance a field of characteristic p for A (this case will be relevant later). Nonetheless one can introduce the algebraic operations $* = +, \cdot$ on $\mathcal{W}(A)$ by means of the structure polynomials S_n^* from Lemma 1:

For $x = (x_0, x_1, \dots)$ and $x = (y_0, y_1, \dots)$ in $\mathcal{W}(A)$, we define $x * y$ by

(38) $$(x * y)_n = S_n^*(x_0, y_0, x_1, y_1, \dots, x_n, y_n) = S_n^*(x, y).$$

This makes $\mathcal{W}(A)$ into a commutative ring with unity.

The validity of the calculation rules for $\mathcal{W}(A)$ can be justified as follows: Consider first the case of a polynomial ring $\tilde{A}$ over $\mathbb{Z}$ in sufficiently many variables. Since in this case the vectors in $\mathcal{W}(\tilde{A})$ are determined by their ghost components, the calculation rules for $\mathcal{W}(\tilde{A})$ follow from those for $\tilde{A}$, and they can be regarded as systems of integer polynomial identities. By substitution one obtains the validity of the calculation rules on $\mathcal{W}(A)$. The vector $0 = (0, 0, 0, \dots)$ is the additive identity of $\mathcal{W}(A)$ and

(39) $$1 = (1, 0, 0, \dots)$$

is the multiplicative identity of $\mathcal{W}(A)$. We call $\mathcal{W}(A)$ the *ring of Witt vectors over A*.

The terminology "ghost components" is justified by the facts expounded above.

Now suppose that

$$k \text{ is a perfect field of characteristic } p.$$

Since the structure polynomials S_n^* have integer coefficients, we have

(40) $$F(x * y) = Fx * Fy$$

in $\mathcal{W}(k)$, where F is the *Frobenius operator* on $\mathcal{W}(k)$:

(41) $$Fx = (x_0^p, x_1^p, \dots).$$

If V is the *shift operator* on $\mathcal{W}(k)$, so that

(42) $$Vx = (0, x_0, x_1, \dots),$$

then the congruence (36) becomes from now on the *equation*

(43) $$px = FVx = (0, x_0^p, x_1^p, \dots),$$

for every $x \in \mathcal{W}(k)$. Since $k = k^p$, therefore, the ideal $p\mathcal{W}(k)$ of $\mathcal{W}(k)$ consists of precisely those vectors of $\mathcal{W}(k)$ whose 0-th components vanish. Thus this ideal is the kernel of the *homomorphism*

(44) $$\begin{aligned} \mathcal{W}(k) &\to k, \\ x &\mapsto x_0. \end{aligned}$$

More generally, the ideal $p^n\mathcal{W}(k)$ consists of those vectors in $\mathcal{W}(k)$ whose n first components equal 0. Thus if we set

(45) $$w(x) = \min\{i \mid x_i \neq 0\},$$

we obtain a map $w : \mathcal{W}(k) \to \mathbb{Z} \cup \{\infty\}$ with $w(x) = \infty$ only for $x = 0$. Then:

(i) *Every $x \neq 0$ in $\mathcal{W}(k)$ has a unique representation of the form*

(46) $$x = p^r u \quad \text{with } u_0 \neq 0;$$

moreover $r = w(x)$.

Taking into account the homomorphism (44), we easily deduce from (i) that

(47) $$w(xy) = w(x) + w(y)$$

for every $x, y \in \mathcal{W}(k)$. As a consequence:

(ii) $\mathcal{W}(k)$ *has no zero-divisors, and is therefore an integral domain.*

The fraction field of $\mathcal{W}(k)$ will be denoted by

(48) $$Q(k).$$

Because of (i) we also have

(49) $$w(x + y) \geq \min(w(x), w(y)).$$

Extending w in the usual way to the fraction field $Q(k)$ of $\mathcal{W}(k)$, we obtain a *discrete valuation on the field* $Q(k)$. As usual we maintain the notation w for this extension. Its valuation group is $\mathbb{Z}$, and p is a *prime element* for w. We don't know yet whether the valuation ring R of w coincides with $\mathcal{W}(k)$. But (i) implies, at any rate, that every element of $Q(k)$ has the form

$$p^n \frac{u}{v}, \quad \text{with } u, v \in \mathcal{W}(k) \text{ and } w(u) = w(v) = 0.$$

Since $\mathcal{W}(k)/p\mathcal{W}(k) \simeq k$ is a *field*, it follows that $\mathcal{W}(k) \subseteq R$ induces an isomorphism from $\mathcal{W}(k)/p\mathcal{W}(k)$ onto the residue field R/pR of w. Now consider the map $k \to \mathcal{W}(k)$ defined by

(50) $$a \mapsto (a, 0, 0, \ldots) =: \{a\}.$$

Its image S is, in view of (44), a full set of representatives for $\mathcal{W}(k)/p\mathcal{W}(k)$, and hence also for the residue field of w. Because of (29), the set S is *multiplicatively closed*. We claim that

(51) $$x = \sum_{i=0}^{\infty} \{x_i^{p^{-i}}\} p^i$$

for every $x = (x_0, x_1, \ldots) \in \mathcal{W}(k)$. Indeed, by (32) and (i), we have

$$x \equiv (x_0, \ldots, x_n, 0, 0, \ldots) \bmod p^{n+1},$$

and repeated application of (32) and (43) yields

$$(x_0, \ldots, x_n, 0, 0, \ldots) = \sum_{i=0}^{n} \{x_i^{p^{-i}}\} p^i.$$

Let $\hat{R}$ be the valuation ring of the w-completion of $Q(k)$. Applying F2 of Chapter 24 we see that each $x \in \hat{R}$ must also have a representation of the form (51). Thus $\hat{R} = \mathcal{W}(k)$, showing that $Q(k)$ is w-*complete* and $R = \mathcal{W}(k)$ is the valuation ring of w.

One more fact before we review what we have accomplished:

(iii) $Q(k)$ *has characteristic zero.*

For suppose $n = 0$ in $\mathcal{W}(k)$ for some prime n. Then $n = 0$ in k; see (44). Since $\operatorname{char} k = p$ and n is prime, it follows that $n = p$.

Theorem 3. *Let a perfect field k of characteristic p be given. There exists, up to isomorphism, a unique complete, unramified, discrete valued field (of characteristic zero) having k as its residue field, namely, the fraction field of the ring $\mathcal{W}(k)$ of Witt vectors over k.*

The existence part of Theorem 3 has already been shown. As for uniqueness, we will soon prove something sharper:

Theorem 4. *Let K be a field of characteristic 0, complete with respect to a discrete valuation w and having a perfect field k of characteristic p as its residue field. Let R be the valuation ring of w and $e = w(p)$ the ramification index. There exists exactly one homomorphism σ from $Q(k) = \operatorname{Frac}\mathcal{W}(k)$ into K that maps $\mathcal{W}(k)$ into R and makes the diagram*

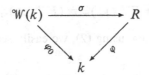

commute; here φ_0 denotes the map (44) and φ the quotient map relative to w. The image K_0 of σ is the unique unramified subfield of K having the same residue field k. The extension K/K_0 is purely ramified of degree e.

Proof. As above, let $\{\ \}: k \to \mathcal{W}(k)$ describe the Teichmüller representatives of $Q(k)$ and $\langle\ \rangle: k \to R$ those of K; let S be the set of the latter. For σ we obviously have no choice but to assign to each element

$$x = (x_0, x_1, \ldots) = \sum_{i=0}^{\infty} \{x_i^{p^{-i}}\} p^i$$

of $\mathcal{W}(k)$ the element

(52)
$$\sigma x = \sum_{i=0}^{\infty} \langle x_i^{p^{-i}}\rangle p^i$$

of R. We must show that σ so defined is a *homomorphism*. Once this is proved, the image K_0 of $Q(k)$ under σ will consist of all elements of K of the form

(53)
$$\sum_{i \gg -\infty} s_i p^i \quad \text{with } s_i \in S,$$

and the set of such elements is a *subfield* of K, which is unramified and has k as its residue field. The remaining assertions of the theorem will also follow (see Theorem 1 in Chapter 24).

Thus the crucial step in the proof is the verification that σ preserves the operations $* = +, \cdot$. Suppose that

(54) $$x * y = z,$$

for x, y, z in $\mathcal{W}(k)$. Because of (40) we have, for every $n \geq 0$,

(55) $$F^{-n}x * F^{-n}y = F^{-n}z.$$

To shorten the notation, set

$$\langle x \rangle := (\langle x_0 \rangle, \langle x_1 \rangle, \dots)$$

and likewise for $\langle y \rangle$ and $\langle z \rangle$. The components of these vectors lie in $S \subseteq R$. Let π be a prime element of K. Since the componentwise validity of (55) is conveyed in terms of integer polynomial identities, one can conclude from (55) that the congruences

$$(F^{-n}\langle x \rangle * F^{-n}\langle y \rangle)_m \equiv (F^{-n}\langle z \rangle)_m \bmod \pi$$

hold for all $m = 0, 1, \dots$. Then, using (2), we easily see that the n-th ghost components satisfy

$$(F^{-n}\langle x \rangle * F^{-n}\langle y \rangle)^{(n)} \equiv (F^{-n}\langle z \rangle)^{(n)} \bmod \pi^{n+1}$$

and therefore also

(56) $$(F^{-n}\langle x \rangle)^{(n)} * (F^{-n}\langle y \rangle)^{(n)} \equiv (F^{-n}\langle z \rangle)^{(n)} \bmod \pi^{n+1}.$$

since $(F^{-n}\langle x \rangle * F^{-n}\langle y \rangle)^{(n)} = (F^{-n}\langle x \rangle)^{(n)} * (F^{-n}\langle y \rangle)^{(n)}$. But, by definition — see (13) and (37) — the n-th partial sum of σx in (52) equals

$$(F^{-n}\langle x \rangle)^{(n)} = \langle x_0 \rangle + p\langle x_1 \rangle^{p^{-1}} + \dots + p^n \langle x_n \rangle^{p^{-n}},$$

and likewise for y and z. One can thus pass to the limit in (56), reaching, as needed,

$$\sigma x * \sigma y = \sigma z.$$

(Of course we also needed here the fact that the set S is multiplicative.)   □

Incidentally, the conclusion of Theorem 2 also follows easily from the arguments just given.

From Theorem 4 it follows in particular that the ring of Witt vectors over the field $\mathbb{F}_p$ coincides with the ring $\mathbb{Z}_p$ of p-adic integers, that is, with the valuation ring of the field $\mathbb{Q}_p$:

(57) $$\mathcal{W}(\mathbb{F}_p) = \mathbb{Z}_p.$$

One can interpret $\mathbb{Z}_p$ as being defined by (57). Then $\mathbb{Q}_p = \operatorname{Frac} \mathbb{Z}_p$ is obtained in a much more constructive way than through the general completion process adopted in Chapter 23.

Remark. Let $q = p^n$ be a power of p. Every $x \in \mathcal{W}(\mathbb{F}_q)$ has a unique representation

$$(58) \qquad x = \sum_{i=0}^{\infty} s_i \, p^i$$

with $s_i \in S = \{0, 1, \zeta, \zeta^2, \ldots, \zeta^{q-2}\}$, where $\zeta = \zeta_{q-1}$ denotes a primitive $(q-1)$-st root of unity. It follows that

$$(59) \qquad \mathcal{W}(\mathbb{F}_q) = \mathbb{Z}_p[\zeta_{q-1}].$$

By Theorem 4 this means that $\mathbb{Z}_p[\zeta_{q-1}]$ is the valuation ring of the *uniquely determined unramified extension* K_n *of degree n over* $\mathbb{Q}_p$. At the same time we recover the equality $K_n = \mathbb{Q}_p(\zeta_{q-1})$.

5. We now turn our attention to another nice application of the *Witt calculus*, namely, the description of the *abelian extensions of exponent* p^n of a field of characteristic p. (This too appeared in the Witt article mentioned earlier).

For now, let A be any commutative ring with unity, and take any $n \in \mathbb{N}$. Letting V be the shift operator on $\mathcal{W}(A)$, consider the set $V^n \mathcal{W}(A)$ of all vectors of the form $(0, \ldots, 0, x_n, x_{n+1}, \ldots)$ in $\mathcal{W}(A)$. Since each polynomial S_n^* in (21) corresponding to one of the operations $* = +, \cdot$ on $\mathcal{W}(A)$ contains only variables of index at most n, the subset $V^n \mathcal{W}(A)$ is an *ideal* of $\mathcal{W}(A)$. Now consider the *quotient ring*

$$(60) \qquad \mathcal{W}_n(A) := \mathcal{W}(A)/V^n \mathcal{W}(A).$$

By (32), the elements of $\mathcal{W}_n(A)$ can be regarded simply as vectors $(x_0, \ldots, x_{n-1})$ of length n; for this reason $\mathcal{W}_n(A)$ is called the *ring of Witt vectors of length n*. (Note that $\mathcal{W}_n(A)$ cannot be taken as a subring of $\mathcal{W}(A)$; this can already be seen in the case $n = 1$, where $\mathcal{W}_1(A) = A$ leads back to the ring A itself.) The shift operator V is *additive* and so gives rise, for $n > 1$, to a well defined additive map

$$(61) \qquad V : \mathcal{W}_{n-1}(A) \to \mathcal{W}_n(A).$$

Now let A be a *field of characteristic* p. The Frobenius operator F defined by (14) is an endomorphism of the ring $\mathcal{W}(A)$; since $FV = VF$, this map passes down to an endomorphism F of the ring $\mathcal{W}_n(A)$. Now consider the map $\wp : \mathcal{W}_n(A) \to \mathcal{W}_n(A)$ defined by

$$(62) \qquad \wp(x) = Fx - x.$$

Clearly $\wp$ is *additive*, and so is an endomorphism of the *abelian group* $\mathcal{W}_n(A)$. For $n = 1$ we have $\wp(x) = x^p - x$ in $A = \mathcal{W}_1(A)$; in this case, therefore, $\wp$ is the familiar map from the *Artin–Schreier theory* of extensions of exponent p (see Chapter 14, Theorem 3 in vol. I). For arbitrary $n \in \mathbb{N}$, the *kernel* of $\wp$ consists of precisely those vectors $(x_0, \ldots, x_{n-1}) \in \mathcal{W}_n(A)$ such that $x_i^p = x_i$, that is, those for which x_i lies in the prime field $\mathbb{F}_p$ of A. Therefore $\ker \wp = \mathcal{W}_n(\mathbb{F}_p)$. In particular, the unit element $e = (1, 0, \ldots, 0)$ of $\mathcal{W}_n(A)$ belongs to $\ker \wp$. Since $px = FVx$, the product $p^n e$ vanishes, but $p^i e$ is nonzero for $i < n$. Thus *the subgroup generated by e has order p^n*, and we see that

$$(63) \qquad \mathcal{W}_n(\mathbb{F}_p) = \ker \wp \simeq \mathbb{Z}/p^n \mathbb{Z}.$$

F2. *Let K be a field of characteristic p, C an algebraic closure of K, and K^s the separable closure of K in C. Then $\wp : \mathcal{W}_n(K^s) \to \mathcal{W}_n(K^s)$ is* **surjective**.

Proof. Take $a = (a_0, \ldots, a_{n-1}) \in \mathcal{W}_n(K^s)$. There exists $x \in C$ such that $\wp(x) = x^p - x = a_0$. Hence $x \in K^s$. Also we have

$$(a_0, \ldots, a_{n-1}) - \wp(x, 0, \ldots, 0) = (0, a'_1, \ldots, a'_{n-1}),$$

for appropriate $a'_i \in K^s$. Since $\wp$ is additive, it suffices to show that every vector of the form $a = (0, a_1, \ldots, a_{n-1})$ lies in $\wp \mathcal{W}_n(K^s)$. By induction on n there exists first an $\alpha \in \mathcal{W}_{n-1}(K^s)$ such that $(a_1, \ldots, a_{n-1}) = \wp \alpha$; applying V we get $a = V \wp(\alpha) = \wp(V\alpha)$. $\qquad\square$

In the situation of F2, let $a \in \mathcal{W}_n(K)$ be given. We claim that any $\alpha \in \mathcal{W}_n(C)$ such that $\wp(\alpha) = a$ lies in $\mathcal{W}_n(K^s)$. Indeed, by F2 there exists $\beta \in \mathcal{W}_n(K^s)$ satisfying $\wp(\beta) = a$. Now $\wp(\alpha - \beta) = 0$, so $\alpha = \beta + \gamma$ with $\gamma \in \ker \wp = \mathcal{W}_n(\mathbb{F}_p) \subseteq \mathcal{W}_n(K)$, and it follows that $\alpha \in \mathcal{W}_n(K^s)$. Thus, if we set $K(\alpha) := K(\alpha_0, \ldots, \alpha_{n-1})$, the extension $K(\alpha)/K$ is algebraic and separable.

More generally, let W be any subset of $\mathcal{W}_n(K)$. Denote by $\wp^{-1}W$ the set of $\alpha \in \mathcal{W}_n(C)$ such that $\wp\alpha \in W$. If $K(\wp^{-1}W)$ is the field obtained from K by adjoining all the $\alpha \in \wp^{-1}W$, the extension $K(\wp^{-1}W)/K$ is *algebraic* and *separable*. It is also *normal*, because for any $\sigma \in G(K^s/K)$ we have $\wp(\sigma\alpha) = \sigma\wp(\alpha) = a$.

If we replace W by the subgroup generated by W and $\wp \mathcal{W}_n(K)$, our extension remains unchanged. For this reason we can assume from now on that W is a *subgroup* of $\mathcal{W}_n(K)$ containing $\wp \mathcal{W}_n(K)$. Let G be the Galois group of $K(\wp^{-1}W)/K$. There is a natural pairing

$$(64) \qquad \begin{aligned} G \times W/\wp\mathcal{W}_n(K) &\to \mathcal{W}_n(\mathbb{F}_p), \\ (\sigma,\ a \bmod \wp\mathcal{W}_n(K)) &\mapsto \chi_a(\sigma), \end{aligned}$$

defined as follows: Given $\sigma \in G$ and $a \in W$, choose $\alpha \in \wp^{-1}W$ and set $\chi_a(\sigma) = \sigma\alpha - \alpha$. Since $\wp(\sigma\alpha) = \sigma\wp(\alpha) = \wp(\alpha)$, the difference $\sigma\alpha - \alpha$ lies in $\ker \wp = \mathcal{W}_n(\mathbb{F}_p)$. If $\wp(\alpha') = a$ for some other α', that is to say if $\alpha' - \alpha \in \ker \wp$, we get $\sigma\alpha' - \sigma\alpha = \sigma(\alpha' - \alpha) = \alpha' - \alpha$. Thus $\chi_a(\sigma)$ is well defined. In case $a \in \wp\mathcal{W}_n(K)$ we have $\chi_a(\sigma) = 0$. For a fixed σ the map $a \mapsto \chi_a(\sigma)$ is clearly additive, so $\chi_a(\sigma)$ only depends on $a \bmod \wp\mathcal{W}_n(K)$. Since $\chi_a(\sigma)$ lies in $\mathcal{W}_n(\mathbb{F}_p)$, and so is invariant under all $\tau \in G$, we have also $\chi_a(\tau\sigma) = \chi_a(\tau) + \chi_a(\sigma)$.

F3. *Let W be a subgroup of $\mathcal{W}_n(K)$ containing $\wp\mathcal{W}_n(K)$, and denote by G the Galois group of the Galois extension $K(\wp^{-1}W)/K$. The canonical group pairing (64) is* **nondegenerate**, *so $K(\wp^{-1}W)/K$ is an abelian extension of exponent p^n.*

Proof. Take $\sigma \in G$ and suppose that $\chi_a(\sigma) = 0$ for all $a \in W$; equivalently, that $\sigma\alpha - \alpha = 0$ for all $\alpha \in \wp^{-1}W$. Then $\sigma = 1$; that is, (64) is nondegenerate with respect to the first variable. Now let $\wp(\alpha) = a \in W$ be given. Then $\sigma\alpha - \alpha = 0$ for every $\sigma \in G$ if and only if α lies in $\mathcal{W}_n(K)$; that is, if and only if $a = \wp(\alpha) \in \wp\mathcal{W}_n(K)$. Thus (64) is also nondegenerate with respect to the second variable. Being a nondegenerate group pairing, (64) gives rise to *injective* group homomorphisms

(65) $$G \to \mathrm{Hom}(W/\wp\mathcal{W}_n(K), \mathcal{W}_n(\mathbb{F}_p)),$$

(66) $$W/\wp\mathcal{W}_n(K) \to \mathrm{Hom}(G, \mathcal{W}_n(\mathbb{F}_p)).$$

Taking the isomorphism $\mathcal{W}_n(\mathbb{F}_p) \simeq \mathbb{Z}/p^n\mathbb{Z}$ into account, this implies that G and $W/\wp\mathcal{W}_n(K)$ are *abelian groups of exponent* p^n. Further: G is *finite* if and only if $W/\wp\mathcal{W}_n(K)$ is, and in this case (65) and (66) are *isomorphisms*. From this we easily deduce that (65) is an isomorphism also when W is arbitrary. The same is true of (66), with the convention that $\mathrm{Hom}(G, \mathcal{W}_n(\mathbb{F}_p))$ there denotes the group of *continuous* homomorphisms from the compact group G into the discrete group $\mathcal{W}_n(\mathbb{F}_p)$. □

In particular, if $W/\wp\mathcal{W}_n(K)$ is *cyclic* (and so finite), the same holds for G. This immediately leads to part (i) of the next theorem:

Theorem 5. *Let K be a field if characteristic p.*

(i) *Take $a = (a_0, \ldots, a_{n-1}) \in \mathcal{W}_n(K)$ and let C be as in F2. If $\alpha \in \mathcal{W}_n(C)$ is a solution of the equation $\wp(\alpha) = a$ and if we set $L = K(\alpha) = K(\alpha_0, \ldots, \alpha_{n-1})$, the extension L/K is cyclic of exponent p^n.*

(ii) *Conversely, every cyclic extension L/K of exponent p^n can be obtained in this way.*

(iii) *For L as in (i), we have $L : K = p^n$ if and only if $a_0 \notin \wp K$.*

Proof. We must show (ii) and (iii). Let L/K be as in (ii) and $G := G(L/K) = \langle\sigma\rangle$. Then $L : K = p^m$, with $m \leq n$. Applying the *trace* $\mathrm{Tr}_{L/K}$ to $c := p^{n-m}e \in \mathcal{W}_n(\mathbb{F}_p) \subseteq \mathcal{W}_n(K)$ we obtain $\mathrm{Tr}_{L/K}(c) = p^n e = 0$. Thus, by F8 in Chapter 13 in vol. I (see also the addendum on p. 108 below), there exists $\alpha \in \mathcal{W}_n(L)$ satisfying $c = \sigma\alpha - \alpha$. Since $\sigma\wp(\alpha) - \wp(\alpha) = \wp(c) = 0$, the image $a := \wp(\alpha)$ lies in $\mathcal{W}_n(K)$. Moreover $\sigma^k\alpha = \alpha + kc$. Since c has order p^m, the equality $\sigma^k\alpha = \alpha$ can only hold when p^m divides k, or equivalently when $\sigma^k = 1$. Thus $L = K(\alpha)$, proving (ii).

By F3, the cyclic groups G and $W/\wp\mathcal{W}_n(K)$ have the same order. Thus $L : K < p^n$ if and only if $p^{n-1}a \in \wp\mathcal{W}_n(K)$. Note that $p^{n-1}a = \wp(p^{n-1}\alpha)$. Since $\ker \wp$ is contained in $\mathcal{W}_n(K)$, we conclude that

$$p^{n-1}a \in \wp\mathcal{W}_n(K) \iff p^{n-1}\alpha \in \mathcal{W}_n(K) \iff \alpha_0^{p^{n-1}} \in K \iff \alpha_0 \in K$$
$$\iff a_0 = \wp(\alpha_0) \in \wp\mathcal{W}_n(K),$$

proving (iii). □

Theorem 6 (Kummer theory of abelian extensions of exponent p^n in characteristic p). *Let K be a field of characteristic p and n a natural number. Then, with the notations above, the map*

(67)
$$W \mapsto K(\wp^{-1}W)$$

is a bijection between the set of subgroups W of $\mathcal{W}_n(K)$ containing $\wp\mathcal{W}_n(K)$ and the set of subfields E of C for which E/K is abelian of exponent p^n. The groups $G = G(K(\wp^{-1}W)/W)$ and $W/\wp\mathcal{W}_n(K)$ stand in full duality; that is, each of them is canonically isomorphic to the (topological) group of characters of the other.

Proof. We need only show that the correspondence (67) is bijective; everything else is covered by F3 and the subsequent remarks. Let E/K be an abelian extension of exponent p^n. Setting $W := \wp\mathcal{W}_n(E) \cap \mathcal{W}_n(K)$ we get $K(\wp^{-1}W) \subseteq E$. To prove equality it suffices to show that $L \subseteq K(\wp^{-1}W)$ for every cyclic subextension L/K of E/K. By part (ii) of Theorem 5 there exists $\alpha \in \mathcal{W}_n(L)$ such that $L = K(\alpha)$ and $\wp(\alpha) \in \mathcal{W}_n(K)$. Hence $\alpha \in \wp^{-1}W$ and therefore $L \subseteq K(\wp^{-1}W)$.

Next we show that (67) is injective. Let E/K be abelian of exponent p^n and set $G = G(E/K)$. For each W such that $E = K(\wp^{-1}W)$ we have the *canonical homomorphism* (66) arising from the map $a \mapsto \chi_a$ derived from (64); but since this is actually an *isomorphism*, W is unique. $\qquad\square$

Addendum. At one point in the proof of Theorem 5 we were a bit too cursory and merely hinted at the justification of the following fact: *If L/K is cyclic, with $G(L/K) = \langle\sigma\rangle$, and if $c \in \mathcal{W}_n(L)$ satisfies $\mathrm{Tr}_{L/K}(c) = 0$, there exists $\alpha \in \mathcal{W}_n(L)$ such that $c = \sigma\alpha - \alpha$.*

Here of course the trace $\mathrm{Tr}_{L/K}(c)$ of an element $c \in \mathcal{W}_n(L)$ is defined as the sum of the elements $\tau c \in \mathcal{W}_n(L)$, where τ runs through $G(L/K)$. To justify the claim, take $c = (c_0, c_1, \ldots, c_{n-1}) \in \mathcal{W}_n(L)$. Then $Sc = (Sc_0, \ldots)$, where $S = \mathrm{Tr}_{L/K}$. Thus the condition $Sc = 0$ implies that $Sc_0 = 0$; that is, there exists (by F8 in Chapter 13 of vol. I) an element $b_0 \in L$ satisfying $c_0 = \sigma b_0 - b_0$. Then

(68)
$$(c_0, c_1, \ldots, c_{n-1}) - (\sigma - 1)(b_0, 0, \ldots, 0) = (0, c_1', \ldots, c_{n-1}')$$

for certain elements $c_i' \in L$. Applying S we get $S(0, c_1', \ldots, c_{n-1}') = 0$. Since $SV = VS$ this implies $S(c_1', \ldots, c_{n-1}') = 0$. By induction, then, there exists $\beta \in \mathcal{W}_n(L)$ satisfying $(c_1', \ldots, c_{n-1}') = (\sigma - 1)\beta$, and we reach the equality

$$(0, c_1', \ldots, c_{n-1}') = V((\sigma - 1)\beta) = (\sigma - 1)V(\beta).$$

Together with (68) this shows that $c \in (\sigma - 1)\mathcal{W}_n(L)$, proving the claim.

27

The Tsen Rank of a Field

1. In a sense, the cornerstone fact of elementary linear algebra is that, over a field, any homogeneous system of linear equations in more unknowns than equations admits a nontrivial solution. We now want to consider more general systems of equations over a field K, of the form

(1)
$$f_1(X_1, \ldots, X_n) = 0,$$
$$f_2(X_1, \ldots, X_n) = 0,$$
$$\vdots$$
$$f_m(X_1, \ldots, X_n) = 0,$$

where there are m equations in n unknowns and $f_1, \ldots, f_m \in K[X_1, \ldots, X_n]$ are polynomials in n variables over K; we also assume these polynomials have no constant term, so (1) always admits the trivial solution $0 = (0, \ldots, 0) \in K^n$. We ask whether there are other solutions to (1). This is of course a fundamental question in algebra, and an extended analysis of it belongs to the vast discipline known as *algebraic geometry*. But it turns out that one can reach very quickly and directly some results that, although elementary, are interesting and remarkable. They are due to C. Tsen, and first appeared in Tsen's article "Zur Stufentheorie und der Quasi-Algebraisch-Abgeschlossenheit kommutativer Körper", *J. Chinese Math. Soc.* **1** (1936), 81–92. Astonishingly, in spite of their elementary nature, these results have hardly ever found their way into algebra textbooks. To make matters worse, even the article just mentioned is seldom cited.

Roughly speaking, what Tsen discovered was the existence of whole series of fields over which a system of the form (1) always has a nontrivial solution, so long as the number of variables is high enough with respect to the degrees of the polynomials involved.

Definition 1. Let $i \geq 0$ be an integer. We call a field K a T_i-*field* if the following condition holds: A system of polynomial equations (1) over K, involving m equations in n variables and whose polynomials have no constant term, always has a nontrivial solution in K^n if the degrees d_k of the polynomials f_k satisfy[1]

$$(2) \qquad\qquad n > d_1^i + d_2^i + \cdots + d_m^i.$$

Since for $i \leq j$ any T_i-field K is trivially also a T_j-field, we can consider the smallest i such that K is a T_i-field. We call this number the *Tsen rank* of K. Here we must allow the value $i = \infty$, of course: for instance, a *formally real* field cannot have a finite Tsen rank, because over such a field the equation

$$\sum_{i=1}^{n} X_i^2 = 0$$

never has a nontrivial solution for any n.

To attest that the preceding definition is not pointless, we exhibit two theorems, whose proof we postpone till Sections 4 and 5. Incidentally, it should be noted that Tsen allowed arbitrary reals to play the role of i in Definition 1, but no fields of noninteger Tsen rank have been found so far.

Theorem 1. *Any **algebraically closed field** has Tsen rank 0. In other words: if $f_1, \ldots, f_m \in K[X_1, \ldots, X_n]$ are polynomials in n variables over an algebraically closed field K, satisfying $f_j(0, \ldots, 0) = 0$ for all $1 \leq j \leq m$, there is a nontrivial common root of $f_1, \ldots, f_m$ in K^n, so long as $n > m$.*

A proof of a weaker version of Theorem 1 — under the assumption that the f_i are all *homogeneous* (of strictly positive degree) — was sketched in problem §19.8 of vol. I. We recall this terminology:

Every polynomial $f \in K[X_1, \ldots, X_n]$ over a commutative ring K has the form

$$(3) \qquad\qquad f = \sum_{v=(v_1,\ldots,v_n)} c_v X_1^{v_1} X_2^{v_2} \ldots X_n^{v_n}.$$

The *degree* of f is defined as the maximum degree of the monomials $X_1^{v_1} X_2^{v_2} \ldots X_n^{v_n}$ occurring in f:

$$(4) \qquad\qquad \deg f = \max \{v_1 + \cdots + v_n \mid c_v \neq 0\}.$$

If all these monomials have the same degree, f is called *homogeneous*. By introducing an additional variable Y, one can obviously state this equivalently by saying that a polynomial $f \in K[X_1, \ldots, X_n]$ is homogeneous of degree d if and only if

$$(5) \qquad\qquad f(YX_1, YX_2, \ldots, YX_n) = Y^d f(X_1, X_2, \ldots, X_n).$$

Remark. The converse of Theorem 1 is also true and easy to prove: *A field K of Tsen rank 0 is algebraically closed.* Indeed, given any polynomial $f(X) =$

---

[1] In this setting we tacitly assume all the f_k to be nonzero, so the case $d_k = -\infty$ is excluded.

$X^n + a_{n-1}X^{n-1} + \cdots + a_0 \in K[X]$ of degree $n \geq 1$, we look at the homogeneous polynomial

(6) $$f^*(X, Y) = X^n + a_{n-1}X^{n-1}Y + \cdots + a_0 Y^n$$

in $K[X, Y]$. By assumption, this has a root $(a, b) \neq 0$ in K^2; then b must be nonzero, and a/b is a root of f in K.

The second promised result confirms a conjecture that goes back to E. *Artin*.

Theorem 2 (Chevalley 1936). *Any finite field K has Tsen rank 1.*

Simple and elegant as their statements become when expressed in terms of the Tsen rank, the preceding results might not by themselves warrant the introduction of Definition 1. But the case is altered when one considers the method used by Tsen to prove Theorem 3 immediately below. Note that unlike Theorem 1, we cannot assert a converse of Theorem 2; Tsen's fecund observation was precisely that, besides finite fields, there are fields of Tsen rank 1 arising in a very different way:

Theorem 3 (Tsen 1933). *Every rational function field $K(X)$ in one variable over an algebraically closed ground field K has Tsen rank 1.*

Tsen's methods actually yield a more general statement (Theorem 4), from which Theorem 3 follows immediately since algebraically closed fields have Tsen rank zero (Theorem 1).

Theorem 4 (Tsen 1936). *If K is a T_i-field, the rational function field $K(X)$ in one variable over K is a T_{i+1}-field. In other words, if K has Tsen rank i, then $K(X)$ has Tsen rank at most $i+1$.*

Proof. Consider a system of m polynomial equations in n unknowns *over $K(X)$*:

(7) $$f_\mu(X_1, \ldots, X_n) = 0, \quad \text{where } 1 \leq \mu \leq m.$$

Assume that all the f_μ vanish at the point $0 \in K^n$, and that

(8) $$n > \sum_{\mu=1}^{m} d_\mu^{i+1},$$

where d_μ is the degree of f_μ. By clearing denominators we can assume also that the coefficients of each f_μ lie in $K[X]$.

The idea is to concoct a system of polynomials *over K* to which we can apply the T_i-property. To this end we restrict our search of solutions of the original system (7) to n-tuples of polynomials in $K[X]$ having degree at most s in X, the choice of s remaining open for now. We can then express each original unknown X_ν as

(9) $$X_\nu = x_{\nu 0} + x_{\nu 1}X + x_{\nu 2}X^2 + \cdots + x_{\nu s}X^s,$$

where $(x_{\nu\sigma})_{\nu\sigma}$ is a family of independent unknowns over $K(X)$. (The easiest way to persuade oneself that this step is valid is through a transcendence degree argument.) By substituting (9) into f_μ and sorting by powers of X, we obtain

$$(10) \qquad f_\mu = f_{\mu 0} + f_{\mu 1} X + f_{\mu 2} X^2 + \cdots + f_{\mu t_\mu} X^{t_\mu},$$

with well defined polynomials $f_{\mu\tau}$ in the $n(s+1)$ variables $x_{\nu\sigma}$ over K, and with powers of X having exponent at most

$$(11) \qquad t_\mu = r + sd_\mu,$$

where X^r is the highest power that occurs in the coefficients of the f_μ. Now, if we can ensure that the system of polynomial equations

$$(12) \qquad f_{\mu\tau} = 0, \quad \text{with } 1 \le \mu \le m \text{ and } 0 \le \tau \le t_\mu$$

(whose coefficients are in K!) admits a nontrivial solution in $K^{n(s+1)}$, we can work our way back, via (9) and (10), to a nontrivial solution in $K[X]^n$ of the given system (7). By assumption, K is a T_i-field; since $f_{\mu\tau}(0) = 0$ and deg $f_{\mu\tau} \le d_\mu$, all that's left to show is that the inequality

$$(13) \qquad n(s+1) > \sum_{\mu=1}^{m}(r+sd_\mu+1)d_\mu^i$$

holds for an appropriate choice of s. After rearrangement, inequality (13) becomes

$$s\left(n - \sum_{\mu=1}^{m} d_\mu^{i+1}\right) > (r+1)\sum_{\mu=1}^{m} d_\mu^i - n.$$

Because of assumption (8), this is obviously true if s is large enough. This proves Theorem 4. $\qquad\square$

It is reasonable to expect that the "at most" in the conclusion of Theorem 4 is superfluous, and that the Tsen rank of a field does go up by 1 upon adjunction of a transcendental element. Later we will indeed establish this, but under an additional assumption (see F4).

When we adjoin an *algebraic* element, by contrast, the Tsen rank can decrease in certain situations; for instance, $\mathbb{R}$ has infinite Tsen rank, yet $\mathbb{C} = \mathbb{R}(i)$, being algebraically closed, has Tsen rank 0. But it's not possible for it to go up, as proved again by Tsen:

Theorem 5. *If a field K has Tsen rank i, any algebraic extension E of K has Tsen rank at most i.*

Proof. Since the coefficients of a given system over E all lie in some finite extension over K, we may as well take E/K to be *finite*. So let a system

$$f_\mu(X_1,\ldots,X_n) = 0, \quad \text{with } 1 \le \mu \le m,$$

be given over E, and suppose that

(14)
$$n > \sum_{\mu=1}^{m} d_{\mu}^{i},$$

where $d_{\mu} = \deg f_{\mu}$. If $b_1, \ldots, b_r$ form a K-basis of E, we can create as in the preceding proof a new family $(x_{\nu\rho})_{\nu,\rho}$ of independent variables over E, in such a way that
$$X_{\nu} = x_{\nu 1} b_1 + x_{\nu 2} b_2 + \cdots + x_{\nu r} b_r$$

for $1 \leq \nu \leq n$. Substituting these expressions into each f_{μ} and collecting terms together, we obtain an equation

(15)
$$f_{\mu} = f_{\mu 1} b_1 + f_{\mu 2} b_2 + \cdots + f_{\mu r} b_r,$$

with well defined polynomials $f_{\mu\rho}$ *over* K in the variables $x_{\nu\rho}$. Then proving the theorem boils down to showing that the system over K given by

$$f_{\mu\rho} = 0, \quad \text{with } 1 \leq \mu \leq m \text{ and } 1 \leq \rho \leq r,$$

with mr equations in nr unknowns, has a nontrivial solution. By assumption, K is a T_i-field; since $f_{\mu\rho}(0) = 0$ and $\deg f_{\mu\rho} \leq d_{\mu}$, it is enough to ensure that

$$nr > \sum_{\mu=1}^{m} r d_{\mu}^{i}.$$

But this is clear because of condition (14).   □

Clearly Theorems 4 and 5 can be combined as follows (in this connection see the remarks after Definition 4 in Chapter 18, vol. I):

Theorem 6. *If K has Tsen rank i and E/K is an extension of transcendence degree j, the Tsen rank of E is at most $i + j$.*

Remark. The Tsen rank is sometimes called the *Tsen level*, but it should not be confused with the *level* of a field, a distinct notion that plays a role in the theory of quadratic forms: this is defined as the smallest natural number s such that -1 is a sum of s squares in the field (or ∞). See F. Lorenz, *Quadratische Formen über Körpern*, Lecture Notes in Mathematics **130**, Springer, 1970.

Speaking of quadratic forms, Theorem 6 has a nice application to them (ibid., §12). As an example of it we mention a theorem: *If F is a function field of transcendence degree n over a real-closed ground field R, every sum of squares in F is a sum of at most 2^n squares.* We thus get a quantitative sharpening of Artin's solution to Hilbert's Seventeenth Problem (Chapter 21, Theorem 1, with $K = R$).

2. We will now look for simple conditions under which the Tsen rank of a field can be bounded *from below*.

Definition 2. Let $i \geq 0$ be an integer, and let $f \in K[X_1, \ldots, X_n]$ be a *homogeneous polynomial of degree* $d \neq 0$ over K. Such an f is also called a *form of degree* d in n variables over K (in the case $d = 2$, a *quadratic form*). We call f a *norm form of level i* if the number of variables n and the degree d of f are related by $n = d^i$ and f only has the trivial root in K^n; we also demand that d be greater than 1, thus excluding forms of the form aX (and only them).

A field K that admits a norm form of level $i+1$ is certainly not a T_i-field — at best it's a T_{i+1}-field. Are there norm forms on any field? The next result addresses the question:

F1. *If K is an extension E of finite degree $n > 1$, the norm map $N_{E/K}$ of E/K is a norm form of level 1 and degree n over K.*

Proof. If $b_1, \ldots, b_n$ is a K-basis of E, we have

$$b_i b_j = \sum_{k=1}^{n} c_{ijk} b_k,$$

with the $c_{ijk} \in K$ uniquely determined. Hence, for an arbitrary element $x = \sum x_i b_i$ in E,

$$x b_j = \sum_k \left(\sum_i c_{ijk} x_i \right) b_k.$$

Thus, in the polynomial ring $K[X_1, \ldots, X_n]$, the homogeneous polynomial defined over K by

$$f(X_1, \ldots, X_n) := \det\left(\left(\sum_i c_{ijk} X_i \right)_{jk} \right),$$

of degree $d = n$, satisfies $f(x_1, \ldots, x_n) = N_{E/K}(x)$ for every $x = \sum x_i b_i$, by the definition of the norm $N_{E/K}$. But $N_{E/K}(x)$ only vanishes for $x = 0$.   □

From F1 we can deduce that a field K that is not algebraically closed admits norm forms of level 1 and arbitrarily high degree: If the algebraic closure of K is an infinite extension over K then K has algebraic extensions of arbitrarily high degree, and if it is finite we are in the case of a real-closed field K (see Chapter 20, Theorem 8), for which — as for any real field — a form with the desired properties is trivially constructed. In a different and wholly elementary way one can actually reach a more general result:

F2. *A field K that has a norm form of level i has norm forms of level i of arbitrarily high degree.*

Proof. Let f, g be norm forms of level i over K, in m and n variables respectively, having degrees e and d. With an array of mn independent variables X_{ij} over K we can consider the form

$$f(g(X_{11}, \ldots, X_{1n}), g(X_{21}, \ldots, X_{2n}), \cdots, g(X_{m1}, \ldots, X_{mn})).$$

We also notate this as

$$f(g \mid g \mid \cdots \mid g).$$

In this way we obtain a form of degree ed in mn variables over K, which is obviously a norm form of level i. Thus by starting with

$$(16) \qquad\qquad f^{(1)} = f(f \mid f \mid \cdots \mid f)$$

and recursively setting $f^{(r)} = f^{(r-1)}(f \mid f \mid \cdots \mid f)$, we obtain norm forms $f^{(r)}$ of level i and increasingly high degrees d^{r+1}.   □

F3. *If a field K admits a norm form of level i, the rational function field $K(X)$ in one variable over K admits a norm form of level $i + 1$.*

Proof. Let f be a norm form of level i over K, and let $n = d^i$ be its number of variables. Taking an array of dn independent variables $X_{\delta\nu}$ over K we can define

$$f_\delta = f(X_{\delta 1}, X_{\delta 2}, \ldots, X_{\delta n}) \quad \text{for } 0 \le \delta \le d - 1,$$

and then consider over $K(X)$ the form

$$f^* = f_0 + f_1 X + f_2 X^2 + \cdots + f_{d-1} X^{d-1}.$$

It has degree d, and $nd = d^{i+1}$ variables. Thus it suffices to show that f^* has no nontrivial root over $K(X)$. Suppose that $(p_{\delta\nu})_{\delta,\nu}$ were such a root. Because f^* is homogeneous, we can suppose that the $p_{\delta\nu}$ lie in $K[X]$ but are not all divisible by X. Because

$$(17) \qquad f_0(p_{01}, \ldots, p_{0n}) + f_1(p_{11}, \ldots, p_{1n})X + \cdots = 0$$

we first conclude that $f_0(p_{01}, \ldots, p_{0n}) \equiv 0 \bmod X$, and hence that

$$(18) \qquad\qquad p_{0\nu} \equiv 0 \bmod X \quad \text{for all } 0 \le \nu \le n,$$

since the root $(p_{01}(0), \ldots, p_{0n}(0))$ of f_0 in K^n is trivial, by our assumption on f. Now, by virtue of (18) — and because f_0 is homogeneous of degree d — we actually have

$$f_0(p_{01}, \ldots, p_{0n}) \equiv 0 \bmod X^d.$$

Thus we obtain, using (17), the congruence $f_1(p_{11}, \ldots, p_{1n}) \equiv 0 \bmod X$, and as above this implies that $p_{1\nu} \equiv 0 \bmod X$. Pushing forward we eventually get the result that *all* the $p_{\delta\nu}$ are divisible by X, a contradiction.   □

F4. *If K has Tsen rank i and admits a norm form of level i, the rational function field $K(X)$ in one variable over K has Tsen rank $i + 1$.*

Proof. We already know from Theorem 4 that $K(X)$ has Tsen rank at most $i + 1$. But F3 says that $K(X)$ is not a T_i-field, so the Tsen rank of $K(X)$ must exceed i. $\square$

F5. *For every integer $i \geq 0$ there are fields of Tsen rank i.*

Proof. By Theorem 1, algebraically closed fields have Tsen rank 0. Now take any field K of rank 1 — say, a finite field (Theorem 2) or the field of rational functions in one variable over an algebraically closed field (Theorem 3). Take a purely transcendental extension

$$(19) \qquad\qquad K_i = K(X_2, X_3, \ldots, X_i)$$

of K with transcendence degree $i-1$. Since K is not algebraically closed (it has Tsen rank 1), it admits a level-1 norm form by F1. We then conclude by induction, using F4 and F3, that K_i has Tsen rank i. $\square$

3. In Definition 1, in formulating the T_i-property of fields, we considered whole *systems* of equations. It might have seemed simpler to start with the following concept instead:

Definition 3 (Lang 1952). We call K a C_i-*field* if any *form* of degree d in n variables over K has a nontrivial root in K^n so long as

$$(20) \qquad\qquad n > d^i.$$

If such a condition is met not only for forms but also for arbitrary polynomials having no constant term, we call K a **strict** C_i-**field** or SC_i-**field**.

The various properties introduced so far are obviously ranked in the order

$$T_i \Rightarrow SC_i \Rightarrow C_i.$$

Since a C_0-field is *algebraically closed* (see right after Theorem 1), we conclude from Theorem 1 that $C_0 \Rightarrow T_0$, that is, $C_0 \Longleftrightarrow T_0$. Whether a similar equivalence holds for some $i \geq 1$ as well is, as far as we know, an open question; counterexamples seem hard to come by. But at least we can say this:

F6 (Tsen). *Let K be a C_i-field admitting a norm form of level i. A system of m forms $f_1, \ldots, f_m$ over K, all of degree d and in n variables, has a nontrivial root over K provided that*

$$n > md^i.$$

A parallel statement holds for polynomials with no constant term, if we start with the assumption that K is an SC_i-field.

Proof. The case $i = 0$ is covered by Theorem 1. Suppose, then, that $i \geq 1$. By assumption, K admits a norm form N of level i. We can assume by F2 that the degree e of N is sufficiently high — specifically, here we want $e^i \geq m$. We form the polynomial

(21) $$N^{(1)} = N(f_1, \ldots, f_m \mid f_1, \ldots, f_m \mid \cdots \mid f_1, \ldots, f_m \mid 0, \ldots, 0),$$

where after each vertical bar n new variables are meant to be used as arguments to the f_μ, as often as a full m-tuple of f_μ's can be accommodated. Thus we obtain a polynomial $N^{(1)}$ in $n[e^i/m]$ variables. If we can show that $N^{(1)}$ has a nontrivial root over K we will be done, because, by the assumption that N is a norm form, the only way the right-hand side of (21) can vanish is if all the arguments vanish, and then *some* n-tuple of coordinates in the root of $N^{(1)}$ will provide a nontrivial common root of all the f_μ.

$N^{(1)}$ has no constant term, and is homogeneous if the f_μ are; moreover its degree is de, because N has no nontrivial roots. So we can apply the defining condition of a C_i-field or SC_i-field if the inequality

$$n \left[\frac{e^i}{m} \right] > (de)^i$$

holds. It certainly does if

$$n \left[\frac{e^i}{m} \right] > md^i \left(\left[\frac{e^i}{m} \right] + 1 \right),$$

since $e^i < m \left([e^i/m] + 1 \right)$. And the last displayed inequality is fulfilled for e large, because $i > 0$ and $n > md^i$ by assumption.   □

F7 (Tsen). *Let K be a C_i-field admitting level-i norm forms of arbitrary degree. A system of m forms $f_1, \ldots, f_m$ of degrees $d_1, \ldots, d_m$ in n variables over K has a nontrivial root over K provided that*

(22) $$n > d_1^i + d_2^i + \cdots + d_m^i.$$

A parallel statement holds for polynomials with no constant term, if we start with the assumption that K is an SC_i-field; in this case, therefore, K has Tsen rank i.

Proof. Set $D = d_1 d_2 \ldots d_m$. For $1 \leq \mu \leq m$, there exists by assumption a norm form N_μ of level i with degree $D/d_\mu =: D_\mu$. (Here we depart from the convention established in Definition 2 and allow the forms aX). Next we fix a set of nD^i independent variables Y_j over K and consider the following m systems of equations, each with the same variables Y_j:

(23)

$$
\begin{array}{ll}
N_1(f_1 \mid f_1 \mid \cdots \mid f_1) = 0, & N_1(f_1 \mid f_1 \mid \cdots \mid f_1) = 0, \quad \ldots\ldots, \\
N_2(f_2 \mid f_2 \mid \cdots \mid f_2) = 0, & N_2(f_2 \mid f_2 \mid \cdots \mid f_2) = 0, \quad \ldots\ldots, \\
\quad \vdots & \quad \vdots \\
N_m(f_m \mid f_m \mid \cdots \mid f_m) = 0, & N_m(f_m \mid f_m \mid \cdots \mid f_m) = 0, \quad \ldots\ldots.
\end{array}
$$

Here the μ-th line contains d_μ^i equations, and the nD^i variables Y_j are apportioned among these equations in order, so that in fact all $nD_\mu^i d_\mu^i = nD^i$ variables do occur. Now consider (23) as a *single* system of equations in nD^i variables; each equation on the μ-th line of (23) has degree $D_\mu d_\mu = D$, so all $d_1^i + d_2^i + \cdots + d_m^i$ equations in (23) have the same degree D. Because of (22) we have

$$nD^i > (d_1^i + d_2^i + \cdots + d_m^i)D^i,$$

and if K is assumed to be a C_i-field or SC_i-field, F6 says that (23) has a nontrivial solution over K. Since the N_i only have trivial roots, we see directly that the system $f_1 = 0$, $f_2 = 0$, ..., $f_m = 0$ also has a nontrivial solution.   □

As mentioned earlier, it would be interesting to know whether in F7 we can make do without the pesky assumption of existence of norm forms. But for F6, at least, this is possible:

F6′ (Lang–Nagata). *The statement of* F6 *is true even without the assumption that* K *has a norm form of level* i.

Proof. The case $i = 0$ is taken care of by the classical Theorem 1, thank goodness. Hence from now on we assume that K is not algebraically closed. By F1 and F2, K admits a norm form N of level $i = 1$ and arbitrarily high degree e. We now start off exactly as in the proof of F6, but by necessity we set $i = 1$ everywhere. Then if we write

$$N^{(1)} = N(f_1, \ldots, f_m \mid f_1, \ldots, f_m \mid \cdots \mid f_1, \ldots, f_m \mid 0, \ldots, 0),$$

the number of variables $n[e/m] =: n_1$ and the degree $ed =: d_1$ of $N^{(1)}$ satisfy

$$n_1 > d_1$$

for e large. If $N^{(1)}$ has a nontrivial root, so does our system $f_1, \ldots, f_m$. Therefore, under the assumption that K is a C_1- or SC_1-field, as the case may be, the conclusion follows.

To deal with the case $i > 1$, we resort to a recursive definition:

$$N^{(r+1)} = N^{(r)(1)}.$$

If n_r is the number of variables of $N^{(r)}$ and d_r its degree, we have

$$(24) \qquad n_{r+1} = n\left[\frac{n_r}{m}\right] \quad \text{and} \quad d_{r+1} = d_r d = ed^{r+1}$$

(since we can assume without loss of generality that $N^{(r)}$ only has the trivial root). So if we can arrange for the inequality

$$n_{r+1} > d_{r+1}^i$$

to hold for e large enough but fixed, we are done.

By (24) we have, first of all,

$$\frac{n_{r+1}}{d_{r+1}^i} = \frac{n[n_r/m]}{d^i d_r^i} \geq \frac{n}{md^i} \frac{n_r}{d_r^i} - \frac{n}{d^i d_r^i},$$

so

$$\frac{n_{r+1}}{d_{r+1}^i} \geq \frac{n}{md^i} \left(\frac{n_r}{d_r^i} - \frac{m}{e^i d^{ir}} \right).$$

By using the corresponding inequalities for $r-1, r-2$, and down to 1, we conclude by recursion, after an easy calculation, that

$$\frac{n_{r+1}}{d_{r+1}^i} \geq \left(\frac{n}{md^i} \right)^r \frac{n_1}{d_1^i} - \frac{m}{e^i} \frac{n}{m} \frac{1}{d^{i(r+1)}} \frac{\left(\frac{n}{m}\right)^r - 1}{\left(\frac{n}{m}\right) - 1}.$$

Now plug in the starting values $n_1 = n[e/m]$ and $d_1 = ed$. Rearranging terms, we are led to

$$\frac{n_{r+1}}{d_{r+1}^i} \geq \left(\frac{n}{md^i} \right)^r \frac{n\left[\frac{e}{m}\right]}{e^i d^i} - \frac{m}{e^i} \frac{n}{m} \frac{1}{d^{i(r+1)}} \frac{m}{m^r} \frac{n^r - m^r}{n - m}$$

$$> \left(\frac{n}{md^i} \right)^{r+1} \frac{e-m}{e^i} - \frac{m}{e^i} \frac{m}{n-m} \frac{n}{md^i} \left(\left(\frac{n}{md^i} \right)^r - \frac{1}{d^{ir}} \right)$$

$$\geq \left(\frac{n}{md^i} \right)^{r+1} \left(\frac{e-m}{e^i} - \frac{m^2}{e^i(n-m)} \right) + 0$$

$$= \left(\frac{n}{md^i} \right)^{r+1} \frac{(e-m)(n-m) - m^2}{e^i(n-m)}.$$

Fix e large enough that $(e-m)(n-m) - m^2 > 0$; the preceding expression then gets arbitrarily large as r grows, because $n/(md^i) > 1$. The proposition follows. $\square$

The result just proved allows us to complement Theorem 6 with the following:

Theorem 7. *If K is a C_i-field and E/K is an extension with transcendence degree j, then E is a C_{i+j}-field.*

Proof. We cannot derive Theorem 7 as a simple logical consequence of Theorem 6, because one of the assumptions has been weakened. But in any case it is enough to consider two cases:

(a) E/K is purely transcendental of transcendence degree 1.

(b) E/K is algebraic.

For case (a) we proceed exactly as in the proof of Theorem 4 (page 116). Moreover now we only have to deal with a single equation $f(X_1, \ldots, X_n) = 0$, and f is assumed *homogeneous* of degree $d > 0$. Then, by homogenity, all (nontrivial) coefficients in (10) are homogeneous of *same* degree $d = \deg f$. Thus instead of the T_i-property of K we need (by F6′) only the C_i-property in this case.

The argument for case (b) also follows along earlier lines; see the proof of Theorem 5. Here what one needs to check is that all the nonzero coefficients in (15) are homogeneous of same degree d. $\qquad\square$

4. In this section and the next we carry out the proof of the two theorems stated at the beginning of the chapter, starting with Theorem 2. So suppose that

(25) $\qquad\qquad\qquad K$ is a finite field with q elements.

Take the ideal

$$(26) \qquad\qquad \mathfrak{a} := (X_1^q - X_1, \; X_2^q - X_2, \; \ldots, \; X_n^q - X_n)$$

of $K[X_1, \ldots, X_n]$. Obviously every polynomial in $\mathfrak{a}$ vanishes at every point of K^n. We will show that, conversely, any $f \in K[X_1, \ldots, X_n]$ vanishing on all of K^n lies in $\mathfrak{a}$. First we introduce a bit of shorthand: A polynomial $g \in K[X_1, \ldots, X_n]$ will be called *reduced* if each variable appears in it only with exponents less than q. Clearly for every $f \in K[X_1, \ldots, X_n]$ there exists a reduced polynomial f_* such that

$$(27) \qquad\qquad f \equiv f_* \bmod \mathfrak{a} \quad \text{and} \quad \deg f_* \leq \deg f.$$

To obtain f_* from f, simply replace factors X_i^q by X_i until this is no longer possible.

Lemma 1. *If a reduced polynomial $g \in K[X_1, \ldots, X_n]$ vanishes at all points of K^n, then $g = 0$.*

Proof. We work by induction on n. For $n = 0$ there is nothing to show. Suppose $n \geq 1$. A reduced polynomial g has a representation

$$g(X_1, \ldots, X_n) = \sum_{i=0}^{q-1} g_i(X_1, \ldots, X_{n-1}) X_n^i,$$

where the $g_i \in K[X_1, \ldots, X_{n-1}]$ are also reduced. Take any $(x_1, \ldots, x_{n-1}) \in K^{n-1}$. By assumption, the element $g(x_1, \ldots, x_{n-1}, X_n)$ of $K[X_n]$ vanishes at every $x \in K$. From degree considerations we then see that

$$g_i(x_1, \ldots, x_{n-1}) = 0 \quad \text{for } 0 \leq i \leq q-1.$$

This holds for every $(x_1, \ldots, x_{n-1}) \in K^{n-1}$, so the induction assumption implies that $g_i(X_1, \ldots, X_{n-1}) = 0$ for $0 \leq i \leq q-1$; that is, g is indeed the zero polynomial. $\qquad\square$

Lemma 2. *Given $f \in K[X_1, \ldots, X_n]$, there exists a **unique** reduced polynomial f^* such that*

$$(28) \qquad\qquad f \equiv f^* \bmod \mathfrak{a}.$$

Moreover f^* satisfies

(29) $$\deg f^* \leq \deg f.$$

If f vanishes on K^n, then f lies in the ideal $\mathfrak{a}$.

Proof. We already know that there exists a reduced polynomial f_* such that $f \equiv f_*$ mod $\mathfrak{a}$ and $\deg f_* \leq \deg f$. Suppose f^* is another reduced polynomial satisfying (28). Then $f^* - f_*$ is also a reduced polynomial, and it lies in $\mathfrak{a}$; thus, it vanishes at all points of K^n. But then Lemma 1 says that $f^* - f_* = 0$, so f^* and f_* are the same.

If f vanishes on K^n, so does f^*, by (28). Again by Lemma 1 we know that $f^* = 0$, and hence that $f \in \mathfrak{a}$.   □

Lemma 3. *If $u \in K[X_1, \ldots, X_n]$ is reduced, vanishes at every point of $K^n \setminus \{0\}$, and takes the value 1 at the point 0, it must equal*

(30) $$u = (1 - X_1^{q-1})(1 - X_2^{q-1}) \ldots (1 - X_n^{q-1}).$$

Proof. The polynomial on the right-hand side of (30) — call it v — satisfies the same assumptions as u. Now apply Lemma 1 to $g = u - v$ to obtain $u = v$.   □

After these preliminary bits of elementary spadework, the proof of Theorem 2 becomes remarkably simple. It all hinges closely on the fact that all nonzero elements of a finite field satisfy

$$x^{q-1} - 1 = 0,$$

if q is the field's cardinality. Let's spell out the statement of Theorem 2:

Theorem 2 (Chevalley). *Let K be a finite field, and let $f_1, \ldots, f_m$ be (nonzero) polynomials in $K[X_1, \ldots, X_n]$, all vanishing at the point $0 \in K^n$. Suppose the sum of their degrees d_μ is less than the number n of variables. Then $f_1, \ldots, f_m$ have a common root in K^n distinct from 0.*

Proof. As before, let q be the cardinality of K. The polynomial

(31) $$f := (1 - f_1^{q-1})(1 - f_2^{q-1}) \ldots (1 - f_m^{q-1})$$

has degree

(32) $$\deg f = (q-1) \sum_{\mu=1}^{m} d_\mu < (q-1)n.$$

Now suppose $f_1, \ldots, f_m$ do not have a common nonzero root in K^n. Then f vanishes at each point of $K^n \setminus \{0\}$, but takes the value 1 at 0. The same is true of the reduced representative f^* of f (see Lemma 2). So by Lemma 3 we must have

$$f^* = (1 - X_1^{q-1})(1 - X_2^{q-1}) \ldots (1 - X_n^{q-1}).$$

It follows that $\deg f^* = (q-1)n$; but this contradicts (32), because $\deg f^* \leq \deg f$ by (29).   □

A remarkable sharpening of Chevalley's Theorem was found by E. Warning ("Bemerkung zur vorstehenden Arbeit von Herrn Chevalley", *Abh. Math. Sém. Univ. Hamburg* **11** (1936), 76–83). Here it what it says:

Theorem 2* (Chevalley–Warning). *Let K be a finite field, and let $f_1, \ldots, f_m$ be (nonzero) polynomials in $K[X_1, \ldots, X_n]$. Suppose the sum of their degrees d_μ is less than the number n of variables. Then the number N of common roots of $f_1, \ldots, f_m$ in K^n satisfies the congruence*

$$(33) \qquad\qquad N \equiv 0 \bmod p,$$

where $p = \operatorname{char} K$. In particular, if $f_1, \ldots, f_m$ do have a common root in K^n, there exist at least $p - 1$ more points in K^n where all m polynomials vanish.

Proof. For $f \in K[X_1, \ldots, X_n]$ arbitrary, set

$$(34) \qquad\qquad S(f) = \sum_{x \in K^n} f(x).$$

Let V be the set of common roots of $f_1, \ldots, f_m$ in K^n. As before, we construct the polynomial

$$(35) \qquad\qquad f := (1 - f_1^{q-1})(1 - f_2^{q-1}) \ldots (1 - f_m^{q-1}).$$

For $x \in V$ we have $f(x) = 1$, whereas $f(x) = 0$ for $x \notin V$. For $x \notin V$, there is at least one nonzero $f_\mu(x)$, so $f(x) = 0$. Thus the polynomial f in (35) evaluates to the *characteristic function of V* on K^n, and consequently the value $S(f)$ in (34) gives the count of elements of V, albeit as an element N_K of K:

$$(36) \qquad\qquad N_K = S(f).$$

Thus the congruence (33) will be proved if we show that $S(f) = 0$. Now, f is a linear combination of monomials of the form

$$X_1^{a_1} X_2^{a_2} \ldots X_n^{a_n},$$

where, by (32),

$$(37) \qquad\qquad a_1 + a_2 + \cdots + a_n < (q-1)n.$$

For each such monomial,

$$S(X_1^{a_1} X_2^{a_2} \ldots X_n^{a_n}) = \left(\sum_{x \in K} x^{a_1} \right) \left(\sum_{x \in K} x^{a_2} \right) \ldots \left(\sum_{x \in K} x^{a_n} \right).$$

Because of (37), there is at least one a_i such that

$$a_i < q - 1.$$

The the proof is completed by resorting to the next lemma:

Lemma 4. *Let K be a finite field with q elements and let $a \geq 0$ be an integer. If a is zero or is not divisible by $q-1$, then*

$$(38) \qquad \sum_{x \in K} x^a = 0.$$

If, contrarywise, a is nonzero and divisible by $q-1$, the sum on the left has value -1.

Proof. For $a = 0$, all terms in the sum equal 1, so the sum is worth $q \cdot 1$, which vanishes since q is a power of the characteristic. Suppose $a \geq 1$. Then $0^a = 0$, and we are left with a sum over the elements of the multiplicative *group* $K^\times$:

$$(39) \qquad S(a) = \sum_{x \in K^\times} x^a.$$

If $q - 1$ divides a, we have $x^a = 1$ for every $x \in K^\times$ and hence $S(a) = q - 1 = -1$. On the other hand, if a is not divisible by $q - 1$, take $z \in K^\times$ of order $q - 1$; then $z^a \neq 1$. But we can write

$$z^a S(a) = \sum_{x \in K^\times} (zx)^a = \sum_{y \in K^\times} y^a = S(a),$$

which, given that $z^a \neq 1$, is only possible if $S(a) = 0$. This completes the proof of Lemma 4 and of the Chevalley–Warning Theorem. $\qquad\square$

Remark. Using rather more sophisticated methods it can be shown that we can replace p in the congruence (33) by the cardinality q of K. In fact something stronger is true: If for some integer $b > 0$ we have $n > b \sum_{\mu=1}^{m} d_\mu$, then

$$(40) \qquad N \equiv 0 \bmod q^b.$$

See J. Ax, "Zeros of polynomials over finite fields", *Amer. J. Math.* **86** (1964), 255–261.

5. We now turn to the proof of Theorem 1. Assume that

$$(41) \qquad K \text{ is an algebraically closed field.}$$

We will draw on the algebraic geometry background and terminology of Chapter 19 (vol. I), rounding it off with some additional observations. In outline, we will prove that adding one polynomial to the system (1) reduces the dimension of the set of common zeros by at most 1: see Theorem 1* on the next page and its corollary.

Let V be a K-variety in K^n (an irreducible algebraic K-set of K^n), and let

$$\mathfrak{a} := \mathfrak{I}(V) \subseteq K[X_1, \ldots, X_n]$$

be its ideal. The affine K-algebra $K[V] = K[X_1, \ldots, X_n]/\mathfrak{a} = K[x_1, \ldots, x_n]$ is an integral domain, because $\mathfrak{a}$ is a prime ideal (V is irreducible). Here $x_1, \ldots, x_n$ are the images of $X_1, \ldots, X_n$ in $K[V]$ under the quotient homomorphism.

Let h be a polynomial in $K[X_1, \ldots, X_n]$ that does not vanish on (all of) V. Since $h \notin \mathfrak{a}$ we can form the extension $K[x_1, \ldots, x_n, 1/h(x_1, \ldots, x_n)]$, which is still an integral domain. Now consider the affine K-algebra

$$K[X_1, \ldots, X_n, X_{n+1}]/(\mathfrak{a}, hX_{n+1} - 1) = K[y_1, \ldots, y_n, y_{n+1}].$$

There is obviously a unique surjective homomorphism of K-algebras

$$(42) \qquad K[y_1, \ldots, y_n, y_{n+1}] \to K[x_1, \ldots, x_n, 1/h(x_1, \ldots, x_n)]$$

with $y_i \mapsto x_i$ for $1 \leq i \leq n$ and $y_{n+1} \mapsto 1/h(x_1, \ldots, x_n)$. From $g(x_1, \ldots, x_n) = 0$ for $g \in K[X_1, \ldots, X_n]$ there follows $g \in \mathfrak{a}$, so $g(y_1, \ldots, y_n) = 0$. Hence (42) is injective on $K[y_1, \ldots, y_n]$ and so injective overall, because every element of $K[y_1, \ldots, y_n, y_{n+1}]$ has the form $g(y_1, \ldots, y_n)/h(y_1, \ldots, y_n)^k$ for some element $g(y_1, \ldots, y_n)$ of $K[y_1, \ldots, y_n]$.

Hence (42) is an *isomorphism*. The zero set

$$(43) \qquad V_h := \mathcal{N}(\mathfrak{a}, hX_{n+1} - 1)$$

of the ideal $(\mathfrak{a}, hX_{n+1} - 1)$ of $K[X_1, \ldots, X_{n+1}]$ is therefore a *K-variety* of K^{n+1}, with affine K-algebra

$$K[V_h] = K[y_1, \ldots, y_n, y_{n+1}] \simeq K[x_1, \ldots, x_n, 1/h(x_1, \ldots, x_n)].$$

In particular we have $\mathrm{TrDeg}(K(V_h)/K) = \mathrm{TrDeg}(K(V)/K)$, and hence

$$(44) \qquad \dim V_h = \dim V.$$

As a subset of K^{n+1}, by definition, V_h consists of all $(a_1, \ldots, a_n, a_{n+1})$ in K^{n+1} such that

$$(45) \qquad (a_1, \ldots, a_n) \in V \quad \text{and} \quad a_{n+1}h(a_1, \ldots, a_n) = 1.$$

We thus have a natural embedding $V_h \to V$, through which V_h can be identified with the set

$$(46) \qquad \{a \in V \mid h(a) \neq 0\} \subset V.$$

After all this preparation we can tackle the proof of a fundamental result:

Theorem 1* (Krull). *Let K be an algebraically closed field and let $V \subset K^n$ be a K-variety. Suppose the polynomial $f \in K[X_1, \ldots, X_n]$ does not vanish at all points of V. Then all the irreducible K-components of the algebraic K-set*

$$(47) \qquad V \cap \mathcal{N}(f) = \{a \in V \mid f(a) = 0\}$$

have dimension $r - 1$, where $r = \dim V$.

Proof. (i) We can assume that $V \cap \mathcal{N}(f) \neq \varnothing$. In fact, we can assume without loss of generality that $V \cap \mathcal{N}(f)$ is irreducible (has a single component), as we now proceed to show. Let W be any component of $V \cap \mathcal{N}(f)$. There is certainly a polynomial h that vanishes on all components of $V \cap \mathcal{N}(f)$ aside from W, but not on W. We then construct V_h and W_h as above and derive from (45) that

$$V_h \cap \mathcal{N}(f) = W_h.$$

Since V_h and W_h are irreducible and since, thanks to (44), the dimensions satisfy $\dim V = \dim V_h$ and $\dim W = \dim W_h$, we have acheived the desired reduction to the case of a single component.

(ii) Let $A = K[V]$ be the affine K-algebra associated to V. We regard f as an element of A. By assumption, $f \neq 0$. Since we are now assuming that $V \cap \mathcal{N}(f)$ is irreducible, the radical of the ideal (f) of A must be prime, say

$$(48) \qquad \sqrt{(f)} = \mathfrak{p}.$$

First we treat the special case $V = K^n$. Then $A = K[X_1, \ldots, X_n]$ is a *unique factorization domain*, so $\mathfrak{p}$ in (48) can only be a principal ideal: $\mathfrak{p} = (f_1)$. But it's easy to check (see proof of Lemma 3 in Chapter 19, vol. I) that the affine K-algebra

$$K[X_1, \ldots, X_n]/(f_1)$$

corresponding to $V \cap \mathcal{N}(f) = \mathcal{N}(f)$ has transcendence degree $n-1$.

(iii) We now reduce the general case to the special case just treated. By Noether's Normalization Theorem (Chapter 18, Theorem 4 in vol. I), there is a subalgebra of A that is a *polynomial algebra*

$$(49) \qquad R = K[x_1, \ldots, x_r]$$

in r variables $x_1, \ldots, x_r$, and such that A/R is *integral*. Being a UFD, the ring R is *integrally closed* in its fraction field $F := \operatorname{Frac} R$. If we set $E := \operatorname{Frac} A$, the field extension E/F is finite, and we can consider the element

$$(50) \qquad f_0 := N_{E/F}(f).$$

This element lies in R (Chapter 16, F9 in vol. I), because A/R is integral, R is integrally closed and $f \in A$. More is true: The minimal polynomial $X^n + a_{n-1}X^{n-1} + \cdots + a_0$ of f over F only has coefficients in R. Taking (48) into consideration we see that $f^n + a_{n-1}f^{n-1} + \cdots + a_0 = 0$ implies $a_0 \in \mathfrak{p}$, and hence also $f_0 \in \mathfrak{p}$ (since $N_{E/K}(f)$ equals a power of a_0, up to a sign). We claim that f_0 has radical

$$(51) \qquad \sqrt{(f_0)} = \mathfrak{p} \cap R$$

in R. As already seen, f_0 lies in $\mathfrak{p} \cap R$; thus we already know that $\sqrt{(f_0)} \subseteq \mathfrak{p} \cap R$. Conversely, let g be any element of $\mathfrak{p} \cap R$. In view of (48), some power g^s of g has the representation

$$g^s = f \cdot h, \quad \text{with } h \in A.$$

Since $g \in R$, an application of the norm $N = N_{E/F}$ yields

$$N(g^s) = N(g)^s = g^{[E:F]s} = N(f)N(h) = f_0 h_0,$$

with $h_0 \in R$. It follows that $g \in \sqrt{(f_0)}$, as claimed.

From (51) the assertion of the theorem follows: Indeed, $\overline{A} := A/\mathfrak{p}$ is integral over $\overline{R} := R/\mathfrak{p} \cap R$, so $\mathrm{TrDeg}(\overline{A}/K) = \mathrm{TrDeg}(\overline{R}/K)$. Taking (48) and (51) into account, we conclude that

$$\dim(V \cap \mathcal{N}(f)) = \dim \mathcal{N}(f_0).$$

But as established earlier, the hypersurface $\mathcal{N}(f_0)$ of K^r has dimension $r - 1$, and r was the dimension of V. □

Corollary. *Suppose that K is an algebraically closed field, and let $f_1, \ldots, f_m \in K[X_1, \ldots, X_n]$ be polynomials in n variables over K. Then each component W of $\mathcal{N}(f_1, \ldots, f_m)$ has dimension at least $n - m$.*

Proof. We apply induction on m. The case $m = 0$ being obvious, assume that $m \geq 1$. Being an irreducible subset of $\mathcal{N}(f_1, \ldots, f_{m-1})$, the set W is contained in an irreducible component V of $\mathcal{N}(f_1, \ldots, f_{m-1})$. Since $W \subseteq V \cap \mathcal{N}(f_m) \subseteq \mathcal{N}(f_1, \ldots, f_m)$, W is an irreducible component of $V \cap \mathcal{N}(f_m)$. By induction we get

$$\dim V \geq n - (m - 1).$$

But by Theorem 1* we have

$$\dim W \geq \dim V - 1.$$

Altogether we get $\dim W \geq n - m$. □

If, in the situation of the corollary, we assume in addition that all the f_i vanish at the point 0 of K^n, we get $\mathcal{N}(f_1, \ldots, f_m) \neq \varnothing$, so there exists an irreducible component W of $\mathcal{N}(f_1, \ldots, f_m)$. If $n > m$, therefore, we have $\dim W \geq 1$, and so W cannot contain only one point. This proves Theorem 1.

Remark. For a more general appreciation of Theorem 1* in the context of commutative algebra, see p. 112 of Atiyah and Macdonald's *Introduction to Commutative Algebra*, Addison-Wesley 1969.

Fundamentals of Modules

1. In this chapter R will always denote a *commutative ring with unity*. We start with some recapitulation:

Definition 1. By an *R-algebra* A we mean a ring A with unity that is at the same time an R-module, so that the equalities

(1) $$\alpha(ab) = (\alpha a)b = a(\alpha b)$$

hold for every $\alpha \in R$ and $a, b \in A$.

Remarks. (i) In the literature a more general definition is sometimes used: namely, an R-algebra is an R-module A endowed with a *bilinear* map

$$A \times A \to A, \qquad (a, b) \mapsto ab.$$

In that case an R-algebra in our sense is called an *associative R-algebra with unity*.

(ii) One generally denotes both the unit element of R and that of A with 1. If necessary one can use the notation 1_R and 1_A. The map

(2) $$\alpha \mapsto \alpha 1_A$$

is a ring homomorphism from R onto the R-subalgebra $R1_A$ of A. This subalgebra is always *central*, that is, it lies in the center

$$Z(A) = \{a \in A \mid ax = xa \text{ for every } x \in A\}$$

of A. The center $Z(A)$ is always an R-subalgebra of A.

(iii) If A is a ring with unity, any ring homomorphism φ from R into the center of A makes A into an R-algebra: $\alpha a = \varphi(\alpha)a$.

(iv) If the map (2) is *injective*, we can identify R with the subalgebra $R1_A$.

(v) In many general statements about R-algebras, the underlying ring R plays no role. As a rule we will forgo the reference to R and use the plain term *algebra*. In this case, when we talk of an *algebra homomorphism* $\varphi : A \to B$, it is tacitly assumed that A and B are algebras over the same R. An algebra homomorphism $A \to B$ is required to carry 1_A to 1_B. Likewise we want a *subalgebra* C of an algebra A to contain the unity of A, so the inclusion $C \to A$ is a homomorphism of algebras.

(vi) It is worth mentioning explicitly that every ring with unity is a $\mathbb{Z}$-algebra, so the study of rings (with unity) is no more general than the study of algebras.

(vii) An *ideal* of an algebra A will always be for us a *two-sided ideal* of the ring A, that is, a subset I of A such that $I + I \subseteq I$ and $AI = I = IA$. Then I is automatically an R-submodule of A, and the quotient map $A \to A/I$ is a homomorphism of R-algebras, with kernel I.

(viii) A subset N of an algebra A satisfying $N + N \subseteq N$ and $AN = N$ will be called a *left ideal* of A. (This collides with other habitual conventions, but we put up with that nonetheless, for the sake of unambiguous terminology.)

(ix) An algebra $A \neq \{0\}$ is called a *division algebra* if every nonzero element in A is invertible in A. When necessary to draw attention to the underlying ring R, one talks of R-*division algebras*. Aside from having an R-module structure, a division algebra satisfies, multiplicatively, all the axioms of a field other than commutativity; for this reason division rings are also called *skew fields*. It is immediately seen that *the center of a division algebra is a field*. If D is a division algebra, a D-module is also called a *vector space over* D.

(x) Many results about algebras later on will require the underlying ring of scalars to be a *field* K. When we talk about a K-*algebra* A, the convention that

(3) K is a field

will always be in effect, and moreover we will assume tacitly that $A \neq \{0\}$, that is, $1_A \neq 0$. Then (2) is injective, so K can be regarded as a subalgebra of A.

A K-algebra is, in particular, a K-vector space. The dimension of a K-vector space V will always be denoted by

$$V : K.$$

The vector space structure of a K-algebra A is fully determined by $A : K$.

(xi) If M is an R-module, the set $\mathrm{End}_R(M)$ of all R-module endomorphisms of M has a natural R-algebra structure. This provides a class of natural examples of algebras.

More generally:

F1. *Let A be an R-algebra and let M, N be A-modules. The set*

$$\mathrm{Hom}_A(M, N)$$

of all A-module homomorphisms $f : M \to N$ from M to N has a natural structure as an R-module. The R-module $\mathrm{End}_A(M) = \mathrm{Hom}_A(M, M)$ of all A-module endomorphisms of M has a natural R-algebra structure; we call this the endomorphism algebra of the A-module M.

Proof. For $f, g \in \mathrm{Hom}_A(M, N)$ and $\alpha \in R$, define $f + g$ and αf by setting

$$(f + g)(x) = f(x) + g(x), \quad (\alpha f)(x) = \alpha f(x).$$

When $M = N$, the product fg is defined by composition:

$$(fg)(x) = f(g(x)).$$

Clearly $f + g$ and αf do lie in $\mathrm{Hom}_A(M, N)$, and fg lies in $\mathrm{End}_A(M)$ if $M = N$. It is easy to check that the operations just defined make $\mathrm{Hom}_A(M, N)$ into an R-module and $\mathrm{End}_A(M)$ into an R-algebra. (The reader should explore why $\mathrm{Hom}_A(M, N)$ is not by the same token an A-module if A is not commutative.) $\square$

Obviously $\mathrm{End}_A(M)$ is a subalgebra of the R-algebra $\mathrm{End}_R(M)$. M can be regarded naturally as a module over the algebra $\mathrm{End}_R(M)$, hence also as a module over any subalgebra of $\mathrm{End}_R(M)$. In particular, M is an $\mathrm{End}_A(M)$-module in a natural way.

Definition 2. Let A be any R-algebra.

(i) By virtue of the multiplication law $A \times A \to A$, we can regard A as an A-module as well; this module is denoted by

$$A_l.$$

The submodules of the A-module A_l are precisely the left ideals of A.

(ii) If we give the R-module A the reverse multiplication law $(a, b) \mapsto a \circ b := ba$, we get an R-algebra, denoted by

$$A^\circ$$

and called the *opposite algebra* of A.

(iii) We denote by A_r the A°-module $(A^\circ)_l$; thus multiplication by scalars in A_r is given by $(a, x) \mapsto xa$. *The submodules of A_r are precisely the right ideals of A.*

(iv) For a given A-module M we have an algebra homomorphism

$$(4) \qquad\qquad A \to \mathrm{End}_R(M)$$

assigning to each $a \in A$ the endomorphism a_M of M defined by *left multiplication by a*: $a_M(x) = ax$.

(v) For $M = A$ this defines a natural homomorphism

$$(5) \qquad\qquad \lambda : A \to \mathrm{End}_R(A)$$

which is an isomorphism onto its image: $A \simeq \lambda(A)$. Similarly, To every $a \in A$ we can associate the endomorphism $_Aa$ of *right multiplication* by a: $_Aa(x) = xa$. We thus obtain an algebra homomorphism

$$(6) \qquad\qquad \rho : A^\circ \to \mathrm{End}_R(A),$$

and again $A^\circ \simeq \rho(A)$ canonically.

F2. $\rho(A)$ *coincides with the subalgebra* $\mathrm{End}_A(A_l)$ *of* $\mathrm{End}_R(A)$; *similarly,* $\lambda(A)$ *is the subalgebra* $\mathrm{End}_{A^\circ}(A_r)$ *of* $\mathrm{End}_R(A)$.

Proof. It is clear that each $_Aa$ lies in $\mathrm{End}_A(A_l)$, because for every $b, x \in A$ we have $_Aa(bx) = (bx)a = b(xa) = b(_Aa(x))$. Conversely, suppose $f \in \mathrm{End}_A(A_l)$. Then

$$f(x) = f(x1) = xf(1) \quad \text{for all } x \in A.$$

Setting $a := f(1)$, we have $f = _Aa$, right multiplication by a. This proves the assertion about $\rho(A)$. The proof for $\lambda(A)$ is completely analogous; alternatively, the equalities $(A^\circ)^\circ = A$ and $\rho(A^\circ) = \lambda(A)$ provide a reduction to the half already proved. □

Trivial as the result in F2 may seem, it is of great importance. We want to stress again what we have proved:

F3. *By associating to each element of an algebra A the endomorphism of right or left multiplication by a, one obtains natural isomorphisms*

$$(7) \qquad\qquad A^\circ \simeq \mathrm{End}_A(A_l), \quad A \simeq \mathrm{End}_{A^\circ}(A_r).$$

Remark. Let A be an R-algebra, and for $n \in \mathbb{N}$ let

$$M_n(A)$$

be the set of $n \times n$ matrices with coefficients in A. Clearly $M_n(A)$ has an R-module structure. The usual multiplication of matrices, obtained by defining the product ab of $a = (a_{ij})_{i,j}$ and $b = (b_{ij})_{i,j}$ as

$$(8) \qquad\qquad ab = \left(\sum_{k=1}^{n} a_{ik}b_{kj} \right)_{i,j},$$

makes $M_n(A)$ into an R-algebra, called the *(full) algebra of $n \times n$-matrices over A*. This will be most easily justified by identifying the elements of $M_n(A)$ with those of the algebra $\mathrm{End}_{A^\circ}(A^n)$ of endomorphisms of the A°-module $A^n = A_r^n$, and checking that the addition and multiplication on $\mathrm{End}_{A^\circ}(A^n)$ are described by matrix addition and multiplication. The matrix $a = (a_{rs})_{r,s}$ associated to a map $f \in \mathrm{End}_{A^\circ}(A^n)$ is the one for which the equations

$$(9) \qquad\qquad fe_i = \sum_j e_j a_{ji}$$

are satisfied, where $e_1, \ldots, e_n$ form the canonical basis of A^n. (An equation $y = fx$ is seen to translate to the coordinate equations

(10) $$y_i = \sum_j a_{ij} x_j,$$

just as in elementary linear algebra; only a certain amount of attention is required since A is not assumed to be commutative.) We obtain in this way a canonical isomorphism

(11) $$\operatorname{End}_{A^\circ}(A^n) \simeq M_n(A).$$

For the endomorphism algebra $\operatorname{End}_A(A^n)$ of the A-module $A^n = A_l^n$ we then get

(12) $$\operatorname{End}_A(A^n) \simeq M_n(A^\circ).$$

The occurence of the opposite algebra A° of A in these relations is part of the nature of things, and cannot be avoided through (say) a cleverer choice of notation.

The first assertion of the next result is a generalization of (12):

F4. *Let N be an A-module and consider the A-module N^n, with $n \in \mathbb{N}$. There is a natural isomorphism*

(13) $$\operatorname{End}_A(N^n) \to M_n(\operatorname{End}_A(N)), \quad f \mapsto (f_{ij})_{i,j},$$

satisfying for every element $x = (x_i)_i \in N^n$ the relation

(14) $$fx = \left(\sum_j f_{ij} x_j \right)_i.$$

With the abbreviations $N' = N^n$, $C = \operatorname{End}_A(N)$, $C' = \operatorname{End}_A(N')$, the natural embedding $\operatorname{End}_C(N) \to \operatorname{End}_C(N')$ gives rise to an isomorphism

(15) $$\operatorname{End}_C(N) \to \operatorname{End}_{C'}(N').$$

Proof. Let $\pi_i : N^n \to N$ be the projections and $\iota_j : N \to N^n$ the injections for the decomposition $N' = N^n$. They satisfy the relations

(16) $$\pi_i \iota_j = \delta_{ij}, \quad \sum_k \iota_k \pi_k = 1,$$

where $\delta_{ij} = 0, 1 \in C = \operatorname{End}_A(N)$ depending on whether $i \neq j$ or $i = j$. For every $f \in \operatorname{End}_A(N')$, the maps

(17) $$f_{ij} := \pi_i f \iota_j$$

are A-endomorphisms of N. Using (16) one confirms easily that the map in (13) is an isomorphism and satisfies (14): just note that $\pi_i f g \iota_j = \pi_i f \left(\sum_k \iota_k \pi_k \right) g \iota_j = \sum_k f_{ik} g_{kj}$ and $\sum_{i,j} \iota_i f_{ij} \pi_j = \sum_{i,j} \iota_i \pi_i f \iota_j \pi_j = f$, while $\pi_r \left(\sum_{i,j} \iota_i f_{ij} \pi_j \right) \iota_s = f_{rs}$.

Now take $g \in \mathrm{End}_C(N)$. Via

$$(18) \qquad\qquad gx = (gx_i)_i,$$

we regard g as an element of $\mathrm{End}_C(N')$, and show first that we do have

$$gf = fg \quad \text{for all } f \in C' = \mathrm{End}_A(N').$$

Indeed, because of (14), $gfx = g\left(\sum_j f_{ij} x_j \right)_i = \left(\sum_j g f_{ij} x_j \right)_i = \left(\sum_j f_{ij} g x_j \right)_i = fgx$. All that remains unproved is the surjectivity of (15). So take $h \in \mathrm{End}_{C'}(N')$; we must show that

$$(19) \qquad\qquad \pi_i h \iota_j = 0 \quad \text{and} \quad \pi_j h \iota_j = \pi_i h \iota_i$$

for every $i \neq j$. By assumption, h commutes (in particular) with $f = \iota_j \pi_i$, so $h \iota_j = h \iota_j (\pi_i \iota_i) = h(\iota_j \pi_i)\iota_i = (\iota_j \pi_i) h \iota_i$, and hence

$$h\iota_j = \iota_j \pi_i h \iota_i.$$

Left multiplication by π_i and π_j leads to (19). $\qquad\qquad\qquad\qquad\square$

We can apply F4 to the A-module A_l and obtain an isomorphism

$$\mathrm{End}_A(A_l^n) \to M_n(\mathrm{End}_A(A_l)).$$

But $\mathrm{End}_A(A_l)$ is canonically isomorphic to A°, by F3; we thus get a canonical isomorphism

$$(20) \qquad\qquad \mathrm{End}_A(A^n) \to M_n(A^\circ).$$

If we consider A^n as an $M_n(A^\circ)$-module by means of this isomorphism, expression (14) gives, for $a = (a_{ij})_{i,j} \in M_n(A^\circ)$ and $x = (x_i)_i \in A^n$,

$$(21) \qquad\qquad ax = \left(\sum_j x_j a_{ij} \right)_i$$

relative to this module structure, and incidentally we recover again the isomorphism (12). If we replace A by A°, we obtain the natural isomorphism

$$(22) \qquad\qquad \mathrm{End}_{A^\circ}(A^n) \to M_n(A).$$

The $M_n(A)$-module structure thus defined on A^n is then given by

$$(23) \qquad\qquad ax = \left(\sum_j a_{ij} x_j \right)_i,$$

which is the usual multiplication rule for an $n \times n$ matrix a by a column vector x.

F5. *For any algebra A, the maps (20) and (15) give rise to natural isomorphisms*

$$\text{(24)} \qquad \text{End}_A(A^n) \simeq M_n(A^\circ),$$

$$\text{(25)} \qquad A \simeq \text{End}_{\text{End}_A(A^n)}(A^n),$$

$$\text{(26)} \qquad A^\circ \simeq \text{End}_{M_n(A)}(A^n).$$

Since the canonical isomorphisms (25) and (26) associate to a $\in A$ respectively left and right multiplication by a, we conclude that the algebras in (24) have centers

$$\text{(27)} \qquad Z(\text{End}_A(A^n)) = Z(A)\,\text{id}_{A^n}, \quad Z(M_n(A)) = Z(A)E_n,$$

where E_n is the $n \times n$ identity matrix in $M_n(A)$.

Proof. The isomorphism (24) has already been explained. We consider (15) in the case $N = A_l$. By F2 we can write $C = \text{End}_A(A_l) = \rho A$, and since $\text{End}_{\rho A}(A_l) = \text{End}_{A^\circ}(A_r) \simeq A$ by F3, we obtain the isomorphism (25) as required. Switching over to A° we obtain (26), taking into account the isomorphism (22).

Finally, suppose $f \in \text{End}_A(A^n)$ commutes with every element of $\text{End}_A(A^n)$. Then f belongs to the algebra in the right-hand side of (25), and so is of the form $f = a_{A^n}$, for $a \in A$. But in view of (25), since f lies in $\text{End}_A(A^n)$ it commutes with c_{A^n} for every $c \in A$. Therefore $ac = ca$ for every $c \in A$. $\qquad\qquad \square$

Remark. As can easily be checked, the map $a \mapsto {}^t a$ taking each matrix into its transpose ${}^t a$ establishes an isomorphism

$$\text{(28)} \qquad M_n(A)^\circ \simeq M_n(A^\circ).$$

2. In this section when we talk about a *module*, we mean an A-module over an R-algebra A.

Definition 3. A module $N \neq 0$ is called *simple* if 0 and N are the only submodules of N. Note that the zero module is excluded.

A module M is called *semisimple* if it is a direct sum of simple submodules of M.

Recall that saying that a module M is a direct sum of submodules M_i ($i \in I$), or that $M = \bigoplus_{i \in I} M_i$, means that the natural homomorphism $\bigoplus_{i \in I} M_i \to M$ is bijective; in other words, M is the sum of the M_i, and a finite sum of elements from distinct submodules M_i can only vanish if all the elements vanish.

Remarks. (1) Simple modules are also called *irreducible*, and semisimple modules *completely reducible*.

(2) A left ideal N of an algebra A is a simple A-module if and only if it is a *minimal left ideal of A* — that is, if N is minimal among nonzero left ideals of A. Of course, an algebra need not have any minimal left ideals: just think of $\mathbb{Z}$.

(3) A submodule L of a module M is *maximal* (among submodules of M distinct from M) if and only if the quotient module M/L is *simple*. Regarding the question whether maximal submodules must exist, we have:

F6. *Let M be a **finitely generated** module. If M_0 is any submodule of M distinct from M, there exists a maximal submodule of M containing M_0.*

Proof. Consider the set X of all submodules N of M such that $M_0 \subseteq N \neq M$. If Y is a nonempty totally ordered subset of X, the union U of all $N \in Y$ is obviously a submodule of M. We show that $U \neq M$. Indeed, M has a finite set of generators $x_1, \ldots, x_n$. If M and U coincided, each x_i would lie in some $N_i \in Y$. But Y is totally ordered, so the generators would lie in some single N_j, contradicting $N_j \neq M$.

The result now follows through an application of *Zorn's Lemma.*   □

F7. *Let N be a nonzero A-module. There is equivalence between:*

(i) *N is simple.*

(ii) *Every element $x \neq 0$ of N generates N; that is, $N = Ax$.*

(iii) *There exists a maximal left ideal L of A such that $N \simeq A/L$.*

Proof. (i) $\Rightarrow$ (ii): For every $x \neq 0$, the set Ax is a nonzero submodule of N.

(ii) $\Rightarrow$ (i): Every nonzero submodule of N contains a submodule of the form Ax, with $x \neq 0$.

(i) $\Rightarrow$ (iii): By assumption, N has a nonzero element x. The map $a \mapsto ax$ is a *surjective* module homomorphism from A into N; its kernel L is a left ideal of A, and A/L is isomorphic to N. Since N is simple, L is maximal, by Remark 3 following Definition 3.

(iii) $\Rightarrow$ (i) follows immediately from Remark 3 to Definition 3.   □

F8. *Let V be an n-dimensional vector space over a division algebra D, and let $A := \mathrm{End}_D(V)$ be its algebra of endomorphisms. We regard V as an A-module.*

(a) *V is a **simple** A-module.*

(b) *$A_l \simeq V^n$. In particular, the A-module A_l is **semisimple**.*

Proof. (a) Every nonzero x in V can be taken as an element of a *basis* of the D-vector space V. Hence, for every $y \in V$, there exists $a \in A = \mathrm{End}_D(V)$ such that $ax = y$. Therefore $V = Ax$; by F7, then, V is a simple A-module.

(b) Let $b_1, \ldots, b_n$ form a basis of V over D. The map $a \mapsto (ab_1, \ldots, ab_n)$ is a homomorphism of A-modules from A_l into V^n. Because of our choice of $b_1, \ldots, b_n$, our map is both injective and surjective.   □

The last assertion in F8 is our cue for the introduction of a fundamental notion, albeit one which we will turn to in earnest only later:

Definition 4. An *algebra* A is called *semisimple* if the A-module A_l is semisimple.

Remark. In analogy with this, one might contemplate defining a *simple algebra A* as one such that A_l is a simple A-module. This is not the practice, and instead the expression is reserved, with good reason, for a more general notion to be introduced in Chapter 29 (Definition 1). In any case it's clear that the A-module A_l is simple if and only if every nonzero $x \in A$ is invertible — in other words, when A is a *division algebra*.

F9 (Schur's Lemma). *Let M and N be A-modules.*

(a) *If M is simple, every nonzero f in $\mathrm{Hom}_A(M, N)$ is injective. If N is simple, every nonzero f in $\mathrm{Hom}_A(M, N)$ is surjective.*

(b) *If M and N are both simple, either $M \simeq N$ or $\mathrm{Hom}_A(M, N) = 0$.*

(c) *If M is simple, $\mathrm{End}_A(M)$ is a division algebra.*

Proof. (a) The kernel of f is a submodule of M; the image of f is a submodule of N. For $f \neq 0$ we have $\ker f \neq M$ and $\mathrm{im}\, f \neq 0$. If M is simple, this implies $\ker f = 0$, and if N is simple, it implies $\mathrm{im}\, f = N$.

(b) It follows from (a) that any nonzero $f \in \mathrm{Hom}_A(M, N)$ is an isomorphism.

(c) Since M is simple, every nonzero $f \in \mathrm{End}_A(M)$ is an isomorphism, and hence has an inverse in $\mathrm{End}_A(M)$. Since $M \neq 0$, we conclude that $\mathrm{id}_M \neq 0$. $\square$

To investigate semisimple modules, we need this result:

Lemma. *Suppose a module M is a sum of a family $(N_i)_{i \in I}$ of simple submodules N_i. If N is any submodule of M, there exists a subset J of I such that*

$$(29) \qquad M = N \oplus \left(\bigoplus_{i \in J} N_i \right).$$

Proof. By Zorn's Lemma, we can choose a maximal subset J of I among those that satisfy

$$N + \sum_{j \in J} N_j = N \oplus \left(\bigoplus_{j \in J} N_j \right).$$

Let M' be this sum for the chosen J. For every $i \in I$, then, the sum $M' + N_i$ is not direct, and so $M' \cap N_i \neq 0$. Since N_i is simple, $M' \cap N_i = N_i$; that is, $N_i \subseteq M'$. Therefore $M = M'$, proving the claim. $\square$

F10. *If M is a module, the follow conditions are equivalent:*

(i) *M is a sum of simple submodules.*

(ii) *M is semisimple.*

(iii) *Every submodule of M is a direct summand in M.*

Proof. (i) $\Rightarrow$ (ii): Just apply the lemma to $N = 0$.

(ii) $\Rightarrow$ (iii): This also follows immediately from the lemma.

(iii) $\Rightarrow$ (i): Given any nonzero $x \in M$, consider the submodule $C = Ax$ generated by x. By F6, C contains a submodule L such that C/L is *simple*. Thanks to (iii), L has a complementary submodule N: $M = L \oplus N$. Since $L \subseteq C$, therefore, $C = L \oplus (N \cap C)$. Hence $N \cap C \simeq C/L$, that is, $N \cap C$ is a *simple* submodule of C. So far we have shown that under assumption (iii), *every nonzero submodule of M contains a* **simple** *submodule*.

Now, again using (iii), take a complement M'' of the sum M' of all simple submodules of M. If M'' were nonzero, it would contain a simple submodule, which would also lie in M' by construction, contradicting the complementarity condition $M' \cap M'' = 0$. Therefore $M'' = 0$ and $M' = M$.   □

F11. *Suppose a module M is the sum of a family $(N_i)_{i \in I}$ of simple submodules.*

(a) *If M' is a submodule of M or a quotient module of M, there exists a subset J of I such that the sum $\sum_{i \in J} N_i$ is direct and isomorphic to M'.*

(b) *Every simple submodule of M is isomorphic to one of the N_i.*

Proof. Part (b) follows immediately from (a). By F10, every submodule of M is a direct summand of M, and so isomorphic to a quotient module of M. Hence it suffices to prove (a) for the case of a quotient $M' = M/N$.

By the lemma, there exists a subset J of I such that the sum $L := \sum_{i \in J} N_i$ is direct and $M = N \oplus L$. Therefore $M' = M/N$ is isomorphic to L.   □

It is worth pointing out explicitly a consequence of part (a) (for the second sentence, consider condition (iii) of F7).

F12. *Any submodule and quotient module of a semisimple module is semisimple. If A is a semisimple algebra, every simple A-module is isomorphic to a submodule of A_l, and so is isomorphic to a minimal left ideal of A.*

F13. *For an algebra A, there is equivalence between:*

(i) *A is semisimple.*

(ii) *Every A-module is semisimple.*

Proof. By definition, A is semisimple if the A-module A_l is semisimple. But *any* A-module M is isomorphic to a quotient of a direct sum of copies of A_l (indexed, say, by some set of generators of M). The result now follows from F12.   □

Definition 5. For a given algebra A, let $T = T(A)$ be the set of isomorphism classes of all simple A-modules (T is not "too large": see property (iii) in F7). Take $\tau \in T$ and let S be a simple A-module *of type τ*. For M an A-module, denote by M_τ the sum of all submodules of M isomorphic to S. We call M_τ the *isogenous component of M of type τ*. (If M has no submodule isomorphic to S, then $M_\tau = 0$.) If $M_\tau = M$ we say that M is *isogenous* (of type τ).

F14. *Let M be a semisimple module.*

(a) *M is the direct sum of its isogenous components M_τ.*

(b) *Every submodule N of M is the direct sum of the modules $N \cap M_\tau$.*

Proof. (a) Since M is semisimple, we necessarily have $M = \sum_\tau M_\tau$. For $\tau \in T$, let M'_τ be the sum of all M_σ such that $\sigma \neq \tau$. We need to show that $M_\tau \cap M'_\tau = 0$. Being a submodule of M, the intersection $M_\tau \cap M'_\tau$ is also semisimple (F12), so if nonzero, it would contain a simple submodule S. By F11(b), then, S would have both type τ and type σ. Contradiction!

(b) N is semisimple because it is a submodule of M. Part (a) then allows us to write $N = \bigoplus_\tau N_\tau$. By definition, $N_\tau \subseteq M_\tau \cap N$. Being a submodule of M_τ, the intersection $M_\tau \cap N$ is the sum of simple modules of type τ (part (a) of F11), and so $M_\tau \cap N \subseteq N_\tau$. $\square$

F15. *If $f : M \to N$ is a homomorphism of A-modules and τ is an isomorphism class of simple A-modules, the isogenous components M_τ and N_τ satisfy $f(M_\tau) \subseteq N_\tau$.*

Proof. Because $f(M_\tau)$ is a homomorphic image of M_τ, it is isogenous of type τ, by part (a) of F11. Hence it is contained in N_τ. $\square$

F16. *Let M be a semisimple A-module. If U is a submodule of M, the following conditions are equivalent:*

(i) $f(U) \subseteq U$ *for every* $f \in \mathrm{End}_A(M)$.

(ii) U *is a (direct) sum of isogenous components of M.*

Proof. The implication (ii) $\Rightarrow$ (i) is clear from F15. Now suppose (i) is satisfied and let S be a simple submodule of U. We must show that every simple submodule S' of M that is isomorphic to S is also contained in U. So let p be a projection from M onto S and let $g : S \to M$ be a homomorphism with image S'. Setting $f = g \circ p \in \mathrm{End}_A(M)$ we get $f(S) = S'$. By condition (i) we then have $S' \subseteq U$, as needed. $\square$

F17. *Let the algebra A be semisimple. If U is a subset of A, the following conditions are equivalent:*

(i) U *is an ideal of A.*

(ii) U *is a (direct) sum of isogenous components of the A-module A_l.*

Proof. The ideals of A are precisely the submodules of A_l that, under right multiplication by any element of A, that is, under any map $f \in \mathrm{End}_A(A_l)$, are mapped to themselves (see F3). With this characterization, F17 is just F16 applied to $M = A_l$. $\square$

F18. *Let A be a semisimple algebra.*

(a) *The isogenous components of A_l are precisely the minimal ideals of A, and every ideal of A is a direct sum of minimal ideals of A.*

(b) *A has only finitely many minimal ideals.*

Proof. (a) follows from F17 in view of F14. Since the A-module A_l is generated by 1, it has only finitely many nonzero isogenous components, again by F14. This leads from (a) to (b).   □

F19 and Definition 6. *Let M be a semisimple A-module, and suppose*

$$(30) \qquad M = \bigoplus_{i \in I} N_i \quad and \quad M = \bigoplus_{j \in J} N'_j,$$

where N_i, N'_j are simple submodules of M. There exists a bijection $\sigma : I \to J$ with $N_i \simeq N'_{\sigma(i)}$ for all $i \in I$; in particular, I and J have the same cardinality. We call this cardinality the *length* of the semisimple module M, and denote it by

$$l(M) = l_A(M).$$

The *length of a semisimple algebra A* is $l(A) := l(A_l)$. For a simple A-module N of type τ,

$$M : N = M : \tau$$

denotes the length of the isogenous component M_τ of M.

Proof. By grouping together simple modules of the same isomorphism type, we recognize that it suffices to work with the case where all the N_i and N'_j are isomorphic to a single simple module N. Let D° be the division algebra of endomorphisms of N. If M is any A-module, $\mathrm{Hom}_A(N, M)$ can be regarded in a natural way as a D-vector space: For $f \in \mathrm{Hom}_A(N, M)$ and $d \in D = \mathrm{End}_A(N)^\circ$, set $df = f \circ d$. If M and M' are isomorphic, $\mathrm{Hom}_A(N, M)$ and $\mathrm{Hom}_A(N, M')$ as isomorphic as D-vector spaces, and for $M = \bigoplus_{i \in I} M_i$ the natural map $\bigoplus_{i \in I} \mathrm{Hom}_A(N, M_i) \to \mathrm{Hom}_A(N, M)$ is an isomorphism (note that N is finitely generated). Hence one gets from assumption (30) the D-isomorphisms $D^{(I)} \simeq \bigoplus_{i \in I} \mathrm{Hom}(N, N_i) \simeq \mathrm{Hom}_A(N, M)$; thus the cardinality $|I|$ of I satisfies

$$(31) \qquad |I| = \dim_D \mathrm{Hom}_A(N, M).$$

The same holds for the cardinality of J, so the length of A is well defined.   □

Remarks. (1) With the notations above, we have

$$(32) \qquad M : N = \dim_D \mathrm{Hom}_A(N, M) = \mathrm{Hom}_A(N, M) : D$$

and

$$(33) \qquad l(M) = \sum_\tau M : \tau.$$

The preceding argument is based, of course, on the well-definedness of the dimension of a D-vector space. We assume this as known (and readers can check it by following the same pattern we used in introducing the transcendence degree: see Section 18.1 in vol. I).

(2) One could also prove F19 more directly, without relying on the D-vector space $\text{Hom}_A(N, M)$ — except that the interpretation of $M : N$ in (32) has its own utility. In any case F19 follows, in the (essential) case that I is finite, from other fundamental theorems in algebra: see Theorems 1 and 2 in the next section.

We conclude this section with the so-called *Jacobson Density Theorem*, which, although perhaps initially overvalued by algebraists, is nonetheless a simple and useful fact in linear algebra.

F20 (Jacobson Density Theorem). *Let M be a semisimple A-module and $C = \text{End}_A(M)$ its algebra of endomorphisms. Consider the natural homomorphism*

$$(34) \qquad A \to \text{End}_C(M).$$

The image of A under (34) *is dense in $\text{End}_C(M)$ in the following sense: Given $f \in \text{End}_C(M)$ and finitely many elements $x_1, \ldots, x_n$ in M, there exists necessarily some $a \in A$ such that*

$$(35) \qquad fx_i = ax_i \quad \text{for } 1 \le i \le n.$$

If M is finitely generated as a C-module, then (34) *is in fact surjective.*

Proof. Consider the A-module $M' = M^n$ and in M' the element $x = (x_i)_i$. Now M' is semisimple because M is, so Ax is a direct summand of M'; let $p : M' \to M'$ be a projector onto Ax. Then p lies in $C' := \text{End}_A(M')$. Looking at f in the standard way as a map from M' into itself, so

$$fy = (fy_i)_i \quad \text{for } y = (y_i)_i \in M' = M^n,$$

we see by F4 that f is a C'-homomorphism of M'. Hence $fp = pf$. There follows $fx = fpx = pfx \in Ax$, so there exists $a \in A$ such as

$$fx = ax,$$

which is (35). □

Remark. Let M be a simple A-module and $D := \text{End}_A(M)$ its division algebra of endomorphisms. Viewing M as a D-vector space, let $x_1, \ldots, x_n \in M$ be linearly independent. Then, for every $y_1, \ldots, y_n$ in M, there exists $a \in A$ such that

$$ax_i = y_i.$$

Indeed, since the x_i are linearly independent, we can find $f \in \text{End}_D(M)$ such that $fx_i = y_i$. Then F20 says there exists $a \in A$ such that $ax_i = fx_i = y_i$.

3. We next introduce two important finiteness properties that a module can possess.

Definition 7. A module M is called *noetherian* if every nonempty set of submodules of M has a *maximal* element. M is called *artinian* if every such set has a *minimal* element. An algebra A is *noetherian* or *artinian* if the A-module A_l is.

Remarks. The validity of statements (1) through (6) below is best checked independently by the reader.

(1) A module M is noetherian if and only if every *increasing sequence* of submodules (the term *ascending chain* is often used) is *stationary*, in the sense that all terms are equal after a certain point. M is artinian if and only if every *decreasing* sequence (or *descending chain*) of submodules is stationary.

(2) For a vector space over a division ring D, the following conditions are equivalent: (i) V is noetherian; (ii) V is finite-dimensional; (iii) V is artinian.

(3) Suppose $M = \bigoplus_{i \in I} M_i$. If I is infinite and all the M_i are nonzero, M is neither artinian nor noetherian.

(4) $\mathbb{Z}$ is noetherian (since it is a principal ideal domain), but not artinian.

(5) If M is a $\mathbb{Z}$-module and p is a prime number, let M_p denote the p-component of M (vol. I, p. 155). Let W be the group of all roots of unity in $\mathbb{C}$. Then the $\mathbb{Z}$-module W_p is artinian but not noetherian.

(6) In view of (2), *a finite-dimensional K-algebra is both noetherian and artinian (and so is its opposite!).*

(7) But it is not hard to find an example of a noetherian and artinian K-algebra whose opposite has neither property; see §28.2 in the Appendix.

(8) We will see in F41 that every artinian algebra is in fact noetherian.

F21. *An A-module M is noetherian if and only if every submodule M is finitely generated.*

Proof. Let N be a submodule of M and let X be the set of finitely generated submodules of N. If M is noetherian, X has a maximal element L. For any $x \in N$ we have $L + Ax \in X$, so $L + Ax = L$ and hence $x \in L$. It follows that $N = L \in X$, so N is finitely generated.

To prove the converse, let $(N_n)_n$ be an increasing sequence of submodules of M. Then $N = \bigcup N_n$ is a submodule of M. If $x_1, \ldots, x_m$ is a finite set of generators of N, there exists $k \in \mathbb{N}$ such that all the x_i lie in N_k. Hence $N = N_k$ and $N_n = N_k$ for every $n \geq k$. $\square$

F22. *Let M be a module and N a submodule. M is noetherian if and only if N and M/N are. M is artinian if and only if N and M/N are.*

Proof. Clearly N inherits either property if M has it. Let $\pi : M \to M/N$ be the quotient map. The correspondence $T \mapsto T' = \pi^{-1}(T)$ between submodules T of M/N and submodules T' of M containing N is one-to-one and preserves inclusion. Hence M/N is noetherian or artinian if M is.

Conversely, assume N and M/N are noetherian (the proof for artinian modules is analogous). Let X be a nonempty set of submodules of M. Since M/N is noetherian, the set $\pi X := \{\pi T \mid T \in X\}$ has a maximal element πT_0. Then the set $Y = \{T \cap N \mid T \in X, \ \pi T = \pi T_0\}$ has a maximal element $T_1 \cap N$, because N is also noetherian. We will show that T_1 is a maximal element of X. Suppose $T_1 \subseteq T$ for $T \in X$. Then $\pi T_0 = \pi T_1 \subseteq \pi T$, and it follows that $\pi T = \pi T_1$. Hence $T \subseteq T_1 + N$; but because $T_1 \subseteq T$, this implies $T \subseteq T_1 + (T \cap N) = T_1 + (T_1 \cap N) = T_1$.   □

F23. *Suppose $M = \bigoplus_{i=1}^{n} M_i$. Then M is noetherian or artinian if and only if each M_i has the same property.*

Proof. Since $M/M_n \simeq M_1 \oplus \cdots \oplus M_{n-1}$, this follows by induction from F22.   □

F24. *If an algebra A is noetherian or artinian, every finitely generated A-module M has the same property.*

Proof. Let $x_1, \ldots, x_n$ be generators of the A-module M, and let N be the kernel of the homomorphism of A-modules $A_l^n \to M$ sending the canonical basis vectors of A_l^n to $x_1, \ldots, x_n$. Then $M \simeq A_l^n/N$, and the assertion follows from F23 and F22.   □

F25. *For a semisimple module M, there is equivalence between:*
  (i) *M is finitely generated.*
  (ii) *M is a direct sum of finitely many simple submodules.*
  (iii) *M is artinian.*
  (iv) *M is noetherian.*

Proof. Since M is semisimple, we can write $M = \bigoplus_{i \in I} N_i$ for appropriate simple submodules N_i of M. If I is infinite, M is neither noetherian nor artinian; see Remark 3 after Definition 7. If I is finite, F23 says that M is both artinian and noetherian, since simple modules are trivially seen to have both properties. This proves (iii) $\Longleftrightarrow$ (ii) $\Longleftrightarrow$ (iv). The implication (iv) $\Rightarrow$ (i) is clear: see F21. To prove (i) $\Rightarrow$ (ii), fix a finite set of generators for M. Each generator lies in a sum of finitely many N_i; hence I is finite.   □

F26. *Every semisimple algebra is artinian and noetherian.*

Proof. This follows trivially from F25, since the A-module A_l is finitely generated: $A_l = A \cdot 1$.   □

F27. *If M is an A-module, there is equivalence between:*
  (i) *M is both artinian and noetherian.*
  (ii) *M has a **composition series**, that is, a chain*

$$0 = M_0 \subseteq M_1 \subseteq \cdots \subseteq M_n = M$$

of submodules M_i, such that each M_i/M_{i-1} is simple.

Proof. The implication (ii) $\Rightarrow$ (i) follows inductively from F22. For the converse, let M be artinian and noetherian, and consider all submodules N of M for which there is a chain

$$N = M_0 \subseteq M_1 \subseteq \cdots \subseteq M_n = M$$

of submodules M_i of M with M_i/M_{i-1} simple for all i. Since M is artinian, there is a minimal submodule N with this property. If $N = 0$, we are done, so assume $N \neq 0$. Since M is noetherian, there is a maximal submodule of N among those distinct from N; call it L. Then N/L is *simple*, and we reach a contradiction with the minimality of N. $\qquad\square$

Definition 8. If M is an artinian and noetherian module, the *length* $l(M)$ of M is the length of a composition series of M (compare F27).

Remarks. (i) The well-definedness of the length is guaranteed by the next theorem, which says that all composition series of M have the same length.

(ii) An artinian and noetherian module is also called a *module of finite length*.

(iii) For a *semisimple* module M (of finite length), the definition of $l(M)$ is clearly in agreement with the earlier Definition 6 (see also F25). Thus, if $M = N_1 \oplus \cdots \oplus N_n$, where the N_i are *simple* submodules, M has length n.

(iv) If N is a submodule of a module M, we have (taking F22 into account)

(36) $$l(M) = l(N) + l(M/N),$$

since one can combine composition series for N and M/N in the obvious way.

Theorem 1 (Jordan–Hölder Theorem for modules). *If*

$$0 = M_0 \subseteq M_1 \subseteq \cdots \subseteq M_n = M \quad and \quad 0 = L_0 \subseteq L_1 \subseteq \cdots \subseteq L_m = M$$

are composition series of the same module M, then $m = n$, and there is a permutation $\sigma \in S_n$ such that $L_i/L_{i-1} \simeq M_{\sigma(i)}/M_{\sigma(i)-1}$ for each i.

Proof. Consider the sequences

(37) $$0 = L_0 \cap M_{n-1} \subseteq \cdots \subseteq L_m \cap M_{n-1} = M_{n-1},$$

(38) $$M_{n-1} = L_0 + M_{n-1} \subseteq \cdots \subseteq L_m + M_{n-1} = M_n.$$

For each j we have an exact sequence

$$0 \to L_j \cap M_{n-1}/L_{j-1} \cap M_{n-1} \to L_j/L_{j-1} \to L_j + M_{n-1}/L_{j-1} + M_{n-1} \to 0.$$

Since L_j/L_{j-1} is simple, this implies that, of the quotient modules

$$L_j \cap M_{n-1}/L_{j-1} \cap M_{n-1} \quad and \quad L_j + M_{n-1}/L_{j-1} + M_{n-1},$$

precisely one is 0, and the other is isomorphic to L_j/L_{j-1}. But because M_n/M_{n-1} is simple, the chain (38) has precisely one "jump"; hence there exists an i such that

$$L_i/L_{i-1} \simeq M_n/M_{n-1},$$
$$L_i \cap M_{n-1} = L_{i-1} \cap M_{n-1},$$
$$L_j \cap M_{n-1}/L_{j-1} \cap M_{n-1} \simeq L_j/L_{j-1} \quad \text{for } j \neq i.$$

In particular,

$$L_0 \cap M_{n-1} \subseteq \cdots \subseteq L_{i-1} \cap M_{n-1} \subseteq L_{i+1} \cap M_{n-1} \subseteq \cdots \subseteq L_m \cap M_{n-1}$$

is a composition series of M_{n-1}. By induction, then, we obtain a bijection σ : $\{1, \ldots, i-1, i+1, \ldots, m\} \to \{1, \ldots, n-1\}$ such that $L_j/L_{j-1} \simeq M_{\sigma(j)}/M_{\sigma(j)-1}$. If we extend σ to a permutation on n elements by setting $\sigma(i) = n$, the theorem is proved.   □

Remark. The Jordan–Hölder Theorem holds, more generally, for *groups with operators* (see, for instance, Bertram Huppert, *Endliche Gruppen I*, pages 55 and 63); in fact, the proof above translates without hindrance to this more general setting.

Definition 9. A module $N \neq 0$ is called *indecomposable* if 0 and N are the only direct summands that N has.

F28. *If a module M is noetherian or artinian, it is a direct sum of finitely many indecomposable submodules.*

Proof. Observe that M, if nonzero, must have a nonzero indecomposable direct summand: for M artinian, any minimal nonzero direct summand will do; for M noetherian, take a complement of any maximal direct summand distinct from M.

For the proof proper, first let M be noetherian and consider a maximal direct summand N among those with the property of being finite direct sums of indecomposable submodules. Write $M = N \oplus M'$. Since M' is noetherian, the preceding observation implies $M' = 0$.

The reasoning for M artinian is analogous: we take a minimal direct summand of M among those with the property that the complement is a finite direct sum of indecomposable modules.   □

Theorem 2 (Krull–Remak–Schmidt). *Let M be an artinian and noetherian A-module. Then M is a direct sum of finitely many indecomposable submodules* (compare F28). *If*

$$(39) \qquad M = N_1 \oplus \cdots \oplus N_m = N_1' \oplus \cdots \oplus N_n',$$

where the N_i and the N_j' are indecomposable, then $m = n$ and, after appropriate reindexing, $N_i \simeq N_i'$ for all $1 \leq i \leq n$.

Proof. Let $\pi_i : M \to N_i$ be the projections and $\iota_i : N_i \to M$ the injections associated with the direct decomposition $M = N_1 \oplus \cdots \oplus N_m$. The endomorphisms $p_i := \iota_i \pi_i$ of M are called the projectors of the decomposition. They satisfy

$$p_i^2 = p_i, \quad p_i p_j = 0 \quad \text{for } i \neq j, \quad \sum_{i=1}^{m} p_i = 1 = \mathrm{id}_M .$$

Let p_i' be the projectors for the decomposition $M = N_1' \oplus \cdots \oplus N_n'$. When restricted to N_1, the sum $p_1 = \sum_{i=1}^{m} p_1 p_i'$ is the identity. Since the restrictions of the $p_1 p_i'$ are endomorphisms of N_1, which is *indecomposable*, at least one of these restrictions must be an *automorphism* of N_1 (see §28.5); it may as well be taken to be that of $p_1 p_1'$. Now consider the endomorphism

$$(40) \qquad f = p_1' p_1 + p_2 + \cdots + p_m = 1 - p_1 + p_1' p_1$$

of M. If $f(x) = 0$, then $0 = p_1 f(x) = (p_1 p_1')(p_1 x) = 0$, so $p_1(x) = 0$, and we get $x = 0$. This shows f is injective, and hence surjective, because an injective endomorphism of a module of finite length is an *automorphism* (§28.3). In view of (40) we have $f(N_1) \subseteq N_1'$, and since $M = fM = fN_1 \oplus \cdots \oplus fN_m$ we see that

$$N_1' = f(N_1) \oplus \left(N_1' \cap \sum_{i \geq 2} f(N_i) \right).$$

But N_1' is *indecomposable*, so $N_1' = f(N_1)$. It follows that $M/N_1 \simeq fM/fN_1 \simeq M/N_1'$, and so from (39)

$$(41) \qquad N_2 \oplus \cdots \oplus N_m \simeq N_2' \oplus \cdots \oplus N_n' .$$

The theorem follows by induction; just observe that (41) implies $N_2' \oplus \cdots \oplus N_n' = N_2'' \oplus \cdots \oplus N_m''$, with $N_i'' \simeq N_i$. □

Remark. The Krull–Remak–Schmidt Theorem, too, can be generalized to the case of groups with operators (see Huppert, *Endliche Gruppen I*, pages 65 and following). See also §28.6.

4. Although later we will be focusing our attention on semisimple algebras, in this section we discuss also the notion of *radical*, because of its importance in algebra.

Definition 10. Let A be an algebra and M an A-module. The intersection $\mathfrak{R}(M)$ of all maximal submodules of M is called the (*Jacobson*) *radical* of M. By Remark 3 following Definition 3, $\mathfrak{R}(M)$ consists of those elements of M that are mapped to 0 under any homomorphism from M into a *simple* module.

The radical of the A-module A_l, that is, the intersection of all maximal left ideals of A, is called the (*Jacobson*) *radical* of A and is denoted by $\mathfrak{R}(A)$. The qualifier "Jacobson" is used when it is necessary to differentiate this notion from other notions of radical, such as that in vol. I, p. 218.

Remarks. We collect here some simple formal properties of the radical:

(a) *For every homomorphism* $f : N \to M$ *of A-modules,* $f(\mathfrak{R}(N)) \subseteq \mathfrak{R}(M)$. Indeed, if $g : M \to S$ is a homomorphism from M into a simple module S, we have $g(fx) = (g \circ f)x = 0$ for every $x \in \mathfrak{R}(N)$.

(b) $\mathfrak{R}(A)$ *is a two-sided ideal of A*. This follows from (a) if we set $N = M = A$ and let f be right multiplication by an arbitrary $a \in A$.

(c) *For every submodule N of M we have* $\mathfrak{R}(N) \subseteq \mathfrak{R}(M)$. This too is a special case of (a), as is the next statement.

(d) *For every submodule N of M we have* $\mathfrak{R}(M/N) \supseteq (\mathfrak{R}(M) + N)/N$. *Hence* $\mathfrak{R}(M/N) = 0$ *implies* $\mathfrak{R}(M) \subseteq N$.

(e) *Given a submodule* $N \subseteq \mathfrak{R}(M)$ *we have* $\mathfrak{R}(M/N) = \mathfrak{R}(M)/N$. *In particular,* $\mathfrak{R}(M/\mathfrak{R}(M)) = 0$. Indeed, if $N \subseteq \mathfrak{R}(M)$, the maximal submodules of M/N are in correspondence with those of M.

(f) $\mathfrak{R}(A/\mathfrak{R}(A)) = 0$, *and if* $\mathfrak{R}(A/I) = 0$ *for an ideal I of A, then* $\mathfrak{R}(A) \subseteq I$. *Thus* $\mathfrak{R}(A)$ *is the smallest ideal I of A such that* $\mathfrak{R}(A/I) = 0$. This follows easily from (e) and (d), because the submodules of the A/I-module $(A/I)_l$ are in correspondence with those of the A-module A_l/I.

(g) $\mathfrak{R}(A)M \subseteq \mathfrak{R}(M)$. For given $x \in M$, the map $a \mapsto ax$ is an A-module homomorphism from A to M, and so (a) implies that $\mathfrak{R}(A)x \subseteq \mathfrak{R}(M)$. It follows that $\mathfrak{R}(A)M \subseteq \mathfrak{R}(M)$.

(h) $\mathfrak{R}(M) = 0$ *if and only if M is isomorphic to a submodule of a direct product of simple modules*. Let N_i, with $i \in X$, be maximal submodules of M such that $\bigcap_{i \in X} N_i = \mathfrak{R}(M)$. The natural homomorphism f of M into the direct product of the simple modules M/N_i has kernel $\bigcap N_i = \mathfrak{R}(M)$, so if $\mathfrak{R}(M) = 0$, this map is injective, proving the forward part. For the converse, let $f : M \to \prod S_i$ be an injective homomorphism from M into a direct product of simple modules S_i, and let p_i be the corresponding projections. For $x \in \mathfrak{R}(M)$ we then have $(p_i \circ f)x = 0$. But $p_i(fx) = 0$ for all i implies $fx = 0$, so, in view of f being injective, we're left with $x = 0$.

(i) *If M is semisimple,* $\mathfrak{R}(M) = 0$. Because then M, being a direct sum of simple modules, is also a submodule of a direct product of simple modules, so (h) applies.

We now show a partial converse to this last result: *Any **artinian** module M with* $\mathfrak{R}(M) = 0$ *is semisimple*. Indeed, we claim that for such an M there is a *finite* family $(N_i)_{i \in X}$ of maximal submodules N_i whose intersection is zero. Assuming this, we see from the first part of the proof of (h) above that M is isomorphic to a submodule of the *semisimple* module $\prod_i M/N_i = \bigoplus_i M/N_i$; but semisimplicity is hereditary (F12).

To prove the claim, consider all intersections of finite families of maximal submodules of M. Since M is artinian, there is a *minimal* such intersection, say D. For any maximal submodule N of M we have $D \subseteq D \cap N$, hence $D \subseteq N$. But since $\mathfrak{R}(M) = 0$, this implies $D = 0$.

Taking F25 into account we can also state:

F29. *M is a finitely generated semisimple module if and only if $\Re(M) = 0$ and M is artinian. In particular, an algebra is semisimple if and only if it is artinian and has zero radical.*

Everything we will say later about the general character of the elements of the radical rests on the following simple fact:

F30. *For a finitely generated module $M \neq 0$, the radical $\Re(M)$ is distinct from M.*

Proof. Indeed, by F6, M has a maximal submodule.   □

F31. *Suppose a submodule N of the module M satisfies*

$$(42) \qquad\qquad N + \Re(M) = M.$$

If M is finitely generated (or if just M/N is), then $N = M$.

Proof. From (42) we have $\Re(M/N) = M/N$; see Remark (d) after Definition 10. If M/N is finitely generated, the result in F30 implies $M/N = 0$.   □

F32 (Nakayama's Lemma). *Suppose a submodule N of the A-module M satisfies*

$$(43) \qquad\qquad N + \Re(A)M = M.$$

If M (or just M/N) is finitely generated, then $N = M$.

Proof. This is an obvious consequence of F31 and Remark (g) after Definition 10.
                                                                            □

Nakayama's Lemma also has a (trivial) converse of sorts:

F33. *If an element x of an A-module M has the property that $N + Ax \neq M$ for all submodules $N \subset M$ distinct from M, then x lies in $\Re(M)$.*

Proof. If $x \notin \Re(M)$, there is a maximal submodule N of M not containing x. For this module, $N + Ax = M$, so x does not have the property in the hypothesis.   □

Taken together, F32 and F33 cast some light on the nature of elements in the radical. We can deduce, for instance:

F34. *Let M be an A-module having a finite set of generators $x_1, \ldots, x_n$. For $x \in M$, the following conditions are equivalent:*

  (i) $x \in \Re(M)$.

  (ii) *For any $a_1, \ldots, a_n \in A$, the elements $x_i + a_i x$ generate M.*

Proof. Suppose (i) holds and let N be the submodule generated by $x_i + a_i x$. Since $x_i = (x_i + a_i x) - a_i x \in N + \Re(M)$, we have $N + \Re(M) = M$, and this immediately gives $N = M$, thanks to F31.

Conversely, suppose (i) does not hold. Then, by F33, there exists a submodule $N \neq M$ such that $N + Ax = M$. In particular, the generators x_i have the form $x_i = y_i - a_i x$, with $y_i \in N$, $a_i \in A$. But then all the y_i lie in N, and so cannot generate M, which is distinct from N.   □

Theorem 3. *The radical of an algebra A consists precisely of the elements x in A such that*

(44) $$1 + ax \in A^\times \quad \text{for all } a \in A.$$

Here $A^\times$ denotes the group of units, or invertible elements, of A.

Proof. Apply F34 to the A-module $M = A_l$, with generator $x_1 = 1$. If (44) is fulfilled for $x \in A$, every element of the form $1 + ax$ is a unit in A, hence a generator of A_l. By F34, then, x belongs to $\mathfrak{R}(A)$.

Conversely, take $x \in \mathfrak{R}(A)$. By F34, every element of the form $1 + ax$ generates A_l, and in particular there exists $b \in A$ such that

$$b(1 + ax) = 1.$$

Therefore $b = 1 - bax$ also generates A_l, by the same argument; that is, b has a left inverse, as well as the right inverse $1 + ax$. This implies that $1 + ax$ is the (two-sided) inverse of b (why?), and so $1 + ax \in A^\times$. □

F35. *Let N be a left ideal of an algebra A. Then*

$$N \subseteq \mathfrak{R}(A) \quad \Longleftrightarrow \quad 1 + x \in A^\times \quad \text{for all } x \in N.$$

Proof. This follows immediately from Theorem 3. □

F36. *For every algebra A we have $\mathfrak{R}(A) = \mathfrak{R}(A^\circ)$, that is, $\mathfrak{R}(A)$ is also the intersection of all maximal **right** ideals of A. (Hence the equivalence in F35 holds for every right ideal N of A as well.)*

Proof. We saw in Remark (b) after Definition 10 that the radical of an algebra is a two-sided ideal. Since A° and A have the same units, the assertion follows from F35. □

F37. *If a left or right ideal N of an algebra A contains only nilpotent elements, it lies in $\mathfrak{R}(A)$.*

Proof. In view of F35 and F36 it suffices to show that if $x \in A$ is nilpotent, $1 + x$ is invertible in A. So assume $x^n = 0$ for some $n \in \mathbb{N}$; then

$$(1 + x)(1 - x + x^2 - \cdots + (-1)^{n-1} x^{n-1}) = 1 + (-1)^n x^n = 1.$$

The inverse is clearly two-sided. □

Remarks. (1) One should beware of reading too much into F37, for instance that every nilpotent element lies in the radical. For if A is not commutative, the nilpotency of $a \in A$ does not imply that all elements of Aa are nilpotent: just take $A = \left(\begin{smallmatrix} 0 & 1 \\ 0 & 0 \end{smallmatrix}\right) \in M_2(K)$. And in fact, we saw in F8 that the matrix algebra $A = M_n(K)$ is semisimple, so $\mathfrak{R}(A) = 0$, whereas A has many nilpotents if $n > 1$.

(2) On the other hand, F37 suggests the question whether $\mathfrak{R}(A)$ perhaps consists only of nilpotent elements. In general this is not the case (see §28.7); but if A is artinian, it is — in fact a much stronger statement is true, given in Theorem 4 just below.

(3) If A is *commutative*, the *nilradical* $\sqrt{0}$ of A (vol. I, p. 218) is always contained in the *Jacobson radical* of A. This is a straightforward consequence of F37, or yet of the characterization of the nilradical as the intersection of all prime ideals (vol. I, §4.14). In general the nilradical does not equal $\mathfrak{R}(A)$ (see again §28.6); but note the result in F39 below.

Theorem 4. *If A is artinian, $\mathfrak{R}(A)$ is nilpotent, that is, there exists $k \in \mathcal{N}$ such that all k-fold products in $\mathfrak{R}(A)$ vanish — in short, $\mathfrak{R}(A)^k = 0$.*

Remark on notation. If I is a left ideal of A and M is an A-module, IM denotes the submodule of M generated by all elements ax with $a \in I$ and $x \in M$. If $M = I'$ is itself a left ideal of A, we call II' the *product ideal* of I and I'. For $k \in \mathbb{N}$ it is then clear what to make of the *power ideal* I^k: it consists of all finite sums of k-fold products of elements of I. In particular I^k contains all k-th powers of elements of I. If I and I' are two-sided ideals, so is II'.

Proof of Theorem 4. Set $I = \mathfrak{R}(A)$. Then $I^0 := A \supseteq I^1 \supseteq I^2 \supseteq \cdots$ is a descending chain of ideals in A. If A is artinian, the chain is stationary after some k, so

$$(45) \qquad\qquad II^k = I^k.$$

If we knew that I^k is finitely generated, Nakayama's Lemma (F32) would imply $I^k = 0$ directly. As we will see later, A is in fact noetherian, but in order to prove this, we need the content of Theorem 4. So we must proceed by a different route, and we do it by contradiction.

Suppose that $I^k \neq 0$, and take the set of all nonzero left ideals N of A such that

$$(46) \qquad\qquad IN = N.$$

Then I^k belongs to this set, and since A is artinian, the set has a minimal element, which we denote by N. At the same time, (46) implies $I^k N = N$, so $I^k x \neq 0$ for some $x \in N$. In view of the minimality of N and since $I(I^k x) = I(I^k Ax) = I^{k+1}Ax = I^k Ax = I^k x$, we get $I^k x = N$, hence $N = Ax$. Thus N is finitely generated, and using Nakayama's Lemma on (46) we obtain $N = 0$. This yields our contradiction. □

F38. *Let N be a left or right ideal of an artinian algebra A. Then*

$$N \subseteq \mathfrak{R}(A) \quad \Longleftrightarrow \quad N \text{ contains only nilpotents.}$$

Proof. Theorem 4 gives the forward implication. The other direction follows from F37 (even absent the assumption that A is artinian). □

F39. *If A is a commutative artinian algebra, $\Re(A)$ is the set of all nilpotent elements of A. In other words, $\Re(A)$ coincides with the nilradical of A.*

Proof. Immediate from F38, since here $a \in \sqrt{0}$ implies $Aa \subseteq \sqrt{0}$ (compare Remark 1 after F37). □

But note that for every affine K-algebra the two radicals coincide (vol. I, §19.3), yet not every such algebra is artinian.

We conclude the section with a remarkable fact:

F40. *Let A be an artinian algebra. The following properties are equivalent for an A-module M:*

(i) *M is artinian*;   (ii) *M is noetherian*;   (iii) *M is finitely generated.*

Proof. The implications (ii) $\Rightarrow$ (iii) and (iii) $\Rightarrow$ (i) are clear from F21 and F24. So we must show that an artinian module over an artinian algebra A is noetherian. Let $I = \Re(A)$ be the radical of A. Since A/I is artinian and its radical is 0 (see Remark (f) after Definition 10), we conclude from F29 that

$$(47) \qquad\qquad A/I \text{ is semisimple.}$$

We claim that, for any $n \in \mathcal{N}$ and any artinian A-module M,

$$(48) \qquad\qquad I^n M = 0 \quad \Longrightarrow \quad M \text{ is noetherian.}$$

With this we are done, because Theorem 4 assures us there is n such that $I^n = 0$, making the left-hand side of (48) true regardless of M.

We prove the claim by induction. Take $n > 1$ and assume the claim true for all natural numbers less than n. Let M be an artinian A-module such that $I^n M = 0$, and take the submodule $N = I^{n-1}M$ of M. Since $IN = 0$, the induction assumption implies that N is noetherian. But M/N is also noetherian by the induction assumption, because $I^{n-1}(M/N) = 0$. So M is noetherian (we have been making liberal use of F22).

There remains to tackle the case $n = 1$. Let M be an artinian A-module with $IM = 0$. We can regard M as an A/I-module as well. As such, M is semisimple, by (47) and F13. It is easy to see that this makes M semisimple also as an A-module. But for semisimple modules we already know from F25 that artinian and noetherian are equivalent properties. □

It is worth spelling out a corollary of F40:

F41. *Every artinian algebra is noetherian. Every finitely generated module over an artinian algebra has a composition series (see F27).*

Wedderburn Theory

1. At the center of this chapter stands the *Wedderburn structure theorem*, according to which every *simple artinian algebra* is isomorphic to a matrix algebra $M_n(D)$ over some division algebra D, with n and (the isomorphism class of) D uniquely determined. A structure result in abstract algebra, and a very satisfying one at that, which one can prove through simple methods of *linear algebra*! (This was first done by E. Artin.) It reduces the study of simple artinian algebras to that of *division algebras* and thus represents not only an achievement but also a starting point for further investigations, in that it leads us to pursue a classification of division algebras. This problem turns out to be tougher than it may appear at first, even after making further restrictions; nonetheless we will be able to deal in Chapter 31 with the case of *local* division algebras.

Because of Wedderburn's theorem it is natural to call two central-simple algebras *similar* if they are isomorphic to matrix algebras over the same division algebra D. The set of such similarity classes is denoted by Br K and constitutes an important invariant of the field K. Also of far-reaching importance is that Br K has a natural *group* structure, with multiplication given by the tensor product of algebras. The tensor product also plays another key role in the theory: it allows one to pass from a central-simple K-algebra $A = M_n(D)$ to a central-simple algebra A_L over a bigger field L, by base change. It is then worth looking in particular for fields L such that A_L is a matrix algebra over L. Such an L is called *splitting field* of A.

We will show that A always has a splitting field L of finite degree over K and that the smallest possible value of this degree coincides with the *Schur index* of A, which is the square root of the dimension of D over K.

These preliminaries will suffice, we hope, to give an idea of what an elegant part of algebra our current subject is.

Definition 1. An algebra $A \neq 0$ is called *simple* if 0 and A are the only ideals of A. (Compare Definition 8 in Chapter 4, and the remark on p. 135 of this volume.)

Without further assumptions, a simple algebra need not be semisimple, as shown by §29.1 in the Appendix. However:

F1. *If A is a simple algebra, there is equivalence between*:

(i) *A is semisimple.*

(ii) *A is artinian.*

(iii) *A possesses a minimal left ideal N.*

Proof. (i) $\Rightarrow$ (ii) is part of F26 in the previous chapter, and the implication (ii) $\Rightarrow$ (iii) is trivial. Hence, suppose that (iii) holds. Then

$$NA = \sum_{a \in A} Na$$

is a (two-sided) nonzero ideal of A. Since A is simple, we obtain $NA = A$. Therefore

$$(1) \qquad\qquad A = \sum_{a \in A} Na.$$

But since N is a simple A-module, so are all nonzero ideals of the form Na, being images of N under the A-module homomorphism $x \mapsto xa$. Thus (1) says that A_l is a sum of simple submodules, hence semisimple (see F10 in previous chapter). $\square$

In fact we can easily say something more: Since A_l is generated by 1, there exist finitely many elements $a_1, \ldots, a_n$ in A such that $A = Na_1 + \cdots + Na_n$. By the argument just given (and F11 in the previous chapter) we then have

$$(2) \qquad\qquad A_l \simeq N^m.$$

for some $m \in \mathbb{N}$. Thus the A-module A_l is *isogenous* (Definition 5 in previous chapter).

F2. *If A is a nonzero semisimple algebra, there is equivalence between*:

(i) *A is simple.*

(ii) *The A-module A_l is isogenous.*

(iii) *All simple A-modules are isomorphic.*

Proof. We have already settled (i) $\Rightarrow$ (ii) above. Any A-module M is isomorphic to a quotient of an A-module of the form $A_l^{(I)}$. If A_l is isogenous of type τ, so is $A_l^{(I)}$ and therefore also M (see F11 in preceding chapter). In particular, every simple A-module has the same type τ. This proves (ii) $\Rightarrow$ (iii). The implication (iii) $\Rightarrow$ (ii) is trivial, so we are left with (ii) $\Rightarrow$ (i). But this follows easily from F17 in the last chapter: if (ii) holds, every nonzero ideal U of A equals A. $\square$

F3. *If D is a division algebra, $M_n(D)$ is a simple artinian algebra for any $n \in \mathcal{N}$.*

Proof. This follows immediately from F2 (and F1), using F8 of the previous chapter. $\square$

Next comes an important property of simple algebras:

F4. *The center of a simple algebra is a field.*

Proof. The center $Z = Z(A)$ of an algebra A consists of all $z \in A$ such that

$$(3) \qquad\qquad za = az \quad \text{for every } a \in A.$$

Clearly Z is a commutative subalgebra of A. Hence we must show that if A is *simple*, every nonzero $z \in Z$ is invertible in A; the inverse will automatically be in Z (why?). So suppose A is simple and $z \in A$ is nonzero and central. Then Az is a nonzero ideal of A, and so equals A; in particular, z has a left (hence right) inverse in A. $\qquad\qquad\qquad\qquad\qquad\qquad\qquad\qquad\qquad\qquad\qquad\qquad\qquad\quad\square$

Theorem 1. *A nonzero semisimple algebra A has only finitely many distinct minimal ideals $A_1, \ldots, A_n$. Each A_i is itself an algebra under the addition and multiplication induced from A. Moreover*

$$(4) \qquad\qquad A = A_1 \times A_2 \times \cdots \times A_n$$

is the direct product of the algebras A_i, and each A_i is a simple artinian algebra. (For this reason the A_i are called the *simple components* of A.)

Conversely, *if $A_1, \ldots, A_n$ are simple artinian algebras, the direct product A of $A_1, \ldots, A_n$ is a semisimple algebra, and the A_i, regarded as subsets of A, are the minimal ideals of A.*

Proof. We use F18 from the last chapter, which says that, as a left A-module, A is the direct sum of all the (finitely many) distinct minimal ideals $A_1, \ldots, A_n$ of A:

$$(5) \qquad\qquad A = A_1 \oplus A_2 \oplus \cdots \oplus A_n.$$

Since the A_i are two-sided ideals of A, we have $A_i A_j \subseteq A_i \cap A_j$. Hence, by (5),

$$(6) \qquad\qquad A_i A_j = 0 \quad \text{for} \quad i \neq j.$$

Again by (5), every $x \in A$ has a *unique* representation of the form

$$x = x_1 + x_2 + \cdots + x_n \quad \text{with } x_i \in A_i.$$

We call the x_i the components of x in A_i. By (6) we have

$$(7) \qquad\qquad xy = x_1 y_1 + x_2 y_2 + \cdots + x_n y_n$$

for any $x, y \in A$; in particular, the components of the unity e of A satisfy

$$e_i x_i = e x_i = x_i = x_i e = x_i e_i.$$

This shows that each A_i is an algebra with unity e_i. So far we have shown that A is the *direct product* of the algebras $A_1, \ldots, A_n$.

Now take a fixed A_i. Every left ideal of A_i is a left ideal of A, and likewise for two-sided ideals. Hence $(A_i)_l$ is artinian (because A_l is), and its only ideals are itself and 0 (since A_i is a minimal ideal of A). Thus A_i is a simple artinian algebra as claimed.

Moving on to the converse: Let $A = A_1 \times A_2 \times \cdots \times A_n$ be a finite direct product of (for now) arbitrary algebras A_i. If T is a subset of A, denote by $T_i = p_i(T)$ the image of T under the i-th projection p_i. If T is a left ideal of A, we can write $T = T_1 \times T_2 \times \cdots \times T_n$, because for any $x \in T$ we have $p_i(x) = e_i x$, where e_i denotes the unity of A_i (and we regard A_i as a subset of A), and therefore $T_i \subseteq T$ for $1 = 1, 2, \ldots, n$. Hence, if the A_i are simple algebras, the minimal ideals of A are the A_i and nothing else. If A_i is a semisimple algebra, we see easily that it is semisimple also as an A-module; but A_l is the direct sum of the A-modules A_i, so A is semisimple.   □

F5. *In the situation of Theorem 1, let* $K_i = Z(A_i)$ *be the center of the algebra* A_i, *for each* i. *The center* $Z = Z(A)$ *of* A *is then given by*

(8) $$Z = K_1 \times K_2 \times \cdots \times K_n,$$

thanks to (4). Thus the center of a semisimple algebra A *is the direct product of the centers* K_i *of the simple components* A_i *of* A. *Each* K_i *is a field (see F4).*

If $n > 1$ in (8), the center of A is obviously not an integral domain, so one can also assert, complementing F2:

F6. *A semisimple algebra is simple if and only if its center is a field.*

Being a field, each K_i in (8) is a simple artinian algebra. Thus, by Theorem 1, $Z = Z(A)$ is itself a semisimple algebra, and its simple components are the fields K_i. In particular, for A commutative, that is, when $Z = A$, we obtain:

F7. *An algebra* A *is a commutative semisimple algebra if and only if* A *is the direct product of finitely many fields*:

(9) $$A = K_1 \times K_2 \times \cdots \times K_n.$$

The fields K_i *satisfying this equality are uniquely determined as subsets of* A.

Putting this together with F39 and F29 of the preceding chapter, we obtain:

Theorem 2. *A commutative artinian algebra having no nonzero nilpotent elements is semisimple* (so F7 applies).

We state explicitly a straightforward but useful consequence of this theorem:

F8. *A finite-dimensional,* **commutative** *K-algebra* A *having no nonzero nilpotents is the direct product of finitely many extension fields* $K_1, \ldots, K_n$ *of* K, *each of finite degree over* K. *The* K_i *satisfying this condition are uniquely determined as subsets of* A.

Remark. In characteristic 0 this result can also be proved more directly; see §29.3. See also the related §23.15.

Theorem 2 can be proved directly as well. One starts by showing that if A is an algebra as in the statement and P is a prime ideal of it, A/P is a field. An application of the Chinese Remainder Theorem (Chapter 4, F16 in vol. 1) and the result in §4.14 conclude the proof.

2. According to F3, any matrix algebra $M_n(D)$ over a division algebra D is an example of a simple artinian algebra. In reality *all* simple artinian algebras arise in this way:

Wedderburn's Theorem. *An artinian algebra A is simple if and only if it is isomorphic to a matrix algebra $M_n(D)$ over a division algebra D: in symbols,*

$$(10) \qquad\qquad A \simeq M_n(D).$$

The number n and the isomorphism class of D are uniquely determined.

As mentioned, F3 takes care of one direction. As for the other, the structural isomorphism (10) is a consequence of Theorem 3 below, and the uniqueness will be dealt with in F9. So let's tackle (10), giving a nice and detailed account:

Theorem 3. *Let A be a simple artinian algebra and let N be a simple A-module with endomorphism algebra D.*

(I) *The natural map $A \to \mathrm{End}_D(N)$ is an isomorphism. The D-vector space N is finite-dimensional, and its dimension coincides with the length r of A. Thus*

$$(11) \qquad\qquad A \simeq M_r(D^\circ).$$

The center of A is a field, isomorphic to the center of D.

(II) *For any nonzero, finitely generated A-module M, the algebra of endomorphisms $B := \mathrm{End}_A(M)$ is simple and artinian, and we have*

$$(12) \qquad\qquad B \simeq M_n(D)$$

with the same D as above and with $n = M:N$. If A is a K-algebra, we have the dimension formula

$$(13) \qquad\qquad (M:K)^2 = (A:K)(B:K),$$

and all these dimensions are finite if one of them is.

Proof. Step 1. A is semisimple because it is a simple artinian algebra (F1). Thus every A-module is a direct sum of simple A-modules (F13 in Chapter 28). But all simple A-modules are isomorphic (F2), so every A-module M is isomorphic to $N^{(I)}$ for some set I. If M is finitely generated, we have

$$(14) \qquad\qquad M \simeq N^n$$

for some $n \in \mathbb{N}$. From (14) together with F4 in the last chapter we immediately get

$$\mathrm{End}_A(M) \simeq M_n(\mathrm{End}_A(N)) = M_n(D),$$

which yields (12). In particular, $B = \mathrm{End}_A(M)$ is a simple artinian algebra.

Step 2. We apply step 1 to the A-module A_l and obtain from the isomorphism

(15) $$A_l \simeq N^r,$$

where $r = A_l : N = l(A)$, the isomorphism of algebras

$$\mathrm{End}_A(A_l) \simeq M_r(D).$$

But we know that $\mathrm{End}_A(A_l) = A^\circ$, so

(16) $$A^\circ \simeq M_r(D).$$

It follows that $A = (A^\circ)^\circ \simeq M_r(D)^\circ \simeq M_r(D^\circ)$; see (28) in Chapter 28.

Step 3. Now fix an isomorphism $\varphi : A_l \to N^r$; it induces a well defined isomorphism φ^* as the bottom row of the following diagram (compare the proof F9 further down):

(17)
$$
\begin{array}{ccc}
A & \longrightarrow & \mathrm{End}_D(N) \\
\downarrow & & \downarrow \\
\mathrm{End}_{\mathrm{End}_A(A_l)}(A_l) & \xrightarrow{\ \varphi^*\ } & \mathrm{End}_{\mathrm{End}_A(N^r)}(N^r)
\end{array}
$$

The remaining maps in (17) are canonical, and the vertical maps are isomorphisms (Chapter 28, F3 and F4). It is easily seen that all maps in (17) take left translations by elements of A into such; hence (17) is commutative. As claimed, therefore, the natural map $A \to \mathrm{End}_D(N)$ is also an isomorphism. But then $\mathrm{End}_D(N)$ is *artinian*, and this in turn obviously implies $N : D < \infty$. The length of $\mathrm{End}_D(N)$ is $N : D$, by F8 in the last chapter, so

(18) $$N : D = l(A) = r.$$

That the center of A is a *field* is true for any simple algebra A, artinian or otherwise, by F4. In the present situation we can also justify it as follows: For $a \in Z = Z(A)$, the element a_N lies in $\mathrm{End}_A(N) = D$, and so is in fact central in D. At the same time, $Z(D)$ is obviously contained in the center of $\mathrm{End}_D(N)$. Therefore the isomorphism

(19) $$A \to \mathrm{End}_D(N)$$

gives rise to an isomorphism from $Z = Z(A)$ onto the center $Z(D)$ of the division algebra D, which is obviously a field.

Step 4. From (14) and (18) we now get $M : K = n(N : K) = n(N : D)(D : K) = nr(D : K)$. On the other hand, (16) and (12) give, respectively, $A : K = r^2(D : K)$ and $B : K = n^2(D : K)$. It follows that $(A : K)(B : K) = r^2n^2(D : K)^2 = (M : K)^2$, which is (13). $\qquad\square$

F9. *Suppose D_1 and D_2 are division algebras such that $M_r(D_1) \simeq M_s(D_2)$. Then $D_1 \simeq D_2$ and $r = s$.*

Proof. Step 1. Let N, N' be A-modules over an algebra A. If N and N' are isomorphic via $\varphi : N \to N'$, we have $\mathrm{End}_A(N) \simeq \mathrm{End}_A(N')$, because conjugation by φ, defined by $f \mapsto \varphi \circ f \circ \varphi^{-1}$, is an isomorphism from $\mathrm{End}_A(N)$ to $\mathrm{End}_A(N')$.

Step 2. Let $f : A \to A'$ be an isomorphism of simple artinian algebras and N, N' simple modules over A, A', respectively. By means of f we can regard N' as an A-module: $ax = f(a)x$. Clearly this A-module is simple. Being a simple artinian algebra, A only admits one type of simple A-module, so $N \simeq N'$ as A-modules, and by step 1 this implies that $\operatorname{End}_A(N) \simeq \operatorname{End}_A(N')$. But by the definition of the A-module structure of N' we have $\operatorname{End}_A(N') = \operatorname{End}_{A'}(N')$; combining both isomorphisms we get $\operatorname{End}_A(N) \simeq \operatorname{End}_{A'}(N')$.

Step 3. If D is a division algebra, D^r is a simple $M_r(D)$-module for every $r \in \mathbb{N}$, and $\operatorname{End}_{M_r(D)}(D^r)$ is isomorphic to D°, by F5 in the preceding chapter. By step 2, then, $M_r(D_1) \simeq M_s(D_2)$ implies $D_1^\circ \simeq D_2^\circ$, hence $D_1 \simeq D_2$. This proves the first part of F9. Since $D_1 \simeq D_2$ we also have $M_s(D_1) \simeq M_s(D_2)$; putting this together with $M_r(D_1) \simeq M_s(D_2)$ we obtain $M_r(D_1) \simeq M_s(D_1)$.

Step 4. There remains to show that if $M_r(D) \simeq M_s(D)$ (where D is a division algebra) then $r = s$. This becomes clear if we recall the notion of length: $M_r(D)$ and $M_s(D)$ have lengths r and s, and isomorphic semisimple algebras obviously have the same length, so $r = s$. □

Remark. We have presented this proof at such length, dotting the i's and crossing the t's, because we have found their counterparts in the literature inadequate, or at best incomplete. In particular, step 4 is often justified simply by concluding from the isomorphism $M_r(D) \simeq M_s(D)$ that the dimensions of the two D-vector spaces are equal. But neither does the isomorphism $M_r(D) \simeq M_s(D)$ have to be a D-morphism, nor do two D-vector space structures on the same additive group have to have the same D-dimension. We mention en passant that certain technical difficulties in the proof of Theorem 3 and F9 do not intervene when one is dealing only with *finite-dimensional K-algebras*. In the applications of the theory in the next few chapters we will always restrict ourselves to this case.

We now spell out some consequences of Wedderburn's Theorem:

F10. (a) *A simple algebra A is artinian if and only if the opposite algebra A° is artinian, in which case $l(A) = l(A^\circ)$.*

(b) *An algebra A is semisimple if and only if the algebra A° is semisimple, in which case $l(A) = l(A^\circ)$.*

Proof. (a) The isomorphism $A \simeq M_n(D)$ implies $A^\circ \simeq M_n(D)^\circ \simeq M_n(D^\circ)$, and moreover $l(M_n(D)) = n = l(M_n(D^\circ))$.

(b) This follows directly from part (a) and Theorem 1. □

From Theorem 1 we also obtain:

Theorem 4. *An algebra A is semisimple if and only if*

(20) $$A \simeq M_{r_1}(D_1) \times M_{r_2}(D_2) \times \cdots \times M_{r_n}(D_n),$$

where the D_i are division algebras. On the right-hand side of (20), the number n of factors and the isomorphism classes of the D_i, as well as the values of $r_1, \ldots, r_n$, are uniquely determined.

For ease of later reference, we wish to state explicitly an important special case of Wedderburn's Theorem:

Theorem 5. *A finite-dimensional K-algebra A is simple if and only if it is isomorphic to a matrix algebra $M_n(D)$ over a K-division algebra D. The number n and the isomorphism class of D are uniquely determined by the condition*

$$(21) \qquad\qquad A \simeq M_n(D),$$

and $D:K$ is finite. K is the center of A if and only if it is the center of D.

Proof. Since any finite-dimensional K-algebra is artinian, Wedderburn's Theorem applies. The K-isomorphism (21) implies $A:K = n^2(D:K)$, so $D:K < \infty$. The centers of A and D are K-isomorphic, by Theorem 3.   □

Theorem 6. *If K is algebraically closed, every finite-dimensional simple K-algebra A is isomorphic to a matrix algebra $M_n(K)$ over the field K; in particular, the center of A is K and the dimension $A:K = n^2$ is a square.*

Proof. This follows from Theorem 5 and the next lemma.   □

Lemma 1. *Let D be a finite-dimensional division algebra over a field K. Every commutative subalgebra E of D is a field. If K is algebraically closed, D coincides with K.*

Proof. E is clearly an integral domain and has finite dimension over K. Therefore E is a field, by F2 in Chapter 2 (vol. 1). If K is algebraically closed, E coincides with K. Applying this to the subalgebra $E = K[a]$ of D generated by an arbitrary $a \in D$, we conclude that $a \in K$.   □

F11. *Let A be an algebra over an algebraically closed field K, and let M be a simple A-module satisfying $M:K < \infty$. Then $K \to \mathrm{End}_A(M)$ is an isomorphism, and $A \to \mathrm{End}_K(M)$ is surjective. If A is commutative we have $M:K = 1$.*

Proof. Since the A-module M is assumed simple, the space $D := \mathrm{End}_A(M)$ is a division K-algebra, by *Schur's Lemma* (F9 in the previous chapter). But $D = \mathrm{End}_A(M)$ is also a subalgebra of the K-algebra $\mathrm{End}_K(M)$, and finite-dimensional at that, because $M:K < \infty$. Now Lemma 1 implies that D indeed equals K, which is the first assertion in F11. The surjectivity of the natural map $A \to \mathrm{End}_K(M)$ comes most easily out of the *Density Theorem* (F20 in the previous chapter). If A is commutative, so is every homomorphic image $\mathrm{End}_K(M) \simeq M_n(K)$; but this is only possible if $n = M:K = 1$.   □

Wedderburn's Theorem leads naturally to the following notion:

Definition 2. Two simple artinian algebras A and B are called *similar* — in symbols, $A \sim B$ — if there exists a division algebra D and natural numbers r, s such

$$(22) \qquad\qquad A \simeq M_r(D), \qquad B \simeq M_s(D).$$

Remark. From Wedderburn's Theorem — especially the part about uniqueness — we get as straightforward consequences the basic properties of the similarity relation among simple artinian algebras:

(a) $A \simeq B$ implies $A \sim B$.

(b) $A \sim B$ and $l(A) = l(B)$ imply $A \simeq B$.

(c) $\sim$ is an equivalence relation.

(d) $A \sim B$ implies $Z(A) \simeq Z(B)$, that is, A and B have isomorphic centers.

Definition 3. A K-algebra A is called *central* if K is the center of A.

Remark. Let A be a *simple* algebra. By F4, the center of A is a *field*, say K. Then A can be regarded naturally as a *central K-algebra*, which we denote by A/K. Of course A/K is simple, just like A.

Wedderburn's Theorem shows that the study of simple artinian algebras is essentially the study of *division algebras*. How should we go about investigating the latter? It is natural to take together all division algebras whose center is isomorphic to one and the same field K, that is, to study all *central* division algebras D over a fixed K. Such a D does not at all need to have finite dimension over K (see §29.16); but we will only be able to make further progress when $D : K$ is in fact finite. For convenience of expression, we will adhere to ingrained custom and make the following definition for this case:

Definition 4. A simple K-algebra A is called *central-simple* if it is central and *finite-dimensional*.

Remark. By Theorem 5, a K-algebra A is central-simple if and only if $A \simeq M_n(D)$, where D is a finite-dimensional and central K-division algebra.

Definition 5. For a given field K, denote by

$$\text{Br } K$$

the set of all *similarity classes of central-simple K-algebras*. This set is called the *Brauer group* of K.

Remarks. (1) First, a word about the set-theoretical admissibility of the definition. Since isomorphism implies similarity (see remark after Definition 2) and we are working in finite dimensions, we need only justify that it is OK to talk about the set of all *isomorphism classes of finite-dimensional K-algebras*. And this is so: If A is an n-dimensional K-algebra, there is a natural monomorphism

$$A \to \text{End}_K(A),$$

given by (5) in Chapter 28. But $\text{End}_K(A)$ and $M_n(K)$ are isomorphic, so A is isomorphic to a subalgebra of a K-algebra $M_n(K)$.

(2) The element of Br K determined by a central-simple K-algebra A will be denoted by $[A]$. If A and B are central-simple K-algebras, we have

$$A \simeq B \implies [A] = [B],$$

$$[A] = [B] \text{ and } A : K = B : K \implies A \simeq B.$$

Every element of Br K is of the form $[D]$, for D a finite-dimensional central division algebra over K, and D is unique up to isomorphism.

3. The main goal of this section is to show that the tensor product $A \otimes_K B$ of two central-simple K-algebras A and B is also a central-simple K-algebra. Then we will see in more detail how this allows us to give Br K a natural group structure, thus vindicating the term *Brauer group* for Br K.

Notation. For a fixed field K, we write the tensor product $A \otimes_K B$ of K-algebras A and B simply as $A \otimes B$, if no misunderstanding is likely. The maps $a \mapsto a \otimes 1$ and $b \mapsto 1 \otimes b$ make A and B into subalgebras of the K-algebra $A \otimes B$ (see vol. I, pp. 61 and following, especially F9; note also §6.14).

If A is a K-algebra, the K-algebra

$$(23) \qquad\qquad A \otimes_K A^\circ$$

is called the *enveloping algebra* of A. We consider the homomorphisms

$$(24) \qquad\qquad A \to \operatorname{End}_K(A) \quad \text{and} \quad A^\circ \to \operatorname{End}_K(A)$$

that associate to each element of A the left and right multiplication by the given element. Since a left multiplication commutes with a right one, we obtain in this way a natural homomorphism

$$(25) \qquad\qquad A \otimes_K A^\circ \to \operatorname{End}_K(A),$$

by means of which A can be regarded as an $A \otimes_K A^\circ$-module. This module structure, therefore, is well defined by the condition

$$(26) \qquad\qquad (a_1 \otimes a_2)x = a_1 x a_2.$$

The $A \otimes_K A^\circ$-module A is simple if and only if the algebra A is simple. Using F3 in the previous chapter, one also checks easily that

$$(27) \qquad\qquad \operatorname{End}_{A \otimes_K A^\circ}(A) = Z(A).$$

Taking *Schur's Lemma* into account we recover from (27) the already proved fact that the center of a simple K-algebra is a field.

If B is a subalgebra of a K-algebra A, the set

$$(28) \qquad\qquad Z_A(B) = \{a \in A \mid ax = xa \text{ for every } x \in B\}$$

is called the *centralizer of B in A*. Clearly $Z_A(B)$ is a subalgebra of A. By restriction, (25) gives rise to a homomorphism

(29) $$B \otimes_K A^\circ \to \mathrm{End}_K(A),$$

and we can regard A also as a $B \otimes A^\circ$-module. As a generalization of (27) we immediately see that

(30) $$\mathrm{End}_{B \otimes A^\circ}(A) = Z_A(B).$$

We call $Z_A(Z_A(B))$ the *bicentralizer of B in A*. By definition, it contains B.

F12. *Given subalgebras A' and B' of K-algebras A and B, respectively, we have*

(31) $$Z_{A \otimes B}(A' \otimes B') = Z_A(A') \otimes Z_B(B'),$$

where $A' \otimes B'$ and $Z_A(A') \otimes Z_B(B')$ are regarded as subalgebras of $A \otimes B$. In particular, the centers are related by

(32) $$Z(A \otimes B) = Z(A) \otimes Z(B).$$

Proof. Let $(a_i)_{i \in I}$ be a K-basis of A, and take $z \in Z_{A \otimes B}(A' \otimes B')$. Like any element of $A \otimes B$, our z has a unique representation

$$z = \sum_i a_i \otimes b_i$$

with elements b_i of B (and $b_i = 0$ for almost all i). We claim that the b_i lie in $Z_B(B')$; indeed, for every $b \in B'$ we have $\sum_i a_i \otimes b_i b = z(1 \otimes b) = (1 \otimes b)z = \sum_i a_i \otimes bb_i$, so $b_i b = bb_i$. Therefore z lies in $A \otimes Z_B(B')$. Analogously, z lies in $Z_A(A') \otimes B$. However, the intersection of $A \otimes Z_B(B')$ and $Z_A(A') \otimes B$ coincides with $Z_A(A') \otimes Z_B(B')$, as one acknowledges easily by choosing for the K-basis of A an extension of a K-basis of $Z_A(A')$. So far, then, we have shown that the left side of (31) is contained in the right side. The reverse inclusion is trivial. $\square$

Remark. Let E and F be extensions of a field K. Then $E \otimes_K F$ is not a field in general (see §6.6 in vol. I). It follows that the tensor product $A \otimes_K B$ of two simple K-algebras need not be simple. Nonetheless:

Theorem 7. *Let A, B be simple K-algebras, and suppose A or B is central over K. Then $A \otimes_K B$ is simple.*

Proof. Let A be a simple and central K-algebra. We will show, more generally, that for any K-algebra B every ideal T of $A \otimes B$ is of the form $A \otimes I$, where we have set $I = T \cap B$. Indeed, any $t \in T$ can be written as

$$t = \sum_{i=1}^n x_i \otimes y_i, \quad \text{with } x_i \in A, \ y_i \in B,$$

where the x_i can be taken to be linearly independent over K. Because of (27) and the assumption $Z(A) = K$, we have

$$\mathrm{End}_{A \otimes A^\circ}(A) = K.$$

An application of the *Jacobson Density Theorem* (F20 in Chapter 28) to the map (25) then yields the existence of elements a_j in the algebra $A \otimes A^\circ$ satisfying

$$a_j x_i = \delta_{ij},$$

where the right-hand side is Kronecker's delta. Now $A \otimes B$ is, like A, an $A \otimes A^\circ$-module, in view of the map $a(x \otimes y) = ax \otimes y$ for $a \in A \otimes A^\circ$, $x \in A$, $y \in B$. Since T is a two-sided ideal of $A \otimes B$, we have

$$aT \subseteq T \quad \text{for every } a \in A \otimes A^\circ.$$

Because $a_j t = \sum_i a_j x_i \otimes y_i = 1 \otimes y_j$, therefore, we have $1 \otimes y_j \in T$. For every j, then, we have $y_j = 1 \otimes y_j \in I = T \cap B$, and altogether we have proved that $T \subseteq A \otimes I$. The reverse is trivial since $I \subseteq T$.   □

Remark. Given K-algebras A, B, the following facts are obvious:

  (a) *If $A \otimes_K B$ is simple, so are A and B.*

  (b) *If $A \otimes_K B$ is artinian, so are A and B.*

In addition:

  (c) *If $A : K < \infty$ and B is artinian, $A \otimes_K B$ is artinian.*

For if $n = A : K$, there is an isomorphism $A \otimes_K B \simeq K^n \otimes_K B \simeq B^n$ of B-modules, and the B-module B^n is artinian (see F23 in the last chapter).

A related result appears in §29.4.

F13. *Let A and B be simple artinian K-algebras. If at least one of A, B is finite-dimensional and at least one is central, then $A \otimes_K B$ is a simple artinian algebra.*

Proof. $A \otimes_K B$ is simple by Theorem 7, and artinian by item (c) just above.   □

We single out a corollary of Theorem 7 and F12:

Theorem 8. *If A and B are central-simple K-algebras, so is $A \otimes_K B$.*

Proof. $A \otimes_K B$ is simple, by Theorem 7. By (32) we have

$$Z(A \otimes_K B) = Z(A) \otimes_K Z(B) = K \otimes_K K = K,$$

so $A \otimes_K B$ is also central. Finally, $(A \otimes_K B) : K = (A : K)(B : K) < \infty$.   □

F14. *If A is a central-simple K-algebra, the natural map $A \otimes_K A^\circ \to \mathrm{End}_K(A)$ of (25) is an isomorphism. By setting $n = A : K$ we obtain*

$$(33) \qquad\qquad A \otimes_K A^\circ \simeq M_n(K).$$

Proof. By Theorem 8, $A \otimes_K A^\circ$ is simple, so the map is injective. For dimensional reasons, then, it is also surjective. □

F15 and Definition 6. *Let A be a K-algebra and L an extension of K. The algebra*

(34) $$A_L := A \otimes_K L$$

can be regarded naturally as an L-algebra. The L-algebra A_L is said to arise from K-algebra A by **base change**. *Base change does not affect the dimension:*

(35) $$A_L : L = A : K.$$

If A is central over K, then A_L is central over L. If the K-algebra A is simple and central, so is the L-algebra A_L. Consequently, A_L is central-simple if A is. (Unless stated otherwise, we will always regard A_L as an L-algebra.)

Proof. As above, we regard L as a subalgebra of $A \otimes L$. Since L is obviously contained in the center of $A \otimes L$, this makes A_L into an L-algebra; multiplication by scalars from L is defined by

$$\alpha(a \otimes \beta) = (1 \otimes \alpha)(a \otimes \beta) = a \otimes \alpha\beta.$$

Any K-basis of A is an L-basis of $A_L = A \otimes L$, so (35) holds. The remaining assertions follow from F12 and Theorem 7. □

It's worth mentioning that over K the dimension of A_L is $A_L : K = (A : K)(L : K)$.

F16. *If A is a central-simple K-algebra, the dimension $A : K$ is a square.*

Proof. Let C be an algebraic closure of K. By F15, $A_C = A \otimes_K C$ is a central-simple C-algebra. But since C is algebraically closed, $A_C : C = A : K$ must be a square, by Theorem 6. □

F17. *If D is a division algebra and K is its center, the dimension of D over K is either infinite or a square.*

Definition 7. If D is a division algebra of finite dimension over its center K, the natural number $s = \sqrt{D : K}$ is called the *Schur index* of D. For A a central-simple K-algebra such that $A \sim D$, we can also say that s is the Schur index of A, or of the element $[A]$ of Br K. We write $s = s(A) = s([A])$ and sometimes abbreviate "Schur index" to "index" (of A or $[A]$).

Remark. Let A be a central-simple K-algebra. The index s of A and the dimension $A : K = n^2$ are related by

(36) $$n = rs,$$

where $r = l(A)$ is the length of A: indeed, if $A = M_r(D)$ we have $r = l(A)$ and $A : K = r^2(D : K) = r^2 s^2$. Hence

$$A : K = s^2 \text{ if and only if } A \text{ is a division algebra.}$$

Otherwise s is strictly less than $n = \sqrt{A : K}$ (and divides n). We call n the *reduced degree* of A.

It is time to officially state our long-heralded result:

Theorem 9. *The set* Br K *of similarity classes of central-simple K-algebras enjoys an abelian group structure provided by the tensor product operation.*

Proof. One must first check that the tensor product is compatible with the similarity equivalence relation:

$$(37) \qquad\qquad A \sim A', \ B \sim B' \ \implies \ A \otimes B \sim A' \otimes B'.$$

Only then is the multiplication in Br K given by

$$(38) \qquad\qquad [A] \cdot [B] = [A \otimes B]$$

well defined; here we rely of course on the all-important Theorem 8. The proof of (37) is by direct calculation, invoking the isomorphisms

$$(39) \qquad M_r(D) \simeq D \otimes M_r(K), \quad M_s(K) \otimes M_r(K) \simeq M_{rs}(K).$$

The first of these holds for any K-algebra D, and with that the second follows by observing that $M_s(K) \otimes M_r(K) \simeq M_r(M_s(K)) \simeq M_{rs}(K)$.

The associativity and commutativity of the multiplication given in (38) are immediate consequences of the same properties for the tensor product. The multiplicative identity is $[K]$, since $A \otimes K \simeq A$. The inverse of $[A]$ is $[A°]$, as can be seen from (33). $\qquad\square$

Remarks. (1) As mentioned in Remark 2 after Definition 5, Br K classifies the finite-dimensional central division algebras over K. But the tensor product of two such algebras need not be a division algebra (see §29.5), so the group structure of Br K cannot be described by considering division algebras alone.

(2) *If K is algebraically closed* (say $K = \mathbb{C}$), *the group* Br K *is trivial.* This follows from Theorem 6 or Lemma 1.

(3) *If K is a finite field,* Br K *is again trivial.* This is an alternate form of another famous theorem of *Wedderburn,* to the effect that every finite division algebra is commutative, that is, a field. This result will come out fairly easily from considerations we will make later.

(4) *If K is real-closed* (say $K = \mathbb{R}$), *then* Br K *is cyclic of order* 2. This follows from a well-known theorem of *Frobenius,* according to which a real-closed field admits only one noncommutative division algebra of finite dimension > 1 over itself, namely the algebra of *quaternions,* introduced by *Hamilton.* This in turn is a straightforward application of the theory of *crossed products,* which we will discuss in the next section; for this reason we refrain from giving a direct proof here.

F18 and Definition 8. *If L is an extension of K, base change to L yields a homomorphism*

$$(40) \qquad\qquad \mathrm{res}_{L/K} : \mathrm{Br}\, K \to \mathrm{Br}\, L$$

from the Brauer group of K into that of L. This is called *restriction* with respect to L/K, for reasons that will become apparent. When there is no likelihood of misunderstanding, we can also write res_L or simply res. If F is an intermediate field of L/K we have

(41) $$\mathrm{res}_{L/F} \circ \mathrm{res}_{F/K} = \mathrm{res}_{L/K}.$$

The *kernel* of the homomorphism $\mathrm{res}_{L/K}$ will be denoted by $\mathrm{Br}(L/K)$, that is,

(42) $$\mathrm{Br}(L/K) = \{[A] \in \mathrm{Br}\, K \mid A_L \sim L\}.$$

If $[A]$ lies in $\mathrm{Br}(L/K)$, we say that L is a *splitting field* of A (or of $[A]$). Thus *L is a splitting field of a central-simple K-algebra A if and only if there is an isomorphism of L-algebras*

(43) $$A \otimes_K L \simeq M_n(L),$$

with $n^2 = A:K$. In this case we also say that A *splits over* L.

Proof. Take $[D] \in \mathrm{Br}\, K$, where D is a division algebra. For any central-simple K-algebra $A \simeq M_r(D)$ similar to D there are L-isomorphisms

$$A \otimes L \simeq D \otimes M_r(K) \otimes L \simeq D \otimes M_r(L) \simeq (D \otimes L) \otimes_L M_r(L),$$

and this last L-algebra is similar to $D \otimes L$. Hence $\mathrm{res}_{L/K}$ is well defined by setting

(44) $$\mathrm{res}_{L/K}([A]) = [A \otimes L] = [A_L].$$

To see that $\mathrm{res}_{L/K}$ is a homomorphism, we write $(A \otimes B) \otimes L \simeq (A \otimes L) \otimes B \simeq ((A \otimes L) \otimes_L L) \otimes B \simeq (A \otimes L) \otimes_L (L \otimes B) \simeq (A \otimes L) \otimes_L (B \otimes L)$. If F is an intermediate field of L/K, we have

(45) $$(A \otimes_K F) \otimes_F L \simeq A \otimes_K L,$$

which yields (41). □

F19. *With the preceding notation, the Schur index $s(A_L)$ divides $s(A)$. The algebra A splits over L if and only if $s(A_L) = 1$.*

Proof. It suffices to prove this for the case when $A = D$ is a division algebra. In view of (36), $s(D_L)^2$ divides $D_L : L = D:K = s(D)^2$. The second assertion is clear: By Definition 7, saying that $s(A_L) = 1$ is the same as saying that $A_L \sim L$. □

Every $[A] \in \mathrm{Br}\, K$ has a splitting field, for instance any algebraic closure of K (Theorem 6). But does any A have a splitting field L of *finite degree* over K? Or perhaps even with $L : K = s(A)$? These questions will be answered in the affirmative in the next section but one.

4. In this section, of merely informational character, we investigate a question that arises of itself in connection with Theorem 7. Let A and B be simple K-algebras. If neither A nor B is central over K, the tensor product $A \otimes_K B$ is not, in general, a simple algebra. But is $A \otimes_K B$ at least semisimple, under an appropriate finiteness assumption? (See statement (c) in the remark on page 162.) This is not always the case, but we do have the following criterion:

Theorem 10. *If two K-algebras A and B are finite direct products of simple algebras, their tensor product $A \otimes_K B$ is a finite direct product of simple algebras if and only if the tensor product $Z(A) \otimes_K Z(B)$ of the centers of A and B has the same property (and therefore is a direct product of finitely many fields). The simple components of $A \otimes_K B$ are then in natural one-to-one correspondence with those of $Z(A \otimes_K B) = Z(A) \otimes_K Z(B)$.*

Theorem 10 represents a generalization of Theorem 7. But it can also be easily derived from Theorem 7 (exercise for the reader). Here we wish to proceed in a slightly different direction and start by pointing out the following fact, whose first part we have already shown in the proof of Theorem 7:

Theorem 11. *Let A be a simple and **central** K-algebra. For any K-algebra B, the ideals T of $A \otimes B$ are in one-to-one correspondence with the ideals I of B, through the assignments $T = A \otimes I$ and $I = T \cap B$.*

Hence, under the same assumptions, $A \otimes B$ is a finite direct product of simple algebras if the same is true of B; and in this case there is a one-to-one correspondence between the simple components of $A \otimes B$ and those of B.

Proof. The second part follows from the first and from the following considerations: If a K-algebra $C = C_1 \times \cdots \times C_n$ is the direct product of simple algebras, each ideal of C is a direct product of some of the C_i. The C_i are precisely the minimal ideals of C. Conversely, if an algebra C has finitely many minimal ideals $C_1, \ldots, C_n$ with the property that C is the direct sum of the C_i, then the C_i are themselves algebras (with the addition and multiplication in C), and C equals the direct product $C_1 \times \cdots \times C_n$ of the algebras C_i; in addition, the C_i are simple (see the proof of Theorem 1). $\square$

Theorem 11'. *Let A be a simple K-algebra. For any K-algebra B, the ideals T of*

$$A \otimes_K B = A \otimes_{Z(A)} (Z(A) \otimes_K B)$$

are in one-to-one correspondence with the ideals I of the K-algebra $Z(A) \otimes_K B$, through the assignment $T = A \otimes_{Z(A)} I$. Thus $A \otimes_K B$ is a finite direct product of simple algebras if and only if the same is true of the algebra $Z(A) \otimes_K B$, and in this case there is a one-to-one correspondence between the simple components of $A \otimes_K B$ and those of $Z(A) \otimes_K B$.

Proof. The center $Z(A)$ of a simple algebra A is a field (by F4), so we just have to apply Theorem 11 to the simple and central $Z(A)$-algebra A and the $Z(A)$-algebra $Z(A) \otimes_K B$. (When the latter is regarded merely as a K-algebra, the ideals don't change.) $\square$

Theorem 12. *Let A and B be simple K-algebras. The ideals of $A \otimes_K B$ are in one-to-one correspondence with those of $Z(A) \otimes_K Z(B)$. Further, $A \otimes_K B$ is a finite direct product of simple algebras if and only if the same is true of the tensor product $Z(A) \otimes_K Z(B)$ of the field extensions $Z(A)$ and $Z(B)$ of K, and in this case there is a one-to-one correspondence between the simple components of $A \otimes_K B$ and those of $Z(A) \otimes_K Z(B) = Z(A \otimes_K B)$.*

Proof. Theorem $11'$ reduces everything to the study of the K-algebra $Z(A) \otimes_K B$. Since B is simple, we can apply Theorem $11'$ to the algebra $B \otimes_K Z(A) = Z(A) \otimes_K B$, thus reducing the problem to the algebra

$$Z(B) \otimes_K Z(A) = Z(A) \otimes_K Z(B). \qquad \square$$

As a result, the one-to-one correspondence between ideals T of $A \otimes_K B$ and ideals H of $Z(A) \otimes_K Z(B)$ is given by

(46)
$$T = A \otimes_{Z(A)} H \otimes_{Z(B)} B.$$

The former correspond to the ideals I of $Z(A) \otimes_K B$ via $T = A \otimes_{Z(A)} I$, and the ideals I, in turn, are in correspondence with the ideals H via $I = H \otimes_{Z(B)} B$.

Proof of Theorem 10. If $A = \prod A_i$ and $B = \prod B_j$, with finitely many simple K-algebras A_i and B_j, we immediately obtain $A \otimes B = \prod_{i,j} A_i \otimes B_j$ and $Z(A \otimes B) = Z(A) \otimes Z(B) = \prod_{i,j} Z(A_i) \otimes Z(B_j)$. Now Theorem 10 follows easily from Theorem 12. $\qquad \square$

In view of the preceding results we should now look into tensor products $E_1 \otimes_K E_2$ of *fields*. First we establish:

F20. *An algebraic field extension E/K is separable if and only if, for every extension F of K, the tensor product $E \otimes_K F$ has no nonzero nilpotent elements.*

Proof. It is easy to see that it suffices to consider the case of E/K finite.

Step 1. Let E/K be finite and separable. By the *Primitive Element Theorem* we have $E \simeq K[X]/f$, where $f \in K[X]$ is an irreducible and separable polynomial. Now obviously $K[X]/f \otimes_K F$ is isomorphic to $F[X]/f$ (compare §6.6 in vol. I). Since f is separable, its prime factorization in $F[X]$ is of the form $f = f_1 f_2 \ldots f_r$, and thus, by the *Chinese Remainder Theorem*,

(47)
$$F[X]/f \simeq K[X]/f_1 \times \cdots \times K[X]/f_r$$

is a direct product of fields, and so has no nilpotent elements other than 0.

Step 2. Let E/K be finite and *inseparable*. There is then some α in E whose minimal equation over K has the form

(48)
$$a_0 + a_1 \alpha^p + a_2 \alpha^{2p} + \cdots + \alpha^{np} = 0,$$

with $p = \operatorname{char} K > 0$ (see F12 in Chapter 7, vol. I). Now there is an extension F of K containing elements b_i such that $b_i^p = a_i$. Looking at degrees we see that $1, \alpha, \alpha^2, \ldots, \alpha^n$ are linearly independent over K, so the sum $t = \sum \alpha^i \otimes b_i$ in $E \otimes_K F$ is nonzero; on the other hand, we have $t^p = \sum \alpha^{ip} \otimes b_i^p = \sum \alpha^{ip} \otimes a_i = \sum a_i \alpha^{ip} \otimes 1 = \left(\sum a_i \alpha^{ip} \right) \otimes 1 = 0$. $\qquad \square$

Thanks to F20, we can extend the notion of *separability* from algebraic field extensions to arbitrary ones:

Definition 9. A field extension E/K is called *separable* if, for *every* field extension F/K, the tensor product $E \otimes_K F$ has no nonzero nilpotents. A *semisimple* K-algebra A is called *separable* if the components K_i of the center of A are all separable over K (compare F5).

Theorem 13. *Let A and B be semisimple K-algebras, of which at least one is finite-dimensional and at least one is separable. Then the K-algebra $A \otimes_K B$ is semisimple and its simple components are in one-to-one correspondence with the simple components of the commutative semisimple algebra $Z(A) \otimes_K Z(B)$.*

Proof. Without loss of generality, we can assume in addition that A and B are simple (compare the proof of Theorem 10). Then $E := Z(A)$ and $F := Z(B)$ are fields. By statement (c) in the remark on page 162, $A \otimes_K B$ is certainly artinian. The theorem's conclusion then follows from Theorem 12, assuming we can prove that $E \otimes_K F$ is semisimple.

Suppose that A, say, is separable. By definition, the field extension E/K is also separable. If E/K is finite, the semisimplicity of $E \otimes_K F$ follows as in step 1 of the proof of F20; see (47). If E/K is not finite — in which case F/K must be finite — one must resort to a stronger argument: in any case $E \otimes_K F$ is artinian, and by definition $E \otimes_K F$ has no nilpotent elements apart from 0. Now an invocation of Theorem 2 shows that $E \otimes_K F$ is semisimple. $\qquad \square$

Remark. We will not go more deeply into this newly extended notion of separable field extensions, but see §29.22–27 in the Appendix for relevant results.

5. We now deal with the answers to the important questions posed at the end of Section 3. It turns out that the key to the solution is the study of the centralizers of simple subalgebras of a central-simple algebra.

Theorem 14 (Centralizer Theorem). *Let A be a simple, artinian and central K-algebra, let B be a simple subalgebra of A such that $B : K < \infty$, and let $C = Z_A(B)$ be its centralizer in A.*

(a) *C is a simple artinian subalgebra of A.*

(b) *C is similar to $B^\circ \otimes_K A$, and in particular $Z(C) = Z(B)$.*

(c) *The centralizer of C in A is once again B; that is, B coincides with its bicentralizer in A:*

(49) $$Z_A(Z_A(B)) = B.$$

(d) *The degree $C : K$ is finite if and only $A : K$ is, and moreover*

(50) $$A : K = (B : K)(C : K).$$

(e) *If L is the center of B, we have*

(51) $$B \otimes_L C \sim A \otimes_K L.$$

(f) *Under the assumption $Z(B) = K$, there is a natural isomorphism $A \simeq B \otimes_K C$.*

Proof. We start from the relation

(52) $$\operatorname{End}_{B \otimes A^\circ}(A) = Z_A(B) = C;$$

see (30). We next apply our Theorem 3, with the role of A (a simple artinian algebra) in that statement being played by $B \otimes A^\circ$, and the role of M by the $B \otimes A^\circ$-module A. In view of (52), conclusions (a), (b) and (d) follow immediately.

Next we prove (e) using (b):

$$B \otimes_L C \sim B \otimes_L (B^\circ \otimes_K A) \simeq (B \otimes_L B^\circ) \otimes_K A \sim L \otimes_K A \simeq A \otimes_K L.$$

We give the proof of (c) only for the case $A : K < \infty$; for the general case, see page 174. Set $B' = Z_A(C)$. In view of (a) we can apply (d) to C instead of B, obtaining

$$A : K = (C : K)(B' : K).$$

But $B \subseteq B'$, so comparing the last equation with (50) yields $B = B'$, proving the assertion.

Part (f) is also proved here only for the case $A : K < \infty$ (but see §29.9). Since the elements of C commute with those of B by definition, we have a natural homomorphism

(53) $$B \otimes_K C \to A.$$

But because of the assumption $Z(B) = K$, the product $B \otimes_K C$ is *simple*, so (53) is injective. On dimension grounds — see (50) — we conclude that (53) is surjective as well. $\qquad\square$

Theorem 15. *Let A be an artinian, simple and central K-algebra, and let L be a finite-dimensional subalgebra L of A that is a field. There is equivalence between:*

(i) *L is a maximal commutative subalgebra of A, that is, $Z_A(L) = L$.*

(ii) *The dimensions of A and L satisfy $A : K = (L : K)^2$.*

If either condition is satisfied, A_L and L are similar, that is, L is a splitting field of A.

Proof. We apply Theorem 14 to $B = L$. If $Z_A(L) = L$ we obtain condition (ii), using (50). Conversely, if $A : K = (L : K)^2 < \infty$, by virtue of (50) the inclusion $L \subseteq Z_A(L)$ must be an equality. If (i) is satisfied, we obtain from (b) the similarity $L \sim L \otimes_K A \simeq A_L$. $\qquad\square$

Not every central-simple K-algebra A has a subfield L satisfying the equivalent conditions in the theorem; just think of $A = M_n(K)$ for K an *algebraically closed* field. However, we do have:

Theorem 16. *Let D be a division algebra having finite dimension over its center K. Then D has maximal subfields, and any of them is a splitting field of D. The degree $L:K$ of every maximal subfield L of D equals $\sqrt{D:K}$, the index of D.*

Proof. A maximal subfield L of D is commutative and obviously contains K. By Lemma 1, every commutative K-subalgebra of D is a subfield of D. Thus being a maximal commutative K-subalgebra and a maximal subfield of D mean the same. Now the theorem follows immediately from Theorem 15. □

Theorem 17. *Every central-simple K-algebra A admits a splitting field L such that $L:K = s(A)$.*

Proof. We have $A \sim D$, with D as Theorem 16. Every maximal subfield L of D is a splitting field of D and hence also of A. Such an L satisfies $L:K = \sqrt{D:K} = s(D) = s(A)$. □

Splitting fields of a central-simple K-Algebra as in Theorem 17 have the lowest possible degree. To show this, we require the following complement to Theorem 15:

Theorem 18. *If L is a splitting field of a central-simple K-algebra A and $L:K$ is finite, there is some algebra A' similar to A that contains L as a maximal commutative subalgebra.*

Proof. Since A splits over L, there is a finite-dimensional L-vector space V such that $A_L = A \otimes_K L$ is isomorphic to the L-algebra $\mathrm{End}_L(V)$. We view A as embedded in $\mathrm{End}_L(V)$ by means of this isomorphism. But $\mathrm{End}_L(V)$ is in turn embedded in $\mathrm{End}_K(V)$, the K-algebra of all K-linear self-maps of V, regarded as a K-vector space. We now apply the *Centralizer Theorem* (page 168) to the simple subalgebra A of the central-simple K-algebra $\mathrm{End}_K(V)$. The centralizer C of A in $\mathrm{End}_K(V)$ then satisfies
$$C \sim A^\circ \quad \text{and} \quad C:K = (V:K)^2/(A:K).$$
Since $A:K = A_L:L = (V:L)^2$, we obtain $C:K = (L:K)^2$. But L is contained in C, so by Theorem 15 it is a maximal commutative subalgebra of C. The conclusion of the theorem is therefore satisfied if we take $A' = C^\circ$. □

Theorem 19. *Let A be a central-simple K-algebra and s the Schur index of A.*

(a) *If L is a splitting field of A, then s divides $L:K$.*

(b) *A has a splitting field L such that $L:K = s$.*

The splitting fields of A whose degree over K equals s are precisely those that are isomorphic as K-algebras to maximal subfields of a division algebra similar to A.

Proof. Let D be the division algebra such that $A \sim D$ (it is well defined up to isomorphism). Let L be a splitting field of A. Assume $L:K$ is finite. By Theorem 18, there exists r such that L is isomorphic to a maximal commutative subalgebra of $M_r(D)$. The K-dimensions of these spaces satisfy $(L:K)^2 = M_r(D):K = r^2(D:K) = r^2 s^2$, by Theorem 15, that is to say

$$(54) \qquad\qquad L:K = r\,s.$$

Clearly $L:K = s$ is equivalent to $r = 1$. In view of Theorem 16, this concludes the proof. $\qquad\qquad\square$

Remarks. (1) We see from Theorem 19 that the *Schur index* of a central-simple K-algebra can be characterized as the unique natural number s such that statements (a) and (b) in the theorem are satisfied.

(2) Note the following generalization of conclusion (a) of Theorem 19: *If L/K is any finite field extension and A is a central-simple K-algebra,*

$$(55) \qquad\qquad s(A) \ \text{divides} \ (L:K)\,s(A_L).$$

Indeed, let E be a splitting field of A_L satisfying $E:L = s(A_L)$. Then E is also a splitting field of A, so $s(A)$ divides $E:K = (E:L)(L:K) = s(A_L)(L:K)$.

Since $s(A_L)$ divides $s(A)$ (by F19), property (55) implies that

$$(56) \qquad\qquad s(A) = q\,s(A_L) \quad \text{for some factor } q \text{ of } L:K.$$

We mention an application of (56):

$$(57) \qquad \textit{If } L:K \textit{ and } s(A) \textit{ are relatively prime, then } s(A_L) = s(A).$$

F21. *Let D be a division algebra having finite dimension over its center K. Among the maximal subfields L of D there are some such that L/K is **separable**. Hence, every central-simple K-algebra A has a separable splitting field L over K of degree $L:K = s(A)$.*

Proof. If char $K = 0$ there is nothing left to show, so assume $p := $ char $K > 0$. Let L be maximal among the subfields of D that contain K and are separable over K. We wish to show that L coincides with its own centralizer C in D; this is enough to prove F21. By Theorem 14, the center of C is L. Clearly $C:L$ is finite. Being a subalgebra of the finite-dimensional division K-algebra D, however, C is itself a division algebra (every nonzero $a \in C$ is invertible in $K[a]$: see Lemma 1). Taking into account our choice of L and the next lemma, we conclude that $C = L$. $\qquad\square$

Lemma 2. *Let D be a division algebra having finite dimension over its center K. If no subfield $E \supsetneq K$ of D is separable over K, then $D = K$.*

Proof. Let L be a maximal subfield of D. Then, by Theorem 16,

$$D:K = n^2, \quad \text{with } n = L:K.$$

In contradiction to the conclusion of the lemma, assume that $D \neq K$, so $n > 1$. By assumption, L/K is *purely inseparable*, and therefore n is a power of the characteristic $p > 0$ of K (see F17 in Chapter 7, vol. I). Now let F be an algebraic closure of K; since D splits over F, we have an isomorphism

$$h : D \otimes_K F \to M_n(F)$$

of F-algebras, with the same n as before. Let x be an element of D. Since $K[x]/K$ is purely inseparable by assumption, and since its degree divides n, we conclude that

$$x^n = a, \quad \text{for some } a \in K.$$

Let α be an n-th root of a in F. Since n is a power of p, we have

$$(hx - \alpha)^n = hx^n - \alpha^n = 0$$

in $M_n(F)$. The matrix $hx - \alpha$ in $M_n(F)$ is therefore nilpotent, and thus its trace $\text{Tr}(hx - \alpha)$ vanishes. But $\text{Tr}(hx - \alpha) = \text{Tr}(hx) - n\alpha = \text{Tr}(hx)$, because n is divisible by p. Thus we have proved that

(58) $$\text{Tr}(y) = 0 \quad \text{for every } y = hx \in hD.$$

In view of the isomorphism h, equation (58) is actually satisfied for all $y \in M_n(F)$. But this is impossible. $\square$

F22. *Every central-simple K-algebra A has a splitting field L such that L/K is a Galois extension of finite degree.*

Proof. By F21 there is a splitting field L of A such that L/K is separable of degree $s(A)$. Let L' be the normal closure of L/K. Then L'/K is Galois of finite degree, and L' is a splitting field of A since it contains L. $\square$

Remark. Whether there are, among the splitting fields L of A whose existence is guaranteed by F22, any such that $L:K = s(A)$ — that is, whether a division algebra D similar to A contains maximal subfields that are Galois over K — is a different and more delicate question. In general the answer is negative (see *Amitsur*, "On central division algebras", *Israel J. Math.* **12**, 408–420), but over certain fields K — including all *local fields*, by F3 in Chapter 31 — it is affirmative.

But in any case F22 implies that the Brauer group $\text{Br } K$ of any field K is the union of the subgroups of $\text{Br } K$ associated to finite Galois extensions:

(59) $$\text{Br } K = \bigcup_{L/K \text{ fin. gal.}} \text{Br}(L/K).$$

(Here we assume the extensions L are subfields of a fixed algebraic closure of K.) The situation described by (59) will be the starting point for the next section and subsequent chapters.

6. A first gateway to the study of the Brauer group, based on (59), is the fundamental result known as the *Skolem–Noether Theorem*:

Theorem 20 (Skolem–Noether). *Let A and B be simple K-algebras, and suppose the center of A coincides with K. Assume further that B has finite dimension over K and A is artinian. Then, for any two K-algebra homomorphisms $f, g : B \to A$, there exists a unit u of A such that*

$$(60) \qquad g(b) = u^{-1} f(b) u \quad \text{for every } b \in B.$$

Proof. Consider the simple artinian K-algebra

$$A \otimes_K B^\circ$$

(see Theorem 7 and statement (c) in the remark thereto). The maps f and g endow A with two $A \otimes_K B^\circ$-module structures:

$$(a \otimes b)x = axf(b) \quad \text{and} \quad (a \otimes b)x = axg(b).$$

We denote these modules by A_f and A_g. Let N be a simple $A \otimes_K B^\circ$-module. Since A_f and A_g are finitely generated, we have

$$(61) \qquad A_f \simeq N^m \quad \text{and} \quad A_g \simeq N^n,$$

with $m, n \in \mathcal{N}$ (see again F1, F2, as well as Chapter 28, F13 and F25). Viewed as A-modules, A_f and A_g are indistinguishable, that is, N^m and N^n are isomorphic A-modules; and since they're also artinian A-modules, a comparison of the lengths (Definition 8 in the previous chapter) yields

$$(62) \qquad m = n.$$

But then (61) implies that A_f and A_g are isomorphic as $A \otimes_K B^\circ$-modules. Let $\varphi : A_f \to A_g$ be an isomorphism of $A \otimes_K B^\circ$-modules. Then

$$(63) \qquad \varphi(a 1 f(b)) = a \varphi(1) g(b) \quad \text{for all } a \in A, \ b \in B.$$

Setting $u = \varphi(1)$, we get for φ the expression

$$\varphi(a) = au \quad \text{for all } a \in A,$$

and in view of (63) we have, for every $b \in B$,

$$(64) \qquad f(b)u = ug(b).$$

But φ is an isomorphism, so u is a unit of A, and (60) follows from (64). □

Definition 10. Let A be an algebra. A map $i : A \to A$ of the form

$$(65) \qquad i(x) = u^{-1}xu,$$

where $u \in A^\times$ is a unit of A, is called an *inner automorphism* of A.

The conclusion of Theorem 20 can be restated to say that, under the assumptions made, two arbitrary K-algebra homomorphisms $f, g : B \to A$ differ only by an inner automorphism i of A: $g = i \circ f$. In particular:

Theorem 20′. *Every automorphism of a central-simple K-algebra A is inner.*

Thus, for every central-simple K-algebra A, the natural map $A^\times \to \mathrm{Aut}_K(A)$ taking each unit u to the inner automorphism of conjugation by u gives rise to a group isomorphism

$$(66) \qquad\qquad A^\times / K^\times \simeq \mathrm{Aut}_K(A)$$

between the quotient $A^\times / K^\times$ and the automorphism group of the K-algebra A.

More generally, Theorem 20 implies:

Theorem 20″. *Let B be a finite-dimensional simple subalgebra of a central, simple and artinian K-algebra A. Denote by N the normalizer of B in A (that is, the set of all $u \in A^\times$ such that $Bu \subseteq uB$) and by $C = Z_A(B)$ the centralizer. Then the map $A^\times \to \mathrm{Aut}_K(A)$ gives rise to a natural group isomorphism*

$$(67) \qquad\qquad N / C^\times \simeq \mathrm{Aut}_K(B).$$

The Skolem–Noether Theorem is a result of great importance; it is the foundation for the *cohomological description of* $\mathrm{Br}\, K$ *by means of crossed products*, which is the subject of the next few chapters and our only source of deeper insights.

But first we want to show how the *Centralizer Theorem* (Theorem 14) follows from Skolem–Noether, which effectively allows us to complete the proof of that theorem by shedding the finiteness assumption $A : K < \infty$ made back on page 169.

Alternate proof of Theorem 14. Let B be a finite-dimensional simple subalgebra of a simple, artinian, central K-algebra A. We apply Theorem 20 to the simple, artinian, central K-algebra

$$A' := A \otimes \mathrm{End}_K(B)$$

and the K-monomorphisms

$$f : b \mapsto b \otimes 1 \quad \text{and} \quad g : b \mapsto 1 \otimes \lambda(b)$$

from B into A', where $\lambda(b)$ is left multiplication by b (in B). The conclusion of Theorem 20 implies in particular that the centralizers of $f(B)$ and $g(B)$ in A' are isomorphic, so

$$(68) \qquad\qquad Z_A(B) \otimes \mathrm{End}_K(B) \simeq A \otimes \rho(B)$$

(see F2 and F3 in the previous chapter). Hence $C := Z_A(B)$ is a simple artinian subalgebra of A similar to $A \otimes B^\circ$. Now (68) yields, in view of (31) and after another application of Theorem 20,

$$Z_A(Z_A(B)) \otimes K \simeq K \otimes \lambda(B),$$

which is to say $Z_A(Z_A(B)) \simeq B$. Since $B \subseteq Z_A(Z_A(B))$ and $B : K < \infty$, we conclude that in fact

$$Z_A(Z_A(B)) = B.$$

The isomorphism (68) also justifies the equality $A : K = (B : K)(C : K)$.   □

As a further application of the Skolem–Noether theorem, we give a proof of the following famous theorem of Wedderburn:

Theorem 21 (Wedderburn). *Every finite division algebra is commutative. In other words, if a field K is finite, its Brauer group is trivial.*

Proof. Let D be a finite division algebra, with center K. Every element of D lies in a maximal subfield of D. By Theorem 16, all maximal subfields of D have the same degree. But for a *finite field* K, all extensions of the same degree are K-isomorphic (Theorem 1′ in Chapter 9, vol. I). So if L is a fixed maximal subfield of D, by the Skolem–Noether Theorem every other maximal subfield of D is of the form $x^{-1}Lx$, for some nonzero $x \in D$. Putting it all together we get for the multiplicative group $D^\times$ of D the equality

(69)
$$D^\times = \bigcup_{x \in D^\times} x^{-1}L^\times x.$$

Now, the number m of distinct maximal subfields of D satisfies

(70)
$$m = D^\times : N,$$

where N denotes the subgroup $\{x \in D^\times \mid x^{-1}L^\times x = L^\times\}$ of $D^\times$. Suppose $D \neq K$, so $m > 1$. From (69) and (70) we obtain for the number of elements of $D^\times$ the bound

(71)
$$(D^\times : 1) < (D^\times : N)(L^\times : 1),$$

the inequality being strict because the multiplicative groups of two subfields of D always intersect. But now we have a contradiction, because the right-hand side of (71) is at most $(D^\times : L^\times)(L^\times : 1) = (D^\times : 1)$.   □

7. Let A be a *central-simple* K-algebra of dimension n^2, and L an extension of K. A K-algebra homomorphism $h : A \to M_n(L)$ is an *L-representation of A of degree* $n = \sqrt{A : K}$, which here we call simply an *L-representation of A*. Such an h can be uniquely extended to an L-algebra homomorphism

(72)
$$h_L : A \otimes L \to M_n(L);$$

we sometimes write just h instead of h_L. Since $A \otimes L$ is simple and both L-algebras in (72) have dimension n^2, the map h_L is an *isomorphism*. Conversely, every L-algebra homomorphism $g : A \otimes L \to M_n(L)$ is of the form $g = h_L$; to see this, just set $h(a) = g(a \otimes 1)$. For a given L, therefore, an L-representation of A exists if and only if L is a splitting field of A.

By means of L-representations the important notions of *trace* and *determinant* can be transferred from the familiar setting of matrices to that of central-simple K-algebras:

F23 and Definition 11. *Let A be a central-simple K-algebra with $A:K = n^2$. For every $a \in K$ there exists a unique normalized polynomial*

$$(73) \qquad\qquad P^0(a) = P^0(a; X) \in K[X]$$

(of degree n) coinciding with the characteristic polynomial of the matrix $h(a) \in M_n(L)$ for any L-representation h of A:

$$(74) \qquad\qquad P^0(a) = P(h(a)).$$

In particular there exist well defined functions $N^0 : A \to K$ and $\mathrm{Tr}^0 : A \to K$ (with values in K!) such that

$$(75) \qquad\qquad N^0(a) = \det h(a) \quad and \quad \mathrm{Tr}^0(a) = \mathrm{Tr}\, h(a),$$

where $\det$ and Tr refer to the determinant and trace of matrices in $M_n(L)$. Hence, if $P^0(a; X) = X^n + c_{n-1}X^{n-1} + \cdots + c_0$, we have

$$(76) \qquad\qquad \mathrm{Tr}^0(a) = -c_{n-1}, \quad N^0(a) = (-1)^n c_0.$$

*$P^0(a)$ is called the **reduced characteristic polynomial**, $N^0(a)$ the **reduced norm** and $\mathrm{Tr}^0(a)$ the **reduced trace** of a. If necessary, we use the more precise notation*

$$(77) \qquad\qquad \mathrm{Tr}^0_{A/K}, \quad N^0_{A/K}, \quad P^0_{A/K}.$$

For any $a \in A$, the reduced characteristic polynomial $P^0_{A/K}(a)$ and the usual characteristic polynomial $P_{A/K}(a)$ (that is, the characteristic polynomial of the K-endomorphism $a_{A/K} : x \mapsto ax$) are linked by

$$(78) \qquad\qquad P_{A/K}(a) = P^0_{A/K}(a)^n.$$

Likewise, the usual normal and trace satisfy

$$(79) \qquad\qquad N_{A/K}(a) = N^0_{A/K}(a)^n, \quad \mathrm{Tr}_{A/K}(a) = n\,\mathrm{Tr}^0_{A/K}(a).$$

If $A = M_n(K)$ is a matrix algebra, $P^0_{A/K}(a)$ is none other than the characteristic polynomial of the matrix $a \in M_n(K)$, and likewise $\mathrm{Tr}^0_{A/K}(a)$ and $N^0_{A/K}(a)$ are just the usual matrix trace and determinant of a.

Proof. Step 1. Let h and g be L-representations of A. The composition $h_L \circ g_L^{-1}$ is an L-automorphism of $M_n(L)$, so the version of Skolem–Noether in Theorem 20$'$ yields $u \in M_n(L)^\times$ conjugating h_L and g_L, that is, $h_L(x) = u^{-1}g_L(x)u$ for $x \in A_L$. Hence $h(a) = u^{-1}g(a)u$ for every $a \in A$, showing that $g(a)$ and $h(a)$ have the same characteristic polynomial.

If L' is an extension of L, every L-representation $h: A \to M_n(L)$ leads naturally to an L'-representation $h' : A \to M_n(L')$, and $h'(a)$ has the same characteristic polynomial as $h(a)$. This shows that the characteristic polynomial is independent of the choice of a splitting field L of A, because any two such fields are contained in a common splitting field L'.

Step 2. Next we have to show that the coefficients of $P^0(a; X)$ lie in K. Fix a splitting field L of A that is a Galois extension of K. Every $\sigma \in G(L/K)$ gives rise (by coefficientwise action) to a K-automorphism of $L[X]$, and likewise to a K-automorphism of $M_n(L)$. We denote both of these by σ still. If $h : A \to M_n(L)$ is an L-representation of A, so is $\sigma \circ h$; hence $P(h(a)) = P(\sigma h(a))$ by step 1, which leads to $P^0(a) = P(\sigma h(a)) = \sigma P(h(a)) = \sigma P^0(a)$. Thus the coefficients of $P^0(a)$ are invariant under any element of $G(L/K)$ and must lie in K.

Step 3. We still need to check that (78) holds. Since

$$P_{A/K}(a) = P_{A \otimes L/L}(a) = P_{M_n(L)/L}(h(a)),$$

we need only show that

$$P_{M_n(L)/L}(x) = P_{L^n/L}(x)^n \quad \text{for any matrix } x \in M_n(L).$$

But this is clear, because as an $M_n(L)$-module,

$$M_n(L) \simeq L^n \oplus \cdots \oplus L^n$$

is the n-fold direct sum of the $M_n(L)$-modules L^n. $\qquad\square$

Remark. The following rules for the reduced norm and trace are obvious:

  (a) $\text{Tr}^0_{A/K}$ *is a K-linear form on A.*

  (b) $N^0_{A/K}$ *is multiplicative.*

  (c) *For all $a, b \in A$ we have $\text{Tr}^0_{A/K}(ab) = \text{Tr}^0_{A/K}(ba)$.*

  (d) *For $a \in K$ we have $N^0_{A/K}(a) = a^n$ and $\text{Tr}^0_{A/K}(a) = na$.*

F24. *Let A be a central-simple K-algebra. For every $a \in A$ we have $P_{A/K}(a) = P_{A^\circ/K}(a)$; that is, the K-endomorphisms $x \mapsto ax$ and $x \mapsto xa$ of A have the same characteristic polynomial. Likewise $\text{Tr}_{A/K} = \text{Tr}_{A^\circ/K}$ and $N_{A/K} = N_{A^\circ/K}$.*

Proof. In view of (78) in F23 it suffices to prove that

(80) $$P^0_{A/K}(a) = P^0_{A^\circ/K}(a).$$

But this is clear, since the transposition map is an isomorphism $M_n(L)^\circ \to M_n(L)$ preserving characteristic polynomials. $\qquad\square$

F25. *Let A be a central-simple K-algebra. An element $a \in A$ is invertible in A if and only if $N^0_{A/K}(a)$ is distinct from 0.*

Proof. We know from F3 in Chapter 13 (vol. I) that every for a finite-dimensional K-algebra A, there is equivalence between $a \in A^\times$ and $N_{A/K}(a) \neq 0$. Now (79) completes the proof. $\qquad\square$

F26. *Let A be a central-simple K-algebra. The reduced trace $\text{Tr}^0_{A/K}$ is nondegenerate, that is, any element $a \in A$ such that $\text{Tr}^0_{A/K}(ax) = 0$ for all $x \in A$ must vanish.*

Proof. Let h be an L-representation of A, and suppose a is such that $\mathrm{Tr}(h(a)y) = 0$ for all $y \in h(A)$. The L-vector space $M_n(L)$ is generated by $h(A)$, so $\mathrm{Tr}(h(a)x) = 0$ for all $x \in M_n(L)$. It follows that $h(a) = 0$, hence $a = 0$.   □

Let A be a central-simple K-algebra of dimension $m = n^2$, and let $e_1, \ldots, e_m$ form a K-basis of A, so each $x \in A$ has a unique representation

(81) $$x = x_1 e_1 + \cdots + x_m e_m$$

with $x_i \in K$. We claim that the coefficients of $P^0(x; X)$ — and so, in particular, $N^0(x)$ and $\mathrm{Tr}^0(x)$ — depend polynomially on the coordinates x_i. More precisely:

Lemma 3. *In the situation just described, there exist homogeneous polynomials $F(t_1, \ldots, t_m)$, $G(t_1, \ldots, t_m)$, $H(t_1, \ldots, t_m, X)$ of respective degrees 1, n, n, with coefficients in K, such that, for every $x = x_1 e_1 + \cdots + x_m e_m$ in A, we have*

$$\mathrm{Tr}^0_{A/K}(x) = F(x_1, \ldots, x_m),$$
$$N^0_{A/K}(x) = G(x_1, \ldots, x_m),$$
$$P^0_{A/K}(x; X) = H(x_1, \ldots, x_m, X).$$

Proof. Take an L-representation $h : A \to M_n(L)$ of A with L/K Galois. Consider the field of rational functions $L(t) := L(t_1, \ldots, t_m)$ in m variables $t_1, \ldots, t_m$ over K. Setting

$$\widehat{h} : A \otimes K(t) \to M_n(L(t))$$
$$x \otimes f(t) \mapsto f(t)h(x)$$

we get an $L(t)$-representation of the central-simple $K(t)$-algebra $A_{K(t)} = A \otimes K(t)$.

The matrix $\widehat{h}(e_1 \otimes t_1 + \cdots + e_m \otimes t_m) = t_1 h(e_1) + \cdots + t_m h(e_m)$ has characteristic polynomial

$$\det\big(XE_n - (t_1 h(e_1) + \cdots + t_m h(e_m))\big).$$

By expanding the determinant one obtains a homogeneous polynomial

$$H(t_1, \ldots, t_m, X) \in L[t, X]$$

of degree n. But H is, by definition, the *reduced characteristic polynomial* of the element $e_1 \otimes t_1 + \cdots + e_m \otimes t_m$ of $A \otimes K(t)$, and so lies in $K(t)[X]$. Therefore

$$H(t_1, \ldots, t_m, X) \in K[t, X].$$

For x as in (81) we then have

$$P^0_{A/K}(x; X) = \det(XE_n - h(x)) = \det\big(XE_n - (x_1 h(e_1) + \cdots + x_m h(e_m))\big)$$
$$= H(x_1, \ldots, x_m, X).$$

Choosing $-F(t)$ and $(-1)^n G(t)$ as the second and last coefficients of $H(t, X)$ we complete the proof.   □

F27. *Suppose the division algebra D is finite-dimensional over its center K, and let $A = M_r(D)$ be any matrix algebra over D. For every matrix $a = (d_{ij}) \in M_r(D)$ we have*

$$(82) \qquad \mathrm{Tr}^0_{A/K}(a) = \sum_i \mathrm{Tr}^0_{D/K}(d_{ii}),$$

and if $a = (d_{ij})$ is a triangular matrix,

$$(83) \qquad N^0_{A/K}(a) = \prod_i N^0_{D/K}(d_{ii}).$$

Proof. Let h be an L-representation of D. By assigning to each matrix $y = (d_{ij})$ in $M_r(D)$ the block matrix $(h(d_{ij}))$, we obtain an L-representation of $A = M_r(D)$. We now use this L-representation to calculate $\mathrm{Tr}^0_{A/K}(a)$ (and $N^0_{A/K}(a)$ in the case of a triangular); this yields (82) and (83). □

Remark. In the situation of F27, formula (82) can be written more elegantly using the trace in the matrix algebra $M_r(D)$:

$$(84) \qquad \mathrm{Tr}^0_{A/K}(a) = \mathrm{Tr}^0_{D/K}(\mathrm{Tr}_D(a)).$$

But $M_r(D)$ also possesses a determinant function $\det_D : M_r(D) \to \bar{D}$ with values in $\bar{D} = (D^\times/D^{\times\prime}) \cup \{0\}$, where $D^\times/D^{\times\prime}$ denotes the quotient by the *commutator group* of $D^\times$. Aside from the codomain change from D to $\bar{D}$, $\det_D$ has all the defining properties of the determinant: columnwise multilinearity, invariance when one column is added to another, and value 1 at the identity matrix. Thus by applying $\det_D$ we obtain

$$(85) \qquad N^0_{A/K}(a) = N^0_{D/K}(\det_D a)$$

for *every* matrix a in $M_r(D)$. Indeed, one can obtain this formula from (83) by applying the usual transformation rules for $\det_D$ and the multiplicativity of N^0. Note also that the right-hand side of (85) really is well defined, because for elements of the form $d = d_1 d_2 d_1^{-1} d_2^{-1}$ we have $N^0_{D/K}(d) = 1$.

Theorem 22. *Let a central-simple K-algebra A contain a field L as a subalgebra. Any element c in the centralizer $C = Z_A(L)$ satisfies*

$$(86) \qquad N^0_{A/K}(c) = N_{L/K} N^0_{C/L}(c), \quad \mathrm{Tr}^0_{A/K}(c) = \mathrm{Tr}_{L/K} \mathrm{Tr}^0_{C/L}(c).$$

Proof. By the Centralizer Theorem (page 168) we have

$$(87) \qquad A:K = (L:K)(C:K).$$

Also, C is a central-simple L-algebra. Consider the C-module A. Since C is simple, we seem from (87) by comparing dimensions that there is an isomorphism of C-modules

$$(88) \qquad A \simeq C^{L:K}.$$

Hence, for any $c \in C$,

$$(89) \qquad P_{A/K}(c) = P_{C/K}(c)^{L:K}.$$

Now apply the equalities in §13.1 (vol. I) to the K-algebra C, to obtain

$$(90) \qquad P_{C/K}(c; X) = N_{L[X]/K[X]} P_{C/L}(c; X).$$

The reduced characteristic polynomial $P_{C/L}^0(c; X)$ therefore satisfies

$$(91) \qquad P_{C/K}(c; X) = N_{L[X]/K[X]} P_{C/L}^0(c; X)^m,$$

with $m^2 = C:L$. Recalling that $n^2 = A:K$ and $P_{A/K}(c) = P_{A/K}^0(c)^n$, we obtain from (89) and (91)

$$(92) \qquad P_{A/K}^0(c; X)^n = N_{L[X]/K[X]} P_{C/L}^0(c; X)^{m(L:K)}.$$

On the other hand, $n^2 = A:K = (L:K)(C:K) = (L:K)^2(C:L) = (L:K)^2 m^2$, so $n = m(L:K)$. Since we have been dealing with normalized polynomials throughout, equation (92) can only be true in the UFD $K[X]$ if

$$(93) \qquad P_{A/K}^0(c; X) = N_{L[X]/K[X]} P_{C/L}^0(c; X).$$

Thus we have found a "nesting formula" for the reduced characteristic polynomial. We now derive the equalities (86) from it. Take arbitrary polynomials

$$f(X) = X^n + c_{n-1} X^{n-1} + \cdots + c_0,$$
$$g(X) = X^m + a_{m-1} X^{m-1} + \cdots + a_0$$

in $K[X]$ and $L[X]$ respectively, such that

$$(94) \qquad f(X) = N_{L[X]/K[X]} g(X).$$

Consider the endomorphism

$$(95) \qquad u = \begin{pmatrix} 0 & & & & -a_0 \\ 1 & 0 & & & -a_1 \\ & 1 & \ddots & & \vdots \\ & & \ddots & 0 & -a_{m-2} \\ & & & 1 & -a_{m-1} \end{pmatrix}$$

of the L-vector space L^m. Its characteristic polynomial is precisely $g(X)$. By (94), therefore, f is the characteristic polynomial of u_K, the K-endomorphism of L^m determined by u:

$$f(X) = P(u_K; X).$$

(See for example LA II, p. 183, Aufgabe 63.) In particular, we see that $c_{n-1} = -\operatorname{Tr}(u_K)$. At the same, because of (95), we have $\operatorname{Tr}(u_K) = -\operatorname{Tr}_{L/K}(a_{m-1})$. Altogether, then,

$$(96) \qquad\qquad c_{n-1} = -\operatorname{Tr}_{L/K}(a_{m-1}).$$

Similarly we obtain $(-1)^n c_0 = \det u_K = N_{L/K}(\det u) = N_{L/K}((-1)^m a_0)$, so

$$(97) \qquad\qquad (-1)^n c_0 = N_{L/K}((-1)^m a_0).$$

With $f = P^0_{A/K}(c; X)$ and $g = P^0_{C/L}(c; X)$, equations (96) and (97) say exactly what we wished to prove in (86). $\qquad\qquad \Box$

Remark. Take Theorem 22 specifically in the case where $A = D$ is a division algebra and L is a *maximal* subfield of D. Then

$$(98) \quad N^0_{D/K}(x) = N_{L/K}(x) \quad\text{and}\quad \operatorname{Tr}^0_{D/K}(x) = \operatorname{Tr}_{L/K}(x) \quad\text{for every } x \in L.$$

Since any element x of D lies in some maximal subfield of D, this brings back the reduced norm and trace in D, in a certain sense, to the norm or trace in *field extensions L*.

Here is a nice application of the notion of reduced norm:

Theorem 23. *If K is a C_1-field* (see Definition 3 in Chapter 27), *then* $\operatorname{Br} K = 1$.

Proof. Take $[D] \in \operatorname{Br} K$, where D is a division algebra with $D:K = m = n^2 > 1$. The reduced norm $N^0_{D/K}$ is then given by a homogeneous polynomial of degree n in $K[X_1, \ldots, X_m]$, by Lemma 3 above. If K is a C_1-field, there exists a nonzero $x \in D$ such that $N^0_{D/K}(x) = 0$. But this cannot be, since x is invertible in D. $\qquad \Box$

Theorem 23 implies Theorem 21 (Wedderburn), because all finite fields are C_1 (Theorem 2 in Chapter 27). But Theorem 23 applies more generally: for instance, when K is a *function field* of transcendence degree 1 *over an algebraic closed field* (see Chapter 27, Theorem 6 together with Theorem 1). Such fields, too, have trivial Brauer group.

30

Crossed Products

1. In this chapter we build up an explicit description of central-simple K-algebras. After choosing a Galois splitting field L for a given central-simple K-algebra A, one can replace A, up to similarity, by a so-called *crossed product*. A crossed product is characterized as a K-algebra by the specification of certain generators and relations (among these generators). The structural description is particularly simple and concise in the case of a *cyclic* Galois extension; although this is a drastic assumption, the study of this special case is of fundamental importance, so we will look at such *cyclic algebras* in detail. The simplest case consists of *quaternion algebras*, which we will exploit as a source of example material.

Assumptions and notation. In this section,

L/K denotes a Galois extension with Galois group G.

For the action of automorphisms $\sigma \in G$ on elements $\lambda \in L$, we adopt *exponential notation*:

$$\lambda^\sigma = \sigma(\lambda), \quad \text{with the convention } \lambda^{\sigma\tau} = (\lambda^\sigma)^\tau.$$

Thus the product $\sigma\tau$ of two maps $\sigma, \tau \in G$ is, contrary to ordinary usage, defined by $\sigma\tau(\lambda) = \tau(\sigma(\lambda))$. Correspondingly, in this context we extend exponential notation to maps in general: the product (concatenation) fg of maps is to be understood so

$$x^{fg} = (x^f)^g.$$

The statements to follow can be regarded as fruit of the work done in the proof of Theorem 18 of the preceding chapter for L/K *Galois*. It is useful to keep this in mind. Now

fix A, a central-simple K-algebra that splits over L.

Thus there exist an n-dimensional L-vector space V and an isomorphism

(1) $$h : A \otimes_K L \to \text{End}_L(V)$$

of L-algebras. If we wish, we can choose V as the standard n-dimensional vector space L^n. Note also that $n^2 = A : K$.

Any $\sigma \in G$ can be extended in a natural way into a K-linear map from $A \otimes_K L$ to itself, again denoted by σ, by setting

$$(2) \qquad\qquad (a \otimes \lambda)^\sigma = a \otimes \lambda^\sigma \quad \text{for } a \in A, \lambda \in L.$$

Via conjugation by the map h from (1), then, σ determines a K-algebra endomorphism p_σ of $\text{End}_L(V)$:

$$(3) \qquad\qquad x^{p_\sigma} = x^{h^{-1}\sigma h}.$$

In other words, p_σ is the one map for which the diagram

$$(4) \qquad\qquad
\begin{array}{ccc}
A \otimes L & \xrightarrow{\;\;h\;\;} & \text{End}_L(V) \\
\sigma \downarrow & & \downarrow p_\sigma \\
A \otimes L & \xrightarrow{\;\;h\;\;} & \text{End}_L(V)
\end{array}$$

commutes. By definition, then, we have for $\lambda \in L$

$$(5) \qquad\qquad \lambda^{p_\sigma} = \lambda^\sigma.$$

For $\sigma, \tau \in G$ we obviously have $p_\sigma p_\tau = p_{\sigma\tau}$. As mentioned, p_σ is merely K-linear; but more precisely, it is σ-*semilinear* with respect to L, in the sense that

$$(6) \qquad\qquad (\lambda x)^{p_\sigma} = \lambda^\sigma x^{p_\sigma}$$

for every $x \in \text{End}_L(V)$ and $\lambda \in L$. Because p_σ is a ring homomorphism, this property follows immediately from (5). But $\text{End}_L(V)$ is, as a K-algebra, a subalgebra of the K-algebra $\text{End}_K(V)$. By the *Skolem–Noether Theorem*, then, there exists $u_\sigma \in \text{End}_K(V)^\times$ such that

$$(7) \qquad\qquad x^{p_\sigma} = u_\sigma^{-1} x u_\sigma \quad \text{for every } x \in \text{End}_L(V).$$

Here u_σ is determined up to a factor from $L^\times$; indeed, the centralizer of $\text{End}_L(V)$ in $\text{End}_K(V)$ is obviously $Z(\text{End}_L(V)) = L$.

Imagine that for each $\sigma \in G$ we *choose* some $u_\sigma \in \text{End}_K(V)^\times$ with property (7). How is $u_{\sigma\tau}$ related to $u_\sigma u_\tau$, for $\sigma, \tau \in G$? Conjugation by each of these two elements of $\text{End}_K(V)^\times$ induces on $\text{End}_L(V)$ the same inner automorphism $p_{\sigma\tau} = p_\sigma p_\tau$, so there exists a well defined element $c_{\sigma,\tau}$ in $L^\times$ such that

$$(8) \qquad\qquad u_\sigma u_\tau = u_{\sigma\tau} c_{\sigma,\tau}.$$

By (5) and (7), conjugation by u_σ induces on L precisely the automorphism σ:

$$u_\sigma^{-1} \lambda u_\sigma = \lambda^\sigma \quad \text{for } \lambda \in L.$$

We can also write this as

(9) $$\lambda u_\sigma = u_\sigma \lambda^\sigma.$$

(As a map from V to itself, therefore, u_σ is σ-semilinear.) For u_1 one can choose $u_1 = 1$, but regardless of this choice we always have

(10) $$u_1 \in L^\times.$$

We now consider the subalgebra Γ of $\text{End}_K(V)$ generated by the elements $\lambda \in L$ and the u_σ. In view of (8), (9) and (10), we have

(11) $$\Gamma = \sum_{\sigma \in G} u_\sigma L.$$

What is the relation between Γ and the algebra A^h, the isomorphic image of A under h? Because of (4), we have $x^{p\sigma} = x$ for all $x \in A^h$, so the maps u_σ commute with all the elements of A^h, by (7). So do all the $\lambda \in L$. Thus Γ lies in the centralizer C of A^h in $\text{End}_K(V)$. By the *Centralizer Theorem*, then, C has dimension $C : K = (L:K)^2$ over K (see the proof of Theorem 18 in the previous chapter). We will see in Theorem 1 below that Γ also has dimension $(L:K)^2$ over K, so

(12) $$\Gamma = C.$$

The elements $c_{\sigma,\tau} \in L^\times$ determine the structure of $\Gamma = C$ fully. Since

(13) $$A \sim C^\circ,$$

the similarity class $[A] \in \text{Br } K$ of the given algebra A is also fully specified by the $c_{\sigma,\tau}$.

On the other hand, the $c_{\sigma,\tau}$ are *not* determined uniquely by A, because, as already mentioned, one can multiply the u_σ by arbitrary factors $a_\sigma \in L^\times$. Set $u'_\sigma = u_\sigma a_\sigma$. The u'_σ then give rise to elements $c'_{\sigma,\tau} \in L^\times$ such that $u'_\sigma u'_\tau = u'_{\sigma\tau} c'_{\sigma,\tau}$. A comparison with (8) leads after a simple computation to the transformation equation

(14) $$c'_{\sigma,\tau} = c_{\sigma,\tau}(a_\sigma^\tau a_\tau a_{\sigma\tau}^{-1}).$$

Thus the $c'_{\sigma,\tau}$ differ from the $c_{\sigma,\tau}$ by factors $d_{\sigma,\tau}$ of the form $d_{\sigma,\tau} = a_\sigma^\tau a_\tau a_{\sigma\tau}^{-1}$. Incidentally, a change in h or in V in (1) also leads at most to a change in the $c_{\sigma,\tau}$ according to (14). The proof of this fact, resorting to the Skolem–Noether Theorem, is left to the reader as an easy exercise.

Theorem 1. *Let L/K be a finite Galois extension with Galois group G, and let Γ be a K-algebra containing L as a subalgebra. Suppose that Γ is generated by the elements $\lambda \in L$ and by certain elements u_σ ($\sigma \in G$), with $u_1 \in L^\times$, these generators satisfying the relations*

$$\lambda u_\sigma = u_\sigma \lambda^\sigma, \tag{15}$$

$$u_\sigma u_\tau = u_{\sigma\tau} c_{\sigma,\tau} \quad \text{with } c_{\sigma,\tau} \in L^\times. \tag{16}$$

Then the u_σ are linearly independent over L; that is, $\Gamma : K = (L : K)^2$. The K-algebra Γ is central-simple and contains L as a maximal commutative subalgebra. The $c_{\sigma,\tau}$ in (16) satisfy, for every ρ, σ, τ in G, the equation

(17)
$$c_{\sigma,\tau}^\rho \, c_{\sigma\tau,\rho} = c_{\sigma,\tau\rho} \, c_{\tau,\rho}.$$

Proof. (a) For $\tau = \sigma^{-1}$ we obtain from (16)

$$u_\sigma u_{\sigma^{-1}} = u_1 c_{\sigma,\sigma^{-1}}.$$

Since u_1 is assumed to be in $L^\times$, therefore, each u_σ is *invertible* in Γ; in particular, $u_\sigma \neq 0$.

(b) For any $\rho, \sigma, \tau \in G$, equations (16) and (15) give on the one hand

$$(u_\sigma u_\tau) u_\rho = u_{\sigma\tau} \, c_{\sigma,\tau} \, u_\rho = u_{\sigma\tau} \, u_\rho \, c_{\sigma,\tau}^\rho = u_{\sigma\tau\rho} \, c_{\sigma\tau,\rho} \, c_{\sigma,\tau}^\rho;$$

on the other hand,

$$u_\sigma(u_\tau u_\rho) = u_\sigma \, u_{\tau\rho} \, c_{\tau,\rho} = u_{\sigma\tau\rho} \, c_{\sigma,\tau\rho} \, c_{\tau,\rho}.$$

Since multiplication in Γ is associative, we get (17) by cancellation: see part (a).

(c) For $\lambda \in L$ we consider the map

$$\lambda_\Gamma : x \mapsto \lambda x$$

from Γ to itself. Clearly, λ_Γ is an endomorphism of the right L-vector space Γ. Because of (15), every u_σ is an eigenvector of λ_Γ, with eigenvalue λ^σ. But there exists some λ for which all the λ^σ, $\sigma \in G$, are distinct (use the Primitive Element Theorem, for example). Hence the claimed linear independence of the u_σ over L follows from the well known fact that eigenvectors belonging to distinct eigenvalues are linearly independent. From the assumptions we have $\Gamma = \sum_\sigma L u_\sigma = \sum_\sigma u_\sigma L$, so $\Gamma : K = (L : K)^2$.

(d) To prove that Γ is *simple*, we show that every surjective algebra homomorphism $\Gamma \to \Gamma'$ has zero kernel. But Γ' is generated by the images u'_σ of the u_σ and the images λ' of the $\lambda \in L$, and these generators satisfy relations completely analogous to (15) and (16). So by part (c) we get $\Gamma : K = \Gamma' : K$.

(e) We prove that Γ is central over K. Let $z = \sum_\sigma u_\sigma \lambda_\sigma$ be any element in the center of Γ. We then have $z\lambda = \lambda z$ for all $\lambda \in L$, and hence, by (15),

$$\sum_\sigma u_\sigma \lambda_\sigma \lambda = \sum_\sigma u_\sigma \lambda^\sigma \lambda_\sigma.$$

Comparing coefficients we obtain $\lambda_\sigma \lambda = \lambda^\sigma \lambda_\sigma$ for all $\lambda \in L$. For $\sigma \neq 1$, then, we must have $\lambda_\sigma = 0$, so z has the form $z = \lambda_1 u_1$ and hence lies in L to begin with. But if some $z \in L$ commutes with all elements of Γ, equation (15) gives $u_\sigma z = z u_\sigma = u_\sigma z^\sigma$, so $z = z^\sigma$ for all $\sigma \in G$. It follows that $z \in K$.

(f) Finally, from the first part of (e) it is clear that L is a maximal commutative subalgebra of Γ. This also follows from Theorem 15 in the previous chapter. $\square$

These considerations give rise to the following notions:

Definition 1. A map $(\sigma, \tau) \mapsto c_{\sigma,\tau}$ from $G \times G$ to $L^\times$ satisfying the functional equation (17) is called a *cocycle* or *factor system* of G (with values in $L^\times$). Cocycles that differ according to (14) are called *equivalent*. Given an assignment $\sigma \mapsto a_\sigma \in L^\times$, the map taking (σ, τ) to

$$(18) \qquad\qquad d_{\sigma,\tau} := a_\sigma^\tau a_\tau a_{\sigma\tau}^{-1}$$

is a cocycle; cocycles of this form are said to *split*. If $b_{\sigma,\tau}$ and $c_{\sigma,\tau}$ are cocycles, so is their product $b_{\sigma,\tau} c_{\sigma,\tau}$. Thus the cocycles of G with values in $L^\times$ form a group, of which the set of split cocycles is a subgroup. The corresponding quotient group is denoted by

$$(19) \qquad\qquad H^2(G, L^\times),$$

and is called the *second cohomology group* of G with coefficients in $L^\times$. A K-algebra Γ as in Theorem 1 is called a *crossed product* of L and G, with cocycle $c = c_{\sigma,\tau}$. Because any two such K-algebras are isomorphic, we use the notation

$$(20) \qquad\qquad \Gamma = (L, G, c).$$

Passing from $c_{\sigma,\tau}$ to an equivalent cocycle $c'_{\sigma,\tau}$ leaves Γ unchanged: it amounts to a change of basis $u'_\sigma = u_\sigma a_\sigma$, where $a_\sigma \in L^\times$. Thus the crossed product Γ depends only on the class $\gamma = [c]$ of c in $H^2(G, L^\times)$, and we also write

$$(21) \qquad\qquad \Gamma = (L, G, \gamma).$$

In this way, every central-simple K-algebra A that splits over L is assigned a well defined element γ in $H^2(G, L^\times)$, which we denote by

$$(22) \qquad\qquad f(A) := \gamma.$$

In view of (13) and (12), then,

$$(23) \qquad\qquad A^\circ \sim (L, G, \gamma).$$

We now state a very satisfying result, which is also of crucial importance:

Lemma (Multiplication theorem). *If A and A' are two central-simple K-algebras that split over L, we have*

$$(24) \qquad\qquad f(A \otimes A') = f(A) f(A').$$

Proof. We first introduce some notation: If M, M', N, N' are vector spaces over L and $\varphi : M \to M', \psi : N \to N'$ are σ-semilinear maps (for the same $\sigma \in G$), we denote by

$$(25) \qquad (\varphi, \psi) : M \otimes_L N \to M' \otimes_L N'$$

the (well defined!) σ-semilinear map given by $x \otimes y \mapsto x^\varphi \otimes y^\psi$.

Now assign to A and A' cocycles $c_{\sigma,\tau}$ and $c'_{\sigma,\tau}$ as above. We obtain from the corresponding commutative diagrams (4) first the commutative diagram

$$
\begin{array}{ccc}
(A \otimes L) \otimes_L (A' \otimes L) & \xrightarrow{(h,h')} & \mathrm{End}_L(V) \otimes_L \mathrm{End}_L(V') \\
{\scriptstyle (\sigma,\sigma)}\downarrow & & \downarrow{\scriptstyle (p_\sigma, p'_\sigma)} \\
(A \otimes L) \otimes_L (A' \otimes L) & \xrightarrow[(h,h')]{} & \mathrm{End}_L(V) \otimes_L \mathrm{End}_L(V')
\end{array}
$$

After making identifications $(A \otimes L) \otimes_L (A' \otimes L) = (A \otimes A') \otimes L$ and

$$(26) \qquad \mathrm{End}_L(V) \otimes_L \mathrm{End}_L(V') = \mathrm{End}_L(V \otimes_L V')$$

the preceding diagram becomes the starting point for the assignment of a cocycle to $A \otimes A'$. Namely, for each $\sigma \in G$ we take the element of $\mathrm{End}_K(V \otimes_L V')^\times$ given by

$$U_\sigma = (u_\sigma, u'_\sigma).$$

The inner automorphism arising from U_σ coincides on $\mathrm{End}_L(V) \otimes_L \mathrm{End}_L(V')$ with (p_σ, p'_σ), because $(u_\sigma, u'_\sigma)^{-1}(x, x')(u_\sigma, u'_\sigma) = (u_\sigma^{-1} x u_\sigma, u_\sigma'^{-1} x' u'_\sigma)$. Hence all that's left to determine is the cocycle corresponding to U_σ. We have $U_\sigma U_\tau = (u_\sigma, u'_\sigma)(u_\tau, u'_\tau) = (u_\sigma u_\tau, u'_\sigma u'_\tau) = (u_{\sigma\tau} c_{\sigma,\tau}, u'_{\sigma\tau} c'_{\sigma,\tau}) = (u_{\sigma\tau}, u'_{\sigma\tau})(c_{\sigma,\tau}, c'_{\sigma,\tau}) = U_{\sigma\tau} c_{\sigma,\tau} c'_{\sigma,\tau}$, so the product $c_{\sigma,\tau} c'_{\sigma,\tau}$ is indeed a cocycle corresponding to $A \otimes A'$. $\square$

Lemma. *Let $[A]$ and $[B]$ be elements of* $\mathrm{Br}(L/K)$.

(a) $A \sim K$ *implies* $f(A) = 1$.

(b) $f(A^\circ) = f(A)^{-1}$.

(c) $A \sim B$ *implies* $f(A) = f(B)$.

Proof. (a) By assumption, A is isomorphic to some $M_n(K)$. We know that

$$A \otimes_K L \simeq M_n(K) \otimes_K L \simeq M_n(L) = \mathrm{End}_L(L^n),$$

so we get an isomorphism $h : A \otimes L \to \mathrm{End}_L(L^n)$ leading to maps p_σ in (4) that just have the form

$$(27) \qquad x^{p_\sigma} = x^\sigma = (x_{ij}^\sigma)_{i,j}.$$

Here we are regarding each $x \in \mathrm{End}_L(L^n)$ as a matrix $x = (x_{ij})_{i,j}$ in $M_n(L)$, and x^σ arises from x by coefficientwise application of $\sigma \in G$. Then, if u_σ denotes the map from L^n to itself defined by $(x_i)_i \mapsto (x_i^\sigma)_i$, we see that $u_\sigma \in \mathrm{End}_K(L^n)^\times$, and for each $x \in M_n(L)$ we have

$$(28) \qquad\qquad u_\sigma^{-1} x u_\sigma = x^\sigma.$$

(To check the equality, just apply both sides to the canonical basis vectors e_i.) But now, for every $\sigma, \tau \in G$, we clearly have $u_\sigma u_\tau = u_{\sigma\tau}$, so $f(A) = 1$ as needed.

(b) We have $A \otimes A^\circ \sim K$. Using (a) and Lemma 1 we conclude that $1 = f(A \otimes A^\circ) = f(A) f(A^\circ)$, so $f(A^\circ) = f(A)^{-1}$.

(c) Suppose $B \sim A$. Then $B \otimes A^\circ \sim A \otimes A^\circ \sim K$, which implies that $f(B \otimes A^\circ) = 1$, by part (a). Then, using Lemma 1 and part (b), we obtain $f(B) f(A)^{-1} = 1$, hence the assertion $f(A) = f(B)$. $\square$

Part (c) of the lemma says that $f(A)$ only depends on the *similarity class* of A; hence we write $f(A) = f([A])$ to obtain a well defined map

$$(29) \qquad\qquad f : \mathrm{Br}(L/K) \to H^2(G, L^\times),$$

also denoted by $f_{L/K}$ when needed.

Theorem 2. *The map* $f : \mathrm{Br}(L/K) \to H^2(G, L^\times)$ *is an isomorphism. For* $[A] \in \mathrm{Br}\, K$ *and* $\gamma \in H^2(G, L^\times)$ *the following conditions are equivalent*:

(i) $[A] \in \mathrm{Br}(L/K)$ *and* $f(A) = \gamma^{-1}$.

(ii) $A \sim (L, G, \gamma)$.

For any $\gamma, \gamma' \in H^2(G, L^\times)$ *we have*

$$(30) \qquad\qquad (L, G, \gamma) \otimes (L, G, \gamma') \sim (L, G, \gamma\gamma').$$

Proof. (a) Suppose $f(A) = 1$. By construction, then, $A^\circ \sim \Gamma = (L, G, c)$ for some *split* cocycle c, and we can even assume that $c_{\sigma,\tau} = 1$ for every $\sigma, \tau \in G$. Then $\Gamma = \sum_\sigma u_\sigma L$, with

$$(31) \qquad\qquad u_\sigma u_\tau = u_{\sigma\tau}.$$

Now consider the K-linear map $\Gamma \to \mathrm{End}_K(L)$ defined by $u_\sigma \lambda \mapsto \sigma\lambda$. By (31), this is in fact a homomorphism of algebras, since the σ's and λ's satisfy relations parallel to those that hold among the u_σ and λ. Hence $A^\circ \sim \Gamma \simeq \mathrm{End}_K(L)$, so A splits over K. This shows that f is *injective*.

(b) Let a given $\gamma \in H^2(G, L^\times)$ be represented by the cocycle $c_{\sigma,\tau}$. Let $V = \sum_\sigma e_\sigma L$ be an n-dimensional vector space, where $n = G : 1 = L : K$. Define $u_\tau \in \mathrm{End}_K(V)$ by

$$(32) \qquad\qquad (e_\sigma \lambda)^{u_\tau} = e_{\sigma\tau} c_{\sigma,\tau} \lambda^\tau$$

for every $\tau \in G$. Finally, let Γ be the subalgebra of $\mathrm{End}_K(V)$ generated by L and the u_τ. It is not too hard to check that Γ satisfies the assumptions of Theorem 1: to show that $u_\sigma u_\tau = u_{\sigma\tau} c_{\sigma,\tau}$, of course, one must use the functional equation (17) for the cocycles $c_{\sigma,\tau}$. By Theorem 1, then, $\Gamma = (L, G, c)$ is the crossed product of L and G with cocycle c.

Now let A be the centralizer of Γ in $\text{End}_K(V)$. By the Centralizer Theorem, A is *simple* with center $Z(A) = Z(\Gamma) = K$, and the various dimensions satisfy $(A : K)(\Gamma : K) = (V : K)^2 = (V : L)^2 (L : K)^2$, from which we get

$$(33) \qquad\qquad A : K = (V : L)^2$$

since $\Gamma : K = (L : K)^2$. Since $L \subseteq \Gamma$, and hence $A \subseteq \text{End}_L(V)$, we obtain a well defined homomorphism

$$h : A \otimes L \to \text{End}_L(V)$$

with $h(a \otimes \lambda) = a\lambda$. This map is injective, because $A \otimes L$ is simple. By (33), h is an isomorphism of L-algebras. With this construction for h, the first step in the determination of $f(A)$ is finished.

The p_σ associated to h are determined by their action on elements of the form $a\lambda$; we have $(a\lambda)^{p_\sigma} = (a\lambda)^{h^{-1}\sigma h} = (a \otimes \lambda)^{\sigma h} = (a \otimes \lambda^\sigma)^h = a\lambda^\sigma$. But at the same time $u_\sigma^{-1}(a\lambda)u_\sigma = a(u_\sigma^{-1}\lambda u_\sigma) = a\lambda^\sigma$, so the u_σ are associated with p_σ in the sense of (7). Therefore the equalities $u_\sigma u_\tau = u_{\sigma\tau} c_{\sigma,\tau}$ imply, by definition,

$$(34) \qquad\qquad f(A) = \gamma,$$

showing that f is *surjective*.

(c) As a consequence of the reasoning in (b), we claim that for each $\gamma \in H^2(G, L^\times)$, the map f associates to the crossed product $\Gamma = (L, G, \gamma)$ the value $f(\Gamma) = \gamma^{-1}$:

$$(35) \qquad\qquad f((L, G, \gamma)) = \gamma^{-1}.$$

Indeed, since $\Gamma \sim A^\circ$ we have $f(\Gamma) = f(A)^{-1}$, which leads via (34) to $f(\Gamma) = \gamma^{-1}$, as desired.

(d) We now prove the equivalence of (i) and (ii). Take $\Gamma = (L, G, \gamma)$. Being a maximal commutative subalgebra of Γ, L is a splitting field of Γ. Every $A \sim \Gamma$ then splits over L as well; that is, A lies in $\text{Br}(L/K)$. Moreover, $A \sim \Gamma$ implies $f(A) = f(\Gamma)$; thus $f(A) = \gamma^{-1}$, by (35). Conversely, suppose $[A] \in \text{Br}(L/K)$ and $f(A) = \gamma^{-1}$. In view of (35) and the injectivity of f, we must have $A \sim (L, G, \gamma)$.

(e) The last statement of the theorem is also clear, being merely an expression of the multiplicativity of f, as shown by the preceding discussion (see Lemma 1). $\square$

2. As in the preceding section, L/K will denote a finite Galois extension with Galois group G. Further, assume that

$$E/K \text{ is a } \mathbf{Galois} \text{ subextension of } L/K.$$

We denote by $\bar{G} = G(E/K)$ and $N = G(L/E)$ the Galois groups of E/K and L/E. Since E/K is Galois, N is a normal subgroup of G. As usual, we identify the quotient group G/N with $\bar{G} = G(E/K)$, so the canonical image $\bar{\sigma}$ of an element $\sigma \in G$ corresponds to the automorphism determined by σ on E: in symbols, $\bar{\sigma} = \sigma_E$.

Since any K-algebra that splits over E surely splits over the extension L of E, we have

(36) $$\mathrm{Br}(E/K) \subseteq \mathrm{Br}(L/K).$$

On account of Theorem 2, this inclusion determines a monomorphism

(37) $$\inf : H^2(\overline{G}, E^\times) \to H^2(G, L^\times),$$

the so-called *inflation map* for $L/E/K$. When necessary, we write it as $\inf_{L/E}$ or even $\inf_{L/E/K}$. By definition, inf is the unique map making the diagram

(38)
$$
\begin{array}{ccc}
\mathrm{Br}(E/K) & \longrightarrow & \mathrm{Br}(L/K) \\
{\scriptstyle f_{E/K}} \downarrow & & \downarrow {\scriptstyle f_{L/K}} \\
H^2(\overline{G}, E^\times) & \xrightarrow[\text{inf}]{} & H^2(G, L^\times)
\end{array}
$$

commute. The next results gives an explicit description of this map.

F1. *Let the situation be as above. If $\gamma \in H^2(\overline{G}, E^\times)$ is represented by a cocycle $c : (\overline{\sigma}, \overline{\tau}) \mapsto c_{\overline{\sigma}, \overline{\tau}}$, then $\inf(\gamma)$ is represented by the cocycle*

(39) $$(\sigma, \tau) \mapsto c_{\overline{\sigma}, \overline{\tau}}.$$

*This cocycle is called the **inflation** of the cocycle c and is denoted by $\inf(c) = \inf_{L/E}(c)$.*

Proof. Take $\gamma = f(A)$, with $[A] \in \mathrm{Br}(E/K)$. Suppose the determination of $f(A) = f_{E/K}(A)$ has already been carried out. For $\sigma \in G$, take the starting diagram in the determination of $p_{\overline{\sigma}}$ and change the base to L over E, obtaining the commutative diagram

$$
\begin{array}{ccc}
(A \otimes E) \otimes_E L & \longrightarrow & \mathrm{End}_E(V) \otimes_E L \\
{\scriptstyle (\overline{\sigma}, \sigma)} \downarrow & & \downarrow {\scriptstyle (p_{\overline{\sigma}}, \sigma)} \\
(A \otimes E) \otimes_E L & \longrightarrow & \mathrm{End}_E(V) \otimes_E L
\end{array}
$$

Here and below we use the notation (25). With the natural isomorphisms

$$(A \otimes E) \otimes_E L \simeq A \otimes L \quad \text{and} \quad \mathrm{End}_E(V) \otimes_E L \simeq \mathrm{End}_L(V \otimes_E L),$$

we then obtain the commutative diagram

$$
\begin{array}{ccc}
A \otimes L & \longrightarrow & \mathrm{End}_L(V \otimes_E L) \\
{\scriptstyle \sigma} \downarrow & & \downarrow {\scriptstyle p_\sigma} \\
A \otimes L & \longrightarrow & \mathrm{End}_L(V \otimes_E L).
\end{array}
$$

Then, with the $u_{\overline{\sigma}}$ in $\mathrm{End}_K(V)$, we have, from the definition,

$$(x, \lambda)^{p_\sigma} = (x^{p_{\overline{\sigma}}}, \lambda^\sigma) = (u_{\overline{\sigma}}^{-1} x u_{\overline{\sigma}}, \lambda^\sigma)$$

for $x \in \text{End}_E(V)$ and $\lambda \in L$. Now define $U_\sigma \in \text{End}_K(V \otimes_E L)$ for $\sigma \in G$ by setting

$$U_\sigma = (u_{\bar{\sigma}}, \sigma).$$

Since $u_{\bar{\sigma}}$ and σ are both $\bar{\sigma}$-semilinear with respect to E, it is the case that U_σ is well defined. Moreover

$$U_\sigma^{-1}(x, \lambda) U_\sigma = (u_{\bar{\sigma}}^{-1} x u_{\bar{\sigma}}, \sigma^{-1} \lambda \sigma) = (u_{\bar{\sigma}}^{-1} x u_{\bar{\sigma}}, \lambda^\sigma),$$

so the inner automorphism on $\text{End}_L(V \otimes_E L)$ of conjugation by $\tilde{U}_\sigma$ agrees with p_σ. Since $U_\sigma U_\tau = (u_{\bar{\sigma}} u_{\bar{\tau}}, \sigma\tau) = (u_{\bar{\sigma}\bar{\tau}} c_{\bar{\sigma}, \bar{\tau}}, \sigma\tau) = (u_{\bar{\sigma}\bar{\tau}}, \sigma\tau)(c_{\bar{\sigma}, \bar{\tau}}, 1) = U_{\sigma\tau} c_{\bar{\sigma}, \bar{\tau}}$, we conclude that $\inf(c)$ is the associated cocycle, as needed. $\square$

We keep the notation L/K for a finite Galois extension with Galois group G. Now suppose

$$K'/K \text{ is any extension.}$$

We denote by $L' = LK'$ the composite of L with K' (in a common extension of L and K'). Consider the diagram of fields

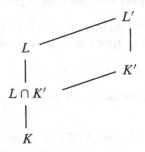

Set $H := G(L/L \cap K')$. Now recall the *Translation Theorem* of Galois Theory (vol. I, Chapter 12, Theorem 1): For every $\sigma \in H$ there is a unique $\sigma' \in G(L'/K')$ inducing σ on L (in symbols, $\sigma'_L = \sigma$), and the map $H \to G(L'/K') =: H'$ taking σ to σ' is an isomorphism. For this reason we will as a rule identify the Galois group of the Galois extension L'/K' with the subgroup H of G.

Because the restriction operation is transitive (F18 in Chapter 29), $\text{res}_{K'/K}$ maps the subgroup $\text{Br}(L/K)$ of $\text{Br}\, K$ into the subgroup $\text{Br}(L'/K')$ of $\text{Br}\, K'$. Therefore, by Theorem 2, the map

(40) $\text{res}_{K'/K} : \text{Br}(L/K) \to \text{Br}(L'/K')$

gives rise to a homomorphism in cohomology, still denoted by $\text{res}_{K'/K}$:

(41) $\text{res}_{K'/K} : H^2(G, L^\times) \to H^2(H', L'^\times),$

this just being the map such that the following diagram commutes:

(42)
$$
\begin{array}{ccc}
\text{Br}(L/K) & \xrightarrow{\text{res}_{K'/K}} & \text{Br}(L'/K') \\
{\scriptstyle f_{L/K}} \downarrow & & \downarrow {\scriptstyle f_{L'/K'}} \\
H^2(G, L^\times) & \xrightarrow[\text{res}_{K'/K}]{} & H^2(H', L'^\times)
\end{array}
$$

The explicit description of the restriction (41) in cohomology is this:

F2. *Let the situation be as above. If $\gamma \in H^2(G, L^\times)$ is represented by a cocycle $c : (\sigma, \tau) \mapsto c_{\sigma,\tau}$, then $\mathrm{res}(\gamma)$ is represented by the cocycle*

$$(43) \qquad (\sigma', \tau') \mapsto c_{\sigma,\tau}.$$

*This is called the **restriction** of the cocycle c and is denoted by $\mathrm{res}(c) = \mathrm{res}_{K'/K}(c)$. Specializing to the case $K' \subseteq L$, where $L' = L$ and $H' = H$, we get for $\mathrm{res}(c)$ literally the restriction of the map $c : G \times G \to L^\times$ to the subset $H \times H$.*

Proof. For $\sigma \in H$, consider the starting diagram (4) for the determination of $\gamma = f(A)$, with $[A] \in \mathrm{Br}(L/K)$. By base change we obtain the commutative diagram

$$
\begin{array}{ccc}
(A \otimes L) \otimes_L L' & \longrightarrow & \mathrm{End}_L(V) \otimes_L L' \\
\big\downarrow{\scriptstyle (\sigma,\sigma')} & & \big\downarrow{\scriptstyle (p_\sigma,\sigma')} \\
(A \otimes L) \otimes_L L' & \longrightarrow & \mathrm{End}_L(V) \otimes_L L'
\end{array}
$$

Via the natural isomorphisms $(A \otimes K') \otimes_{K'} L' \simeq (A \otimes L) \otimes_L L'$ and $\mathrm{End}_L(V) \otimes_L L' \simeq \mathrm{End}_{L'}(V \otimes_L L')$, one gets from this the commutative diagram

$$
\begin{array}{ccc}
(A \otimes K') \otimes_{K'} L' & \longrightarrow & \mathrm{End}_{L'}(V \otimes_L L') \\
\big\downarrow{\scriptstyle \sigma'} & & \big\downarrow{\scriptstyle p_{\sigma'}} \\
(A \otimes K') \otimes_{K'} L' & \longrightarrow & \mathrm{End}_{L'}(V \otimes_L L')
\end{array}
$$

With the u_σ in $\mathrm{End}_K(V)$, then, we have from the definition

$$(x, \alpha)^{p_{\sigma'}} = (u_\sigma^{-1} x u_\sigma, \alpha^{\sigma'})$$

for $x \in \mathrm{End}_L(V)$ and $\alpha \in L'$. Thus, if we define $v_{\sigma'} \in \mathrm{End}_{K'}(V \otimes_L L')$ by setting $v_{\sigma'} = (u_\sigma, \sigma')$, the inner automorphism induced on $\mathrm{End}_{L'}(V \otimes_L L')$ by $v_{\sigma'}$ coincides with $p_{\sigma'}$. The corresponding cocycle is then the restriction of the cocycle c as desired, because $v_{\sigma'} v_{\tau'} = (u_\sigma u_\tau, \sigma' \tau') = (u_{\sigma\tau} c_{\sigma,\tau}, \sigma' \tau') = (u_{\sigma\tau}, \sigma' \tau')(c_{\sigma,\tau}, 1) = v_{\sigma' \tau'} c_{\sigma,\tau}.$ □

Remark 1. By means of Theorem 2, the conclusions of F1 and F2 can be rephrased as follows:

$$(44) \qquad (E, \overline{G}, c) \sim (L, G, \mathrm{inf}(c)),$$

$$(45) \qquad (L, G, c) \otimes_K K' \sim (L', G(L'/K'), \mathrm{res}(c)).$$

Moreover, if $L' : K' = L : K$, the latter similarity is actually an isomorphisms.

Remark 2. Let L/K be a Galois extension with *finite* Galois group G, as before, and let N be a *normal subgroup* of G with fixed field

$$E := L^N.$$

As usual, we identify G/N with the Galois group of the Galois extension E/K. We write inf for the inclusion map $\mathrm{Br}(E/K) \to \mathrm{Br}(L/K)$, too; this map and the restriction res $= \mathrm{res}_{E/K}$ fit into a sequence

$$(46) \qquad 1 \longrightarrow \mathrm{Br}(E/K) \overset{\mathrm{inf}}{\longrightarrow} \mathrm{Br}(L/K) \overset{\mathrm{res}}{\longrightarrow} \mathrm{Br}(L/E),$$

which is obviously *exact* from the definitions. From this we obtain, using F1 and F2, an *exact sequence* of cohomology groups:

$$(47) \qquad 1 \longrightarrow H^2(G/N, L^{\times N}) \overset{\mathrm{inf}}{\longrightarrow} H^2(G, L^{\times}) \overset{\mathrm{res}}{\longrightarrow} H^2(N, L^{\times}).$$

For didactic reasons, we have written the multiplicative group $E^{\times}$ of the intermediate field E as $L^{\times N}$, the fixed module of $L^{\times}$ under N.

3. As an important application of the methodology developed in Section 1, we now prove a theorem that is far from obvious:

Theorem 3. *For any field* K, *the Brauer group* $\mathrm{Br}\, K$ *is a torsion group. More precisely, every* $[A] \in \mathrm{Br}\, K$ *having Schur index* $s = s(A)$ *satisfies*

$$[A]^s = 1;$$

that is, the s-fold tensor product $A \otimes A \otimes \cdots \otimes A$ *is isomorphic to a matrix algebra* $M_m(K)$ *over* K.

Proof. In view of Theorem 2 (and (59) in the previous chapter) what we must show is this: If $\Gamma = (L, G, c)$ is any *crossed product* with Schur index $s = s(\Gamma)$, the class γ of the cocycle $c = c_{\sigma,\tau}$ in $H^2(G, L^{\times})$ satisfies

$$(48) \qquad\qquad\qquad \gamma^s = 1.$$

Let N be a simple Γ°-module (hence a simple right Γ-module). We claim that N has dimension

$$(49) \qquad\qquad\qquad N : L = s$$

as a vector space over L. Indeed, if $\Gamma \simeq M_r(D)$, for D a division algebra, the maximal subfield L of Γ has degree $L : K = rs$ (see (54) in previous chapter). Since $\Gamma^{\circ} \simeq N^r$ we have $\Gamma : L = r(N : L)$, so (49) follows from the equalities $r(N : L) = \Gamma : L = L : K = rs$.

Now let $b_1, \ldots, b_s$ be an L-basis of N. For each u_{σ} in Γ, there is a unique representation

$$b_i u_{\sigma} = \sum_k b_k u_{ki}(\sigma), \quad \text{where } u_{ki}(\sigma) \in L.$$

In this way every $\sigma \in G$ gets assigned an $s \times s$-matrix

$$U(\sigma) = (u_{ij}(\sigma))_{i,j} \in M_s(L).$$

Now,

$$b_i u_\sigma u_\tau = \sum_k b_k u_\tau u_{ki}(\sigma)^\tau = \sum_k \sum_l b_l u_{lk}(\tau) u_{ki}(\sigma)^\tau,$$

so, using the equality $u_\sigma u_\tau = u_{\sigma\tau} c_{\sigma,\tau}$, we see that the matrix equation

(50) $$U(\sigma\tau) c_{\sigma,\tau} = U(\tau) U(\sigma)^\tau$$

holds. For all determinants $a_\sigma := \det(U(\sigma)) \in L^\times$ of (nonsingular) $s \times s$-matrices $U(\sigma)$ we obtain from (50) the equation

$$a_{\sigma\tau} c_{\sigma,\tau}^s = a_\tau a_\sigma^\tau.$$

Thus the s-th power of the cocycle $c_{\sigma,\tau}$ has the form $c_{\sigma,\tau}^s = a_\sigma^\tau a_\tau a_{\sigma\tau}^{-1}$, and is therefore a *split* cocycle. But this says the same as equation (48).   □

Definition 2. The order of an element $[A]$ in the group Br K is called the *exponent* of $[A]$ (or of A), and is denoted by

$$e(A) = e([A]).$$

F3. *For every* $[A] \in$ Br K *the exponent of A divides the Schur index of A:*

(51) $$e(A) \mid s(A).$$

Conversely, every prime factor of $s(A)$, at least, divides $e(A)$.

Proof. That $e(A)$ divides $s(A)$ follows directly from Theorem 3. On the other hand, let p be a prime not dividing the exponent of A:

(52) $$p \nmid e(A).$$

We must show that p does not divide $s(A)$ either. Choose a finite Galois extension L/K with $[A] \in \mathrm{Br}(L/K)$ and consider the fixed field F of a Sylow p-group H of $G = G(L/K)$. Then $L:F$ is a power of p, yet

(53) $$F:K \not\equiv 0 \bmod p.$$

Next take the restriction map

$$\mathrm{res} : \mathrm{Br}(L/K) \to \mathrm{Br}(L/F).$$

On the one hand, $e(\mathrm{res}[A])$ divides $e(A)$, because res is a group homomorphism; on the other, $e(\mathrm{res}[A])$, being a divisor of $L:F$, is a p-power (apply Theorem 3 to res$[A]$, together with Theorem 19 in the last chapter). This conflicts with (52) unless $e(\mathrm{res}[A]) = 1$. Hence res$[A] = 1$ in Br F and so $s(A)$ is a divisor of $F:K$ (again by Theorem 19 in Chapter 29). By (53), therefore, $s(A)$ is relatively prime to p.   □

Remark 1. It is not always the case that $s(A) = e(A)$, as we will see with an example — see (108) and paragraphs following. However the equality does hold over *local* and *global* fields; in the local case we will show this in full (Chapter 31, Theorem 5).

Remark 2. We mention a generalization of (51). *Let L/K be an **arbitrary** finite field extension. For any $[A] \in \operatorname{Br} K$,*

$$(54) \qquad\qquad e(A) \quad divides \quad (L:K)\, e(A_L).$$

Proof. For brevity, set $k = e(A_L)$, $\alpha = [A]$. Since $\operatorname{res}_{L/K}(\alpha^k) = (\operatorname{res}_{L/K} \alpha)^k = 1$, L is a splitting field of α^k, so $s(\alpha^k)$ divides $L:K$. By F3 then also

$$(55) \qquad\qquad e(\alpha^k) \quad divides \quad (L:K).$$

But k is a divisor of $e(A) = e(\alpha) = \operatorname{ord}(\alpha)$; hence $e(\alpha) = k e(\alpha^k)$. Multiplying (55) by k we obtain (54).   □

As an application of (54), we have: *If $L:K$ and $e(A)$ are relatively prime,* $e(A_L) = e(A)$.

Remark 3. For $[A] \in \operatorname{Br} K$, let $s(A) = mn$, where m, n are *relatively prime*. Then there is a unique decomposition in $\operatorname{Br} K$ of the form

$$(56) \qquad\qquad [A] = [B][C], \quad \text{with } s(B) = m,\ s(C) = n.$$

If A is a *division algebra*, there exist division algebras B and C such that

$$(57) \qquad\qquad A \simeq B \otimes C \quad \text{and} \quad s(B) = m,\ s(C) = n.$$

B and C are uniquely determined up to K-isomorphism.

Proof. Since $e(A)$ divides $s(A)$, it has the form $e(A) = m_1 n_1$ with m_1 dividing m and n_1 dividing n. Thus, in the abelian group $\operatorname{Br} K$, we have a *unique* decomposition

$$[A] = [B][C], \quad \text{with } e(B) = m_1,\ e(C) = n_1.$$

Since m_1 and n_1 are relatively prime, so are $s(B)$ and $s(C)$, by F3. By §29.12, then, we have $s(A) = s(B \otimes C) = s(B)s(C)$. But since $s(A) = mn$ and $\gcd(m, n) = 1$, the only remaining possibility, by F3, is that $s(B) = m$, $s(C) = n$. This shows the existence of the decomposition (56). Its uniqueness is clear: since $e(B)$ and $e(C)$ divide $s(B)$ and $s(C)$, respectively, these two divisors are relatively prime.

We can assume without restriction that B and C are division algebras. Again by §29.12, $B \otimes C$ is also a division algebra. Thus, if A is a division algebra, the representation (56) implies (57). Uniqueness in (57) follows from uniqueness in (56).   □

4. We now investigate $\operatorname{Br}(L/K)$ more closely in the case that L/K is *cyclic*. Although this restriction is severe, the study of this special case is nonetheless of fundamental significance.

Let L/K be a *cyclic* extension, that is, a finite Galois extension with *cyclic* Galois group G. Let τ be a generator of G:

$$G = \langle \tau \rangle.$$

We denote by $n = \mathrm{ord}(\tau) = G:1 = L:K$ the degree of L/K.

Fix a crossed product

$$(58) \qquad\qquad \Gamma = (L, G, c)$$

of L with cyclic group G and cocycle c; an algebra of this form is also called a *cyclic algebra*. We abbreviate as

$$(59) \qquad\qquad u := u_\tau$$

the representative u_τ in Γ corresponding to the generator τ. Since conjugation by u^i induces the automorphism τ^i on L, we can choose the remaining u_σ by the rule

$$(60) \qquad\qquad u_{\tau^i} = u^i \quad \text{for } 0 \le i \le n-1.$$

Since u^n yields the automorphism $\tau^n = 1$ on L, and since it commutes with all the u^i, it lies in the center of Γ. Thus

$$(61) \qquad\qquad a := u^n \in K^\times.$$

After a choice of u, then, the structure of Γ can be fully described by the specification of the element $a \in K^\times$. For $0 \le i, j < n$ we have

$$u_{\tau^i} u_{\tau^j} = u^i u^j = u^{i+j} = \begin{cases} u_{\tau^{i+j}} = u_{\tau^i \tau^j} & \text{for } i+j < n, \\ a u^{i+j-n} = a u_{\tau^i \tau^j} & \text{for } i+j \ge n. \end{cases}$$

Thus the cocycle $\hat{c}$ corresponding to the u_σ as in (60) has the form

$$(62) \qquad\qquad \hat{c}_{\tau^i, \tau^j} = \begin{cases} 1 & \text{for } i+j < n, \\ a & \text{for } i+j \ge n, \end{cases}$$

where we assume that $0 \le i, j < n$. Such a cocycle is called *uniformized*. For a fixed τ, the structure of Γ is fully determined by the element $a \in K^\times$; for this reason we can denote the cyclic algebra Γ by

$$(63) \qquad\qquad \Gamma = (L, \tau, a).$$

The element a, in contrast, is *not* determined uniquely by Γ, because we can multiply the representative $u = u_\tau$ of τ by any factor $\lambda \in L^\times$. Set

$$(64) \qquad\qquad u' = u\lambda.$$

Then the element

$$a' = u'^n$$

associated with u' is given by

$$a' = (u\lambda)^n = (u\lambda)(u\lambda)(u\lambda)\ldots(u\lambda) = (uu\lambda^\tau \lambda)(u\lambda)\ldots(u\lambda)$$
$$= u^n \lambda^{\tau^{n-1}} \ldots \lambda^\tau \lambda = a\lambda^{\tau^{n-1}} \ldots \lambda^\tau \lambda = a N_{L/K}(\lambda).$$

This leads to the transformation equation

(65) $$a' = a N_{L/K}(\lambda).$$

Conversely, if (65) holds, equation (64) yields a representative u' of τ such that $u'^n = a'$.

The upshot is that every cyclic algebra is associated to a well defined element α of the *norm residue group* $K^\times/N_{L/K}(L^\times)$ of the extension L/K, namely the residue class $\alpha = a\, N_{L/K}(L^\times)$ of a modulo the subgroup $N_{L/K}(L^\times)$ of $K^\times$. Every element of $K^\times/N_{L/K}(L^\times)$ is obtained by means of this correspondence, because if we take any $a \in K^\times$ and use (62) to define a function $\hat{c} : G \times G \to K^\times$, it is easy to check that this function is a cocycle. This, together with the fundamental result stated in Theorem 2, leads to:

Theorem 4. *Let L/K be cyclic with Galois group $G = \langle \tau \rangle$. There exist natural (but τ-dependent) isomorphisms*

(66) $$\mathrm{Br}(L/K) \simeq K^\times/N_{L/K}L^\times \simeq H^2(G, L^\times),$$

and every element of $\mathrm{Br}(L/K)$ is represented by a cyclic crossed product (L, τ, a), with a uniquely determined modulo $N_{L/K}L^\times$.

Remarks. (1) If α is the residue class of $a \in K^\times$ in $K^\times/N_{L/K}L^\times$, we denote the cyclic algebra (L, τ, a) also by

(67) $$(L, \tau, \alpha).$$

(2) For $a, b \in K^\times$, Theorem 4 implies that

(68) $$(L, \tau, a) \otimes_K (L, \tau, b) \sim (L, \tau, ab).$$

Further: *A cyclic algebra (L, τ, a) splits if and only if $a = N_{L/K}(\lambda)$ is the norm of an element in the extension L/K.*

(3) What about the dependence on the choice of a generator τ for G? Any other generator has the form τ^k, with k relatively prime to $n = G : 1$, and it is easy to check that

(69) $$(L, \tau, a) \simeq (L, \tau^k, a^k).$$

(4) The norm residue group $K^\times/N_{L/K}L^\times$ is defined, of course, for *any finite* field extension L/K. When, in addition, L/K is *Galois* with arbitrary (finite) Galois group, we call $K^\times/N_{L/K}L^\times$ the *zeroth cohomology group* of G with coefficients in $L^\times$, and denote it by

$$H^0(G, L^\times).$$

By Theorem 4 we have

(70) $$H^0(G, L^\times) \simeq H^2(G, L^\times) \quad \text{for cyclic } G,$$

but for arbitrary G the two groups generally diverge.

(5) In the situation of Theorem 4, given an *arbitrary* cocycle c of $G = \langle \tau \rangle$ in $L^\times$, the structure element a of the corresponding *uniformized* cocycle $\hat{c}$ in (62) can be expressed in terms of c as

$$(71) \qquad a = \prod_{j=0}^{n-1} c_{\tau, \tau^j}.$$

One sees this easily by computing the powers u_τ^i inductively by means of c. For $a = u_\tau^n$ we get (71).

Simple as the arguments leading to Theorem 4 were, the theorem is nonetheless a powerful tool. A feeling for this can be obtained from some examples:

Example Application 1 (Frobenius' Theorem). Since $\mathbb{C}$ is algebraically closed, we have $\mathrm{Br}\,\mathbb{R} = \mathrm{Br}(\mathbb{C}/\mathbb{R})$. But $\mathbb{C}/\mathbb{R}$ is cyclic, with $\tau : z \mapsto \bar{z}$ the only nontrivial element of $G(\mathbb{C}/\mathbb{R})$. Since $N = N_{\mathbb{C}/\mathbb{R}}$ is given by $Nz = z\bar{z} = |z|^2$, we have $N\mathbb{C}^\times = \mathbb{R}_{>0}$. Thus $\mathbb{R}^\times / N\mathbb{C}^\times = \mathbb{R}^\times / \mathbb{R}_{>0}$ is of order 2, represented by the elements 1 and -1 of $\mathbb{R}^\times$. In view of (66), we conclude that $\mathrm{Br}\,\mathbb{R}$ has a single element distinct from 1, namely the class of the cyclic algebra

$$(72) \qquad \mathbb{H} := (\mathbb{C}, \tau, -1).$$

($\mathbb{H}$ is necessarily a *division algebra*.) If we denote $u = u_\tau$ by j, we get $\mathbb{H} = \mathbb{C} + \mathbb{C}j = \mathbb{R} + \mathbb{R}i + \mathbb{R}j + \mathbb{R}ij$ with an $\mathbb{R}$-basis $1, i, j, ij$, and the elements of this basis satisfy, by definition,

$$(73) \qquad ij = -ji, \quad i^2 = -1, \quad j^2 = -1.$$

Thus $\mathbb{H}$ is the well known *quaternion algebra* introduced by Hamilton. Of course, one gets the same results for any *real closed* field R, instead of $\mathbb{R}$.

Example Application 2 (Wedderburn's Theorem). Let K be a *finite* field. To prove that $\mathrm{Br}\,K = 1$, we must show that $\mathrm{Br}(L/K) = 1$ for every finite extension L/K. Since L/K is cyclic, we need only use Theorem 4 and the fact that for an extension L/K of finite fields we have

$$(74) \qquad K^\times = N_{L/K} L^\times,$$

that is, the norm map $N_{L/K}$ is always *surjective*. We saw this result in F10 of Chapter 13 (vol. 1).

Conversely, if we suppose Wedderburn's Theorem to be known by other means — say by Theorem 21 of the last chapter — the field-theoretical statement in (74) will follow as an application of the theory of algebras: an example of the uses of "noncommutative tools" in a "commutative theory".

Example Application 3 (Quaternion algebras). Suppose a field K satisfies

$$(75) \qquad \mathrm{char}\,K \neq 2.$$

Consider an element $[A] \in \mathrm{Br}\, K$ of Schur index $s(A) = 2$. (By F3 this equality implies that $e(A) = 2$, but the converse is not always true.) Then A has a splitting field L of degree 2 over K, by Theorem 19 in the previous chapter. By (75), L/K is automatically Galois, so A is similar to a cyclic crossed product

$$(76) \qquad \qquad \Gamma = (L, \tau, a),$$

where τ denotes the nontrivial automorphism of L/K. Since $\Gamma : K = 4$ and $s(\Gamma) = 2$, the algebra Γ is necessarily a *division algebra*. Being a quadratic extension of K, the field L has the form

$$(77) \qquad \qquad L = K[v], \quad \text{with} \quad v^2 = b \in K^\times \smallsetminus K^{\times 2}.$$

Since $\Gamma = L + Lu = K + Kv + Ku + Kvu$, the elements $1, u, v, uv$ form a K-basis for Γ, and by definition they satisfy

$$(78) \qquad \qquad u^2 = a, \quad v^2 = b, \quad uv = -vu$$

(since $u^{-1}vu = v^\tau = -v$ in Γ). The algebra Γ in (76) is therefore an example of the following notion:

Definition. A *quaternion algebra over K*, where K is a field of characteristic other than 2, is any K-algebra Q generated by two elements u and v satisfying (78) for certain fixed

$$(79) \qquad \qquad a, b \in K^\times.$$

Clearly a quaternion algebra Q over K consists of K-linear combinations of

$$(80) \qquad \qquad 1, \ u, \ v, \ uv.$$

We show these elements are linearly independent: For any $x = a_1 + a_2 u + a_3 v + a_4 uv$ in Q we have $xu + ux = 2a_1 u + 2a_2 a =: y$, hence $yv - vy = 4a_1 uv$. Thus $x = 0$ implies $4a_1 uv = 0$, hence $a_1 = 0$, since $4uv$ is a unit in Q, by (75) and (78). The vanishing of the other coefficient follows. This shows that

$$(81) \qquad \qquad Q : K = 4.$$

But then Q is necessarily a *simple* K-algebra, since every homomorphic image of Q is again a quaternion algebra and so again 4-dimensional over K. Since Q is noncommutative—again by (78) and (75)—its center can only be K, or the dimensions don't work out. We conclude, then, that Q *is a central-simple K-algebra of dimension* 4, and its structure is fully determined by the constants $a, b \in K$. We use the notation

$$(82) \qquad \qquad Q = \left(\frac{a, b}{K} \right).$$

Because the relations (78) are completely symmetric, we have

(83)
$$\left(\frac{a,b}{K}\right) = \left(\frac{b,a}{K}\right).$$

Thus there is a hidden symmetry in the cyclic algebra (76) that is not yet apparent in the notation of (76). *Hamilton's quaternion algebra* $\mathbb{H}$ can be written as

(84)
$$\mathbb{H} = \left(\frac{-1,-1}{\mathbb{R}}\right).$$

A dimensional argument shows that any quaternion algebra Q over K either splits over K or is a division algebra. In the first case Q is isomorphic to $M_2(K)$, whereas in the second it has Schur index 2 and is therefore expressible in the form (76) as a cyclic algebra, besides being of the form (82).

We can summarize the discussion as follows: *Quaternion algebras over K are precisely the four-dimensional central-simple K-algebras. If $[A] \in \mathrm{Br}\, K$ has Schur index 2, then A is similar to a quaternion algebra, unique up to isomorphism, and this is a division algebra. The condition*

(85)
$$\left(\frac{a,b}{K}\right) \sim 1,$$

which expresses the splitting of the quaternion algebra in (82), is equivalent to the condition that

(86)
$$a \text{ is a norm in the extension } K(\sqrt{b})/K.$$

Here all that remains to show is that (85) and (86) are equivalent. Assume first that $K(\sqrt{b}) : K$ equals 2, so Q can be written as the cyclic algebra $(K(\sqrt{b}), \tau, a)$; by the remark after Theorem 4, the latter algebra splits if and only if (86) is satisfied. On the other hand, if $K(\sqrt{b}) = K$, so that $b = x^2$ is a square in $K^\times$, then Q is not a division algebra, because it has zero divisors: $(v+x)(v-x) = v^2 - x^2 = 0$. This leaves only the possibility that $Q \simeq M_2(K)$, that is, Q splits. And of course (86) is trivially satisfied if $K(\sqrt{b}) = K$.

We can also check easily that *for any a, b in $K^\times$ there exists precisely one quaternion algebra Q satisfying* (82). If $\sqrt{b} \notin K$ we take the crossed product $\Gamma = (K(\sqrt{b}), \tau, a)$, where τ is the nontrivial automorphism of $K(\sqrt{b})/K$ (recall the assumption that char $K \neq 2$). As we saw earlier, we then have $\Gamma = \left(\frac{a,b}{K}\right)$. In the opposite case where $\sqrt{b} \in K$, our only choice is to take $\Gamma = M_2(K)$. And indeed, the matrices

$$u = \begin{pmatrix} 0 & a \\ 1 & 0 \end{pmatrix}, \qquad v = \sqrt{b}\begin{pmatrix} 1 & 0 \\ 0 & -1 \end{pmatrix}$$

in $M_2(K)$ satisfy $u^2 = a$, $v^2 = b$, $vu = -uv$, and the subalgebra Q of $M_2(K)$ they generate coincides with $M_2(K)$, by (81).

We now turn to the behavior of *cyclic algebras* under *inflation* and *restriction*. Making use, of course, of the rules (44) and (45) for *arbitrary* crossed products, we will reach the following rules for *cyclic algebras*:

Theorem 5. *Let L/K be cyclic with Galois group $G = \langle \tau \rangle$, and let a be any element of $K^\times$.*

(a) *If E is an intermediate field of L/K, then E/K is also cyclic with Galois group $G(E/K) = \langle \tau_E \rangle$, and*

$$(87) \qquad (E, \tau_E, a) \sim (L, \tau, a^{L:E}).$$

(b) *For any extension K'/K, let $L' = LK'$ be the composite of L with K'. If we define $r = L \cap K' : K$ and identify $G(L'/K')$ in the usual way with the subgroup $G(L/L \cap K')$ of G, the element τ^r is then a generator of $G(L'/K')$, and*

$$(88) \qquad (L, \tau, a) \otimes_K K' \sim (L', \tau^r, a).$$

In particular, for any intermediate field F of L/K,

$$(89) \qquad (L, \tau, a) \otimes_K F \sim (L, \tau^{F:K}, a).$$

Proof. (a) Since $G = G(L/K)$ is abelian, every intermediate field E is Galois over K. Hence $\bar{G} = G(E/K)$, a homomorphic image of G, is cyclic, and generated by the automorphism $\bar{\tau} = \tau_E$ arising from E. Let c denote a *uniformized* cocycle associated to (E, τ_E, a). From (44) we get

$$(E, \tau_E, a) = (E, \bar{G}, c) \sim (L, G, \inf(c)).$$

For the crossed product on the right, let $u = u_\tau$ be a representative of τ. According to our procedure at the beginning of the section, there remains to determine $u^{L:K}$. By the definition of c and $\inf(c)$, we first get

$$u^i = u_{\tau^i} \quad \text{for } 1 \le i \le n := E : K,$$

while $u^n = u^{n-1} u = u_{\tau^{n-1}} u_\tau = u_{\tau^n} c_{\bar{\tau}^{n-1}, \bar{\tau}} = u_{\tau^n} a$. But then

$$u^{L:K} = u^{n(L:E)} = u_{\tau^n}^{L:E} a^{L:E} = a^{L:E},$$

because a recursion shows that $u_{\tau^n}^j = u_{\tau^{nj}}$ for every $j \in \mathbb{N}$, and for $j = L:E$ we have $u_{\tau^{nj}} = u_1 = c_{1,1} = 1$.

(b) After the warmup of part (a), the reader will have no difficulty tackling the (easier) proof of this part. □

Remark. As an application of Theorem 5(a) we give yet another proof of *Wedderburn's Theorem* (Chapter 29, Theorem 21) — or rather, this time the theorem almost proves itself. Let K be a finite field with q elements. Every finite extension of K is cyclic, so we need only show that every cyclic algebra (E, ρ, a) splits. As a finite field, E has an extension L of any given degree $n \in \mathbb{N}$. Let τ be a generator of $G(L/K)$ such that $\tau_E = \rho$. For $n = q - 1$, formula (87) implies what we need:

$$(E, \rho, a) \sim (L, \tau, a^{q-1}) = (L, \tau, 1) \sim 1.$$

We have repeatedly applied the algebra-theoretic methods developed so far to the task of proving that there are no *finite* noncommutative division algebras. But as far as positive information on proper division algebras (of finite dimension over their center) is concerned, we have so far only Hamilton's quaternion algebra (72), together with the fact that there are no others over a real-closed field. Of course this is already interesting; it tells us that the quaternion algebra

$$\left(\frac{-1,-1}{\mathbb{Q}}\right) = (\mathbb{Q}(\sqrt{-1}), \tau, -1)$$

is an example of a division algebra, with center $\mathbb{Q}$. But Theorem 4 would seem to afford a reliable method to ensure the existence of further nontrivial division algebras having a field K as their center. One need only consider *cyclic* extensions L/K and look for elements $a \in K^\times$ that *are not norms* in L/K. Then $\Gamma = (L, \tau, a)$ is a nonsplitting central-simple K-algebra, and thus has an associated nontrivial division algebra with center K (and when a has order $L:K$ in $K^\times/N_{L/K}L^\times$, it's Γ itself that is a division algebra). However, deciding whether a given $a \in K^\times$ is a norm is a difficult problem in general. The simplest nontrivial case is that of an extension L/K of degree 2. Here the question is to decide whether or not a given quaternion algebra

$$(90) \qquad Q = \left(\frac{a,b}{K}\right)$$

splits (where char $K \neq 2$). For $K = \mathbb{Q}$ this question in fact leads into the deeper realms of *number theory*, where it admits a full — and splendid — solution in a very general framework. (See, for example, F. Lorenz, *Algebraische Zahlentheorie*, BI-Verlag, 1993, Chapters 10 and 12.)

We want in any case to push a bit further our analysis of quaternion algebras, with the goal of amassing additional nontrivial examples. We start with a *splitting criterion for quaternion algebras*:

F4. *Let Q be a quaternion algebra over K of the form* (90), *still with* char $K \neq 2$. *The following conditions are equivalent*:

(i) *Q splits, that is, $Q \simeq M_2(K)$.*

(ii) *b is a norm of the extension $K(\sqrt{a})/K$.*

(ii') *a is a norm of the extension $K(\sqrt{b})/K$.*

(iii) *The quadratic form $q = X_1^2 - aX_2^2 - bX_3^2$ is isotropic over K.*

Proof. The equivalence of (i) and (ii') has already been shown: it is that of (85) and (86). The equivalence of (ii) and (ii') is clear from symmetry; see (83).

If $K(\sqrt{a}) = K$, both (ii) and (iii) hold — note that $q(\sqrt{a}, 1, 0) = 0$. So assume instead that $K(\sqrt{a}) : K = 2$. If (ii) holds, there exists x_1, x_2 in K such that $b = N(x_1 + x_2\sqrt{a}) = x_1^2 - ax_2^2$; it follows that $x_1^2 - ax_2^2 - b1^2 = 0$, giving (iii). Conversely, if $x_1^2 - ax_2^2 - bx_3^2 = 0$, where $x_1, x_2, x_3 \in K$ don't all vanish, the element $bx_3^2 = x_1^2 - ax_2^2$ is a norm in $K(\sqrt{a})/K$. But x_3 is nonzero since $\sqrt{a}$ is not in K; hence b is a norm, proving (ii). $\qquad\square$

F5. *Still assuming that* char $K \neq 2$, *we have the following calculation rules for quaternion algebras*:

(a) $\left(\dfrac{a,b}{K}\right) = \left(\dfrac{b,a}{K}\right)$.

(b) $\left(\dfrac{x^2 a, y^2 b}{K}\right) = \left(\dfrac{a,b}{K}\right)$.

(c) $\left(\dfrac{a,-a}{K}\right) \sim 1$.

(d) $\left(\dfrac{a, 1-a}{K}\right) \sim 1$.

(e) $\left(\dfrac{a_1 a_2, b}{K}\right) \sim \left(\dfrac{a_1, b}{K}\right) \otimes \left(\dfrac{a_2, b}{K}\right)$, $\quad \left(\dfrac{a, b_1 b_2}{K}\right) \sim \left(\dfrac{a, b_1}{K}\right) \otimes \left(\dfrac{a, b_2}{K}\right)$.

(f) $\left(\dfrac{a, b_1}{K}\right) \simeq \left(\dfrac{a, b_2}{K}\right) \iff \left(\dfrac{a, b_1 b_2}{K}\right) \sim 1$.

Proof. The validity of (a) and (b) can be checked by replacing u, v with v, u and xu, yu, respectively. Items (c) and (d) can be easily read out from F4. If b is not a square in K, (e) follows from the corresponding rule (68) for cyclic algebras. Since a quaternion algebra Q always satisfies $e(Q) = s(Q) \leq 2$, we have $[Q] = [Q]^{-1}$ in Br K. For arbitrary quaternion algebras Q_1, Q_2 over K, therefore, $Q_1 \otimes Q_2 \sim 1$ is equivalent to $Q_1 \sim Q_2$, hence also to $Q_1 \simeq Q_2$. This gives (f) simply as a consequence of (e). $\qquad\square$

F6. *Let Q be a quaternion division algebra over K of the form* (90), *with* char $K \neq 2$, *and consider the quadratic form* $q = a X_1^2 + b X_2^2 - ab X_3^2$. *Every field $E = K(\sqrt{q(x)})$ with nonzero $x \in K^3$ is K-isomorphic to a maximal subfield of Q, and conversely every maximal subfield of Q has this form.*

Proof. The maximal subfields of Q are the fields $K(w)$ where

(91) $$w \in Q, \quad w \notin K, \quad w^2 \in K.$$

Every w in Q has a unique representation

$$w = \lambda + u\mu, \quad \text{with } \lambda, \mu \in L = K(v).$$

When does w^2 lie in K? Using primes to denote the action of the nontrivial automorphism of L/K, we have $(\lambda + u\mu)^2 = \lambda^2 + \lambda u\mu + u\mu\lambda + (u\mu)(u\mu) = \lambda^2 + u(\lambda'\mu + \lambda\mu) + u^2\mu\mu'$, so

$$(\lambda + u\mu)^2 = (\lambda^2 + a\mu\mu') + u(\lambda'\mu + \lambda\mu).$$

Hence $c := w^2 = (\lambda + u\mu)^2$ lies in K if and only if

(91') $$(\lambda + \lambda')\mu = 0 \quad \text{and} \quad \lambda^2 + a\mu\mu' \in K.$$

We now distinguish two cases. If $\mu \neq 0$, the first condition in (91') is then equivalent to $\lambda' = -\lambda$, that is, $\lambda = x_2 v$, while the second is automatically satisfied if the first is. For c we obtain the value $c = x_2^2 v^2 + a(x_1^2 - x_3^2 v^2)$, so

$$c = w^2 = ax_1^2 + bx_2^2 - abx_3^2.$$

If, contrariwise, μ vanishes, (91') and the condition $w \notin K$ are equivalent to $\lambda^2 \in K$ and $\lambda \notin K$, hence to $\lambda = x_2 v$, and we obtain in this case

$$c = w^2 = bx_2^2.$$

Thus, putting it all together, the elements characterized by (91) are precisely the square roots in Q of elements of the form

$$q(x) = ax_1^2 + bx_2^2 - abx_3^2,$$

where $x = (x_1, x_2, x_3)$ runs through all nonzero triples in K^3. This proves F6. $\square$

For the case $K = \mathbb{Q}$ we now investigate when a given quaternion algebra

$$\left(\frac{a, b}{\mathbb{Q}} \right)$$

splits. In view of F5(b) we can assume in addition that a and b are square-free integers. By F4, what we have to find out is whether the equation

(92) $$aX_1^2 + bX_2^2 - X_3^2 = 0$$

has a nontrivial solution over $\mathbb{Z}$ (a solution over $\mathbb{Q}$ implies one over $\mathbb{Z}$, by homogeneity.) Let d be the greatest common divisor of a and b, and set $a' = a/d$, $b' = b/d$. Solving (92) over $\mathbb{Z}$ amounts to solving

(93) $$a'X_1^2 + b'X_2^2 - dX_3^2 = 0,$$

and for this situation — a', b', d pairwise relatively prime, square-free integers — we can invoke a number-theoretic result:

Theorem 6 (Legendre 1798). *If r, s, t are pairwise relatively prime, square-free integers, the equation*

(94) $$rX_1^2 + sX_2^2 + tX_3^2 = 0$$

has a nontrivial integer solution if and only if r, s, t don't all have the same sign and each of the quadratic congruences

(95) $$X^2 \equiv -st \bmod r, \quad Y^2 \equiv -rt \bmod s, \quad Z^2 \equiv -rs \bmod t$$

is solvable in integers.

Proof. The forward implication is easy to check; we prove the converse. Multiplying (94) by $-t$, we pass to an equivalent equation having the form (92) with square-free integers a and b; the assumptions are then equivalent to the numbers a and b not being both negative and the congruences

$$(96) \qquad X^2 \equiv a \bmod b, \quad Y^2 \equiv b \bmod a, \quad Z^2 \equiv -a'b' \bmod d$$

being all solvable, where $d := \gcd(a, b)$, $a' := a/d$, $b' := b/d$. For $a = 1$, (92) always has a nontrivial solution, so from now on we assume $a \neq 1$. We work by induction on the value of the sum $|a| + |b|$. In the starting case, $|a| + |b| = 2$, the only possibility is $a = -1$ and $b = 1$, since $a \neq 1$: then (92) certainly has a nontrivial solution. Now take $|a| + |b| > 2$; without loss of generality we can assume

$$(97) \qquad |a| \leq |b| \geq 2.$$

Given the solvability of the first congruence in (96), there exist integers x and b_1 such that

$$(98) \qquad a = x^2 - b_1 b,$$

where we also require that

$$(99) \qquad |x| \leq |b|/2.$$

Since $a \neq 1$ must be square-free, b_1 is nonzero. Equation (98), in the form

$$(100) \qquad bb_* c^2 = x^2 - a \quad \text{with } b_* \text{ square-free,}$$

is the crux of the proof. By (100),

$$bb_* \text{ is a norm of } \mathbb{Q}(\sqrt{a}).$$

By F4 and F5(f), then,

$$\left(\frac{a, b}{\mathbb{Q}} \right) \sim \left(\frac{a, b_*}{\mathbb{Q}} \right).$$

Again by F4, this shows that the starting equation (92) has a nontrivial solution (still over $\mathbb{Q}$ or $\mathbb{Z}$) if and only if the same is true of

$$(101) \qquad aX_1^2 + b_* X_2^2 - X_3^2 = 0.$$

From (100), (99) and (97) there now follows the estimate $|b_* b| \leq |x|^2 + |a| \leq |b|^2/4 + |b|$, and hence

$$|b_*| \leq |b|/4 + 1 < |b|.$$

The induction assumption then completes the proof, provided we can show that the coefficients of (101) satisfy analogous conditions as those of (92). First one reads off (100) that a and b_* are not both negative, since a and b are not both negative by assumption. There remains to show that the congruences

$$(102) \qquad X^2 \equiv a \bmod b_*, \quad Y^2 \equiv b_* \bmod a, \quad Z^2 \equiv -a'_* b'_* \bmod d_*$$

are solvable, where $d_* := \gcd(a, b_*)$, $a'_* := a/d_*$, $b'_* := b_*/d_*$. The solvability of the first congruence in (102) can be immediately inferred from (100). Now let

(103)
$$p \mid a$$

be any prime divisor of a. In view of (100), p cannot divide c, since a is square-free. For the same reason, p cannot divide simultaneously b and b_*. Hence

(104)
$$p \mid a \Longrightarrow p \nmid c \quad \text{and} \quad p \mid d \Longrightarrow p \nmid d_*.$$

Under assumption (103), we now consider first the case $p \nmid b$. By (100), bb_*c^2 is a square mod p; since by assumption b is one as well, and since $p \nmid c$ and $p \nmid b$, the congruence

(105)
$$Y^2 \equiv b_* \bmod p$$

also has a solution. To verify the solvability of the second congruence in (102), there remains to show that (105) is solvable also in the case where p divides both a and b, that is,

$$p \mid d.$$

Dividing (100) by d, we obtain

$$b'b_*c^2 = dy^2 - a',$$

for some $y \in \mathbb{Z}$. Multiplication by b' yields

$$b'^2 b_* c^2 \equiv -a'b' \bmod d.$$

But $-a'b'$ is by assumption a square mod d, and *a fortiori* mod p; since $p \nmid b'$ and $p \nmid c$, this shows that b_* is a square mod p. There remains to verify that the third congruence in (102) has a solution. Dividing (100) by d_*, we obtain

$$bb'_* c^2 = d_* z^2 - a'_*,$$

with $z \in \mathbb{Z}$. Multiplying this equation by a'_* yields

$$-a'_* b'_* bc^2 \equiv a'^2_* \bmod d_*,$$

so $-a'_* b'_* bc^2$ is a square mod d_*. By assumption, b is a square mod a, and *a fortiori* mod d_*; because b and c are relatively prime to d_* — see (104) — this implies that $-a'_* b'_*$ too is a square mod d_*. $\square$

The problem of deciding whether a quaternion algebra over $\mathbb{Q}$ splits can be regarded as essentially solved by the theorem of Legendre just proved. To cap this elegant line of argumentation there is the *quadratic reciprocity law* (vol. I, pp. 107 and following), which affords a simple and effective decision procedure for whether the congruences (95) in Theorem 6 are solvable.

In any case, thanks to Theorem 6, we are in a position to exhibit an infinite family of nonisomorphic division algebras of finite dimension having center $\mathbb{Q}$:

Example. For every prime $p \equiv 3 \bmod 4$,

$$(106) \qquad\qquad Q(p) := \left(\frac{p, -1}{\mathbb{Q}}\right)$$

is a division algebra of Schur index 2. If $p, q \equiv 3 \bmod 4$ are distinct primes, $Q(p)$ and $Q(q)$ are not isomorphic. (Using a slight variation of Euclid's famous proof, one can show the existence of infinitely many primes congruent to 3 mod 4; see also §11.7(d).)

Proof. First let p be any prime at all. The quaternion algebra (106) splits if and only if $pX_1^2 - X_2^2 - X_3^2 = 0$ admits a nontrivial solution in integers (see F4). By Theorem 6 applied to $t = -1$ — a special case that incidentally is much easier to prove than the full-fledged theorem — there is such a solution if and only if the congruence $X^2 \equiv -1 \bmod p$ has a solution in $\mathbb{Z}$. This never happens when $p \equiv 3 \bmod 4$ (see vol. I, p. 107). Next, if $Q(p)$ and $Q(q)$ were isomorphic for primes $p \neq q$ with $p \equiv 3 \bmod 4$, it would follow by F5(f) that

$$\left(\frac{pq, -1}{\mathbb{Q}}\right) \sim 1.$$

By F4 and Theorem 6 the congruence $X^2 \equiv -1 \bmod pq$ would then have a solution in $\mathbb{Z}$, contradicting the absence of solutions to $X^2 \equiv -1 \bmod p$.   □

In searching for further examples of division algebras, it is natural to look into tensor products $Q_1 \otimes Q_2$ of quaternion algebras Q_1 and Q_2 over a given field. Dimensionally speaking, the only possibilities for the Schur index s of $Q_1 \otimes Q_2$ are $1, 2$ and 4. Hence, $Q_1 \otimes Q_2$ is either similar to a quaternion algebra, or it is a division algebra. The second case can happen at best when Q_1 and Q_2 are themselves division algebras. The question then is: When is the tensor product of two quaternion division algebras also a division algebra? For $K = \mathbb{Q}$ this is never the case — a fact that cannot easily be proved without more sophisticated number-theoretical tools, so we will not go into it. But it can happen over other fields K. The following criterion enables us to give explicit examples:

F7. *Let*

$$Q_1 = \left(\frac{a, b}{K}\right), \qquad Q_2 = \left(\frac{c, d}{K}\right)$$

be quaternion algebras over a field K with char $K \neq 2$. *The following conditions are equivalent:*

 (i) *$Q_1 \otimes Q_2$ is a division algebra.*

 (ii) *Every quadratic extension E/K satisfies $Q_1 \otimes E \not\cong Q_2 \otimes E$ (over E).*

 (iii) *The quadratic form*

$$(aX_1^2 + bX_2^2 - abX_3^2) - (cX_4^2 + dX_5^2 - cdX_6^2)$$

 is anisotropic over K.

(iv) *No maximal subfields L_1 of Q_1 and L_2 of Q_2 are isomorphic over K.*

(v) *For any maximal subfields L_1 of Q_1 and L_2 of Q_2 the algebras $Q_2 \otimes L_1$ and $Q_1 \otimes L_2$ are division algebras.*

Proof. Let E/K be a quadratic extension and consider res $= \mathrm{res}_{E/K}$. In Br E we have $\mathrm{res}[Q_1 \otimes Q_2] = \mathrm{res}[Q_1]\,\mathrm{res}[Q_2] = \mathrm{res}[Q_1]\,\mathrm{res}[Q_2]^{-1}$, so $Q_1 \otimes Q_2$ splits over E if and only if $\mathrm{res}[Q_1] = \mathrm{res}[Q_2]$. This shows the equivalence of (i) and (ii).

Suppose (iv) holds. We claim that Q_1 and Q_2 are division algebras. For suppose $Q_1 \simeq M_2(K)$, for instance. Surely Q_2 does not split, otherwise $Q_2 \simeq Q_1$, contrary to (iv). Every maximal subfield L_2 of Q_2 has degree 2 over K, and so can be isomorphically embedded into $M_2(K) \simeq Q_1$. This contradicts (iv).

Suppose (iii) holds. Then the "partial forms"

$$aX_1^2 + bX_2^2 - abX_3^2 \quad \text{and} \quad cX_4^2 + dX_5^2 - cdX_6^2$$

are anisotropic over K. Thus F4(iii) implies (after multiplication with ab and cd) that Q_1 and Q_2 cannot both split.

The equivalence of (iii) and (iv) is now an easy consequence of F6.

Suppose (v) does not hold, and $Q_2 \otimes L_1$ is not a division algebra. Then Q_2 splits over L_1, so L_1 is isomorphic to a maximal subfield of Q_2 (see Theorem 19 in the previous chapter). This proves that (iv) implies (v). The converse is obvious, as is the implication (ii) $\Rightarrow$ (v).

There remains therefore the much more delicate implication (v) $\Rightarrow$ (i). What we must show is that, under (v), every nonzero $x \in Q_1 \otimes Q_2$ is invertible. We take Q_1 to be of the form (78), for given u, v. In the algebra $Q_1 \otimes Q_2$, the element x has a unique decomposition

$$x = y_1 + uy_2, \quad \text{with } y_i \in L_1 \otimes Q_2,$$

where $L_1 = K[v]$. If y_2 vanishes we are done, since $L_1 \otimes Q_2$ is a division algebra by assumption. Hence we may assume that $y_2 \neq 0$. After multiplying with y_2^{-1}, we can then write x in the form

$$x = y + u, \quad \text{with } y \in L_1 \otimes Q_2.$$

Now we apply the inner automorphism of conjugation by u, using exponential notation to write it: we get

$$(y + u)(y^u - u) = yy^u - u^2 = yy^u - a \in L_1 \otimes Q_2.$$

If $yy^u - a \neq 0$ there is nothing else to show. We therefore suppose the opposite, and write $y = z_1 + vz_2$, with $z_i \in Q_2$; then

$$0 = yy^u - a = z_1^2 + v(z_2 z_1 - z_1 z_2) - bz_2^2 - a,$$

that is, $z_1^2 - bz_2^2 - a = 0$ and $z_2 z_1 - z_1 z_2 = 0$. If $z_2 \in K$, the element

(107) $$x = y + u = z_1 + vz_2 + u$$

lies in $K[vz_2+u] \otimes Q_2$, which is a division algebra by assumption, since $K[vz_2+u]$ is a maximal subfield of Q_1. Hence suppose $z_2 \notin K$. Since $z_1 z_2 = z_2 z_1$ we must have $z_1 \in K[z_2]$. But then, by (107), x lies in $Q_1 \otimes K[z_2]$; this product, too, is by assumption a division algebra, so we are done.   □

Example. Over a *real* ground field F, consider the field of rational functions

$$(108) \qquad\qquad K = F(X, Y)$$

in two variables. K is obviously real as well. Consider over K the quaternion algebras

$$(109) \qquad\qquad D_1 = \left(\frac{X, -1}{K}\right), \qquad D_2 = \left(\frac{-X, Y}{K}\right),$$

and their tensor product $A = D_1 \otimes D_2$. We claim:

(i) *A is a division algebra (and hence so are D_1 and D_2). Thus A affords an example of a central-simple K-algebra for which*

$$(110) \qquad\qquad s(A) \neq e(A),$$

since $s(A) = 4$ and $e(A) = 2$. Even more is true:

(ii) *For every extension of the form $L = K(\sqrt{a^2 + b^2})$, with $a, b \in K$, the algebra $A \otimes L$ is a division algebra.* This generalization of (i) is significant because it lends itself to the derivation of the following property of A:

(iii) *A is an example of a **noncyclic** division algebra of Schur index 4.*

Indeed, (iii) follows from (ii) and from a purely field-theoretical fact:

(iv) *If K is any field such that $\sqrt{-1} \notin K$, any cyclic extension E/K of degree 4 has an intermediate quadratic field of the form $L = K(\sqrt{a^2 + b^2})$, with $a, b \in K$.*

The demonstration of (iv) is a charming exercise in field theory: see §13.2 in vol. I.

To prove (ii), we apply F7 to the quaternion algebras

$$Q_1 := D_1 \otimes L = \left(\frac{X, -1}{L}\right), \qquad Q_2 := D_2 \otimes L = \left(\frac{-X, Y}{L}\right).$$

What we need to show is that the quadratic form in F7(iii)—call it β—is anisotropic over L. Its coefficients are

$$X, \ -1, \ X, \ X, \ -Y, \ -XY,$$

Thus, suppose $\beta(z) = 0$ for some $z \in L^6$. Setting $d := a^2 + b^2$ we can write

$$z = x + y\sqrt{d},$$

with $x = (x_1, \ldots, x_6)$ and $y = (y_1, \ldots, y_6)$ in K^6. Denoting the symmetric bilinear form corresponding to β by the same symbol, we have

$$0 = \beta(x + y\sqrt{d}) = \beta(x) + d\beta(y) + 2\beta(x, y)\sqrt{d}.$$

If $\sqrt{d} \notin K$, we obtain

$$(111) \qquad \qquad \beta(x) + d\beta(y) = 0.$$

We now show that (111), for any $d = a^2 + b^2$, can only have a solution if $x = dy = 0$. This will take care of the case $L = K$ as well, since we can take $d = 0$. By homogeneity, we can also assume that the x_i and y_i are polynomials, that is, elements of $F[X, Y]$. The same applies to a and b. Defining the sums of squares

$$(112) \qquad \qquad q_i = x_i^2 + a^2 y_i^2 + b^2 y_i^2 \quad \text{for } 1 \le i \le 6$$

in K, we can write (111) as $Xq_1 - q_2 + Xq_3 + Xq_4 - Yq_5 - XYq_6 = 0$, or yet

$$(113) \qquad \qquad X(q_1 + q_3 + q_4) = Y(q_5 + Xq_6) + q_2.$$

Since F is *real*, the polynomial on the left-hand side of (113) has even degree in Y (unless it vanishes), whereas the right-hand side has odd degree in Y, again unless $q_5 + Xq_6 = 0$, which (as we see by comparing degrees in X) can only happen if $q_5 = q_6 = 0$. Thus the equation becomes

$$X(q_1 + q_3 + q_4) = q_2.$$

Again by a degree comparison in X, and taking into account that K is a real field, one obtains $q_1 = q_3 = q_4 = q_2 = 0$. The upshot is that equation (113) can only be satisfied if all the q_i vanish. But by (112) this means that $x = 0$ and $dy = 0$.

Remark. If K is any field of characteristic other than 2, *every central-simple K-algebra of exponent 2 is similar to a tensor product of finitely many quaternion algebras*. This arresting theorem, which solved a long-standing conjecture, was proved in 1981 by *Merkurjev*. He also showed that every relation between quaternion algebras in $\mathrm{Br}\,K$ is a consequence of the fundamental relations listed in F5. The proof of this theorem, and of its generalization by Merkurjev and Suslin to algebras of higher exponents, lies beyond the scope of this text; a good source is I. Kersten, *Brauergruppen von Körpern*, Vieweg, 1990 (see also F. Lorenz, *K_2 of fields and the Theorem of Merkurjev*, Report, Univ. of South Africa, 1986).

In the case of an *algebraic number field* K we have an additional result: *The tensor product of two quaternion algebras over K is always similar to a quaternion algebra over K.* This follows from F7 by an application of *Meyer's Theorem*, mentioned at the end of Chapter 22. Using tools from number theory one can also show right away that over a number field K every central-simple K-algebra of exponent 2 is similar to a quaternion algebra: see F. Lorenz, *Algebraische Zahlentheorie*, 10.6 or 10.10.

Appendix: Cohomology Groups

The terminology of *cocycles* introduced at the beginning of this chapter will have cued the reader to the fact that we're dealing with an aspect of the *cohomology* of algebraic structures — a vast subject of which this is but the tip of an iceberg. Although we cannot explore this theory systematically, we would like to place some of the definitions and results of the chapter in the context of group cohomology, adopting a more or less naïve approach.

Let G be a group. The starting point is the notion of a *G-module*, that is, an abelian group M, on which G acts by automorphisms. For consistency with the parallel material in the body of the chapter, we write the group operation on M multiplicatively and the action of G exponentially. Thus the G-module structure on M is given by a map

$$G \times M \to M$$
(1*)
$$(\sigma, x) \mapsto x^\sigma$$

satisfying

(2*)
$$x^1 = x, \quad x^{\sigma\tau} = (x^\sigma)^\tau, \quad (xy)^\sigma = x^\sigma y^\sigma.$$

If M is a G-module, we call

(3*)
$$M^G = \{x \in M \mid x^\sigma = x \text{ for all } \sigma \in G\}$$

the *fixed module* of M under G; it consists of the elements of M left invariant by G. If $M^G = M$ we say that M is a *trivial G-module*.

If G is a *profinite group*, that is, a projective limit of finite groups (see vol. I, p. 126), we also require that the map (1) be *continuous*, where M is given the discrete topology. It is easy to check that this amounts to demanding that

(4*)
$$M = \bigcup_H M^H,$$

where H runs over a neighborhood basis of open subgroups of G.

Example. If L/K is a Galois extension with Galois group G, the multiplicative group $M = L^\times$ of L is naturally a G-module. Condition (4*) is satisfied by Galois theory; see Theorem 4 in Chapter 12 (vol. I). The fixed module of M under G is $L^{\times G} = K^\times$.

Given G-modules M, N, we denote by $\mathrm{Hom}_G(M, N)$ the abelian group of G-homomorphisms from M into N, that is, homomorphisms $f : M \to N$ such that

(5*)
$$f(x^\sigma) = f(x)^\sigma \quad \text{for every } \sigma \in G, x \in M.$$

If G is *finite*, one can procure invariant elements as follows: For any $x \in M$, define the *norm of x with respect to G* as

(6*)
$$N_G(x) = \prod_\sigma x^\sigma.$$

Then

(7*)
$$N_G(x^\rho) = N_G(x)^\rho = N_G(x),$$

for every $\rho \in G$, and in particular $N_G(x) \in M^G$. But in general, not all invariant elements of M are norms; the departure is encoded by the quotient group

(8*)
$$H^0(G, M) := M^G / N_G(M),$$

which we call the *zeroth cohomology group of G with coefficients in M*. Let n be the order of G. Since

(9*)
$$N_G x = x^n \quad \text{for } x \in M^G,$$

all the elements α of $H^0(G, M)$ satisfy $\alpha^n = 1$.

We now show that the construction (3*) of the fixed module inexorably leads to a certain formalism. (We must of course forgo an in-depth study of this formalism, which is the scaffolding for *group cohomology theory*). The starting point is an exact sequence of G-modules

(10*)
$$1 \to X \xrightarrow{i} Y \xrightarrow{p} Z \to 1;$$

exactness means that the G-homomorphism $i : X \to Y$ is injective, the G-homomorphism $p : Y \to Z$ is surjective, and $\ker p = \operatorname{im} i$. If we identify X with its image under i in Y, we obtain from p a G-isomorphism $Y/X \simeq Z$. *We now ask whether the corresponding sequence of fixed modules is also exact.* The only doubtful point is whether p leads to a surjective homomorphism $Y^G \to Z^G$, so take $z \in Z^G$. Then

(11*)
$$z = p(y) \quad \text{for some } y \in Y.$$

Since $p(y^\sigma) = z^\sigma = z = p(y)$ we have

(12*)
$$x_\sigma := y^\sigma y^{-1} \in X$$

for every $\sigma \in G$. A straightforward computation then shows that

(13*)
$$x_{\sigma\tau} = x_\sigma^\tau x_\tau \quad \text{for every } \sigma, \tau \in G.$$

We see that for any $z \in Z^G$, after choosing $y \in Y$ with $p(y) = z$, we obtain a map $\sigma \mapsto x_\sigma$ from G into X satisfying the functional equation (13*). If a different choice of y is made, say y' with $p(y') = z$, we can write $y' = xy$ for some $x \in X$, so

(14*)
$$x'_\sigma := y'^\sigma y'^{-1} = x^\sigma x_\sigma x^{-1} = x_\sigma(x^\sigma x^{-1}).$$

Definition 1*. A map $\sigma \mapsto x_\sigma$ from G to X satisfying (13*) is called a *1-cocycle* or *crossed homomorphism* of G in X. For every $x \in X$, the map

(15*)
$$\sigma \mapsto x^{\sigma-1} = x^\sigma x^{-1}$$

is a 1-cocycle of G in X; a cocycle of this form is said to *split*, and is called a
1-*coboundary* of G in X. The 1-cocycles of G in X form a group $C^1(G, X)$ under
the obvious multiplication. The 1-coboundaries of G in X make up a subgroup
$B^1(G, X)$ of $C^1(G, X)$. The quotient group

$$(16^*) \qquad H^1(G, X) := C^1(G, X)/B^1(G, X)$$

is called the *first cohomology group of G with coefficients in X*.

To summarize, we can say that every $z \in Z^G$ can be assigned a well defined
element $\delta_1 z$ in $H^1(G, X)$. Thus the exact sequence of G-modules (10^*) gives rise
to a homomorphism

$$(17^*) \qquad \delta_1 : Z^G \to H^1(G, X),$$

called the associated *connecting homomorphism*.

Clearly $H^1(G, X)$ behaves functorially in X; that is, a given G-homomorphism
$f : X \to Y$ leads naturally to a homomorphism $H^1(G, X) \to H^1(G, Y)$. When no
misunderstanding is to be feared, we denote this derived homomorphism also by f,
or suppress entirely its dependence on f.

Now the question posed a short while ago can be answered:

F1*. *An exact sequence (10^*) of G-modules gives rise to an exact sequence of
abelian groups*

$$(18^*) \qquad 1 \to X^G \to Y^G \to Z^G \xrightarrow{\delta_1} H^1(G, X) \to H^1(G, Y) \to H^1(G, Z).$$

Proof. Exactness at Z^G, that is, the equality $p(Y^G) = \ker \delta_1$, is proved as follows.
Let $z = py$ for some $y \in Y^G$; then $x_\sigma = 1$ in (12^*), so $\delta_1 z = 1$. Conversely,
assuming $\delta_1 z = 1$, we have $x_\sigma = x^{\sigma\ -1}$ for some $x \in X$. Using (12^*) we then see
that $y^{\sigma\ -1} = x^{\sigma\ -1}$, hence $(yx^{-1})^{\sigma\ -1} = 1$ and so $(yx^{-1})^\sigma = yx^{-1}$ for all σ. This
shows that yx^{-1} lies in Y^G and hence that $z = p(yx^{-1})$ in $p(Y^G)$.

The proof of exactness at the other spots is left to the reader. □

The exact sequence (10^*) motivates a question: When does an element of
$H^1(G, Z)$ lie in the image of the map $H^1(G, Y) \to H^1(G, Z)$? Let the given
class in $H^1(G, Z)$ be represented by a 1-cocycle z_σ. Choose $y_\sigma \in Y$ such that
$p(y_\sigma) = z_\sigma$; since $z_{\sigma\tau} = z_\sigma^\tau z_\tau$, we must have

$$(19^*) \qquad x_{\sigma,\tau} := y_\sigma^\tau y_\tau y_{\sigma\tau}^{-1} \in X.$$

Now it is easy to check that, for every σ, τ, ρ in G,

$$(20^*) \qquad x_{\sigma,\tau}^\rho x_{\sigma\tau,\rho} = x_{\sigma,\tau\rho} x_{\tau,\rho}.$$

A different choice of inverse images for the z_σ, say $y_\sigma' = y_\sigma x_\sigma$, leads to

$$(21^*) \qquad x_{\sigma,\tau}' = x_{\sigma,\tau}(x_\sigma^\tau x_\tau x_{\sigma\tau}^{-1}).$$

If one changes z_σ to an equivalent 1-cocycle, the $x_{\sigma,\tau}$ can be made to remain the
same (for an appropriate choice of inverse images).

Definition 2*. A map $(\sigma, \tau) \mapsto x_{\sigma,\tau}$ from $G \times G$ into X satisfying (20*) is called a 2-*cocycle* or *factor system* of G in X. For any map $\sigma \mapsto x_\sigma \in X$, the assignment

$$(22^*) \qquad\qquad (\sigma, \tau) \mapsto x_\sigma^\tau x_\tau x_{\sigma\tau}^{-1}$$

is a 2-cocycle; a cocycle of this form is said to *split*, and is called a 2-*coboundary*. The 2-cocycles of G in X form a group $C^2(G, X)$, of which the set $B^2(G, X)$ of 2-coboundaries is a subgroup. The quotient

$$(23^*) \qquad\qquad H^2(G, X) := C^2(G, X)/B^2(G, X)$$

Compare this with Definition 1 on page 187; what we have here is a more general version of the same notion introduced — from a very different direction — in our study of algebras. In terms of the present discussion, the construction above gives rise to a well defined *connecting homomorphism* $\delta_2 : H^1(G, Z) \to H^2(G, X)$. One checks without difficulty that the sequence

$$(24^*) \quad H^1(G, Y) \to H^1(G, Z) \xrightarrow{\delta_2} H^2(G, X) \to H^2(G, Y) \to H^2(G, Z)$$

of abelian groups is exact. Thus the sequence (18*) can the extended to a longer exact sequence. And as one might expect, the game does not end there; it is possible to define cohomology groups

$$H^n(G, X) = C^n(G, X)/B^n(G, X)$$

and connecting homomorphisms $\delta_n : H^{n-1}(G, Z) \to H^n(G, X)$ for *every* $n \geq 1$, which, starting from a short exact sequence (10*), form the so-called *long exact sequence in cohomology*, extending (18*) endlessly. We shall not go into this, however.

If the group G is *finite*, it is natural enough to replace the fixed module M^G by the zeroth cohomology group $H^0(G, M) = M^G/N_G M$. Then the exact sequence (10*) gives rise to a variant of (18*), the exact sequence

$$H^0(G, X) \to H^0(G, Y) \to H^0(G, Z) \xrightarrow{\delta_1} H^1(G, X) \to H^1(G, Y) \to H^1(G, Z).$$

One checks easily that the map δ_1 in (17*) takes norms in Z to 1-coboundaries of X; moreover the sequence above is obviously exact at $H^0(G, Y)$. There is however an important difference relative to (18*): unlike the map $X^G \to Y^G$, the derived map $H^0(G, X) \to H^0(G, Y)$ is generally not injective. We therefore investigate its kernel.

Let $x \in X^G$ be such that $x = N_G y$ for some $y \in Y$. Then $z := py$ obviously satisfies $N_G z = 1$. Thus the elements in the kernel of $H^0(G, X) \to H^0(G, Y)$ are connected with the elements of norm 1 in Z. This motivates the following:

Definition 3*. If G is a *finite* group and M is a G-module, we denote by

$$(25^*) \qquad\qquad C^{-1}(G, M) = \{z \in M \mid N_G z = 1\}$$

the set of elements of norm 1 in M, and by

$$(26^*) \qquad\qquad B^{-1}(G, M) = \langle z^{\sigma - 1} \mid z \in M, \sigma \in G \rangle$$

the $\mathbb{Z}$-submodule of $C^{-1}(G, M)$ generated by elements of the form $z^{\sigma-1} = z^{\sigma} z^{-1}$. The quotient

$$(27^*) \qquad\qquad H^{-1}(G, M) = C^{-1}(G, M)/B^{-1}(G, M)$$

is called the (-1)-*st cohomology group of G with coefficients in M.*

From this definition it is easy to see that if G is a finite group, the exact sequence (10^*) gives rise to a well defined connecting homomorphism

$$\delta_0 : H^{-1}(G, Z) \to H^0(G, X)$$

and to an exact sequence

$$H^{-1}(G, X) \to H^{-1}(G, Y) \to H^{-1}(G, Z) \xrightarrow{\delta_0} H^0(G, X) \to H^0(G, Y) \to \cdots$$

Summarizing:

F2*. *If G is a finite group, any exact sequence (10^*) of G-modules yields an exact sequence in cohomology*

$$H^i(G, X) \to H^i(G, Y) \to H^i(G, Z) \xrightarrow{\delta_{i+1}} H^{i+1}(G, X) \to H^{i+1}(G, Y) \to H^{i+1}(G, Z),$$

where $i = -1, 0, 1$. For $i = 1$ we do not need G to be finite.

Remark. If G is a *finite group* and M is an G-module, one can define cohomology groups $H^i(G, M)$ for *all* $i \in \mathbb{Z}$ and extend the statement of F2* to all $i \in \mathbb{Z}$ without exception. See, for instance, J.-P. Serre, *Local fields*, Chapter VII.

For the rest of this appendix we let U be a *subgroup* and N a *normal subgroup* of G. Any G-module X is also a U-module in a natural way. If $U = N$ is normal, the fixed module

$$X^N \text{ is an } G/N\text{-module in a natural way.}$$

For a function $x : (\sigma_1, \dots, \sigma_n) \mapsto x_{\sigma_1, \dots, \sigma_n}$ from G^n in X, denote by $\mathrm{res}(x)$ the set-theoretical restriction of x to U^n; for a function $x : (G/N)^n \to X$, denote by $\mathrm{inf}(x)$ the function from G^n to X defined by $(\sigma_1, \dots, \sigma_n) \mapsto x_{\sigma_1 N, \dots, \sigma_n N}$. Of course the homomorphisms res and inf take n-cocycles to n-cocycles and n-coboundaries to n-coboundaries, for each n (however these may be defined in the case $n \geq 3$). Thus res and inf lead to homomorphisms of the same name in cohomology:

$$\mathrm{res} : H^n(G, X) \to H^n(U, X) \qquad \text{(restriction)}, \qquad (28^*)$$
$$\mathrm{inf} : H^n(G/N, X^N) \to H^n(G, X) \qquad \text{(inflation)}. \qquad (29^*)$$

As to fixed modules, let

(30*) $\text{res}: X^G \to X^U$ and $\text{inf}: (X^N)^{G/N} \to X^G$

be respectively the inclusion of X^G in X^U and the identity on $(X^N)^{G/N} = X^G$. In the case of *finite* groups, $\text{res}: X^G \to X^U$ gives rise to a homomorphism

(31*) $\text{res}: H^0(G, X) \to H^0(U, X),$

but this does not work for inf. For $n = 1$, incidentally,

(32*) $1 \to H^1(G/N, X^N) \xrightarrow{\text{inf}} H^1(G, X) \xrightarrow{\text{res}} H^1(N, X)$

is *exact*, as can be checked without much effort. No such statement holds for H^2 in general, but we remark that *under the assumption* $H^1(N, X) = 1$, the sequence

(33*) $1 \to H^2(G/N, X^N) \xrightarrow{\text{inf}} H^2(G, X) \xrightarrow{\text{res}} H^2(N, X)$

is *exact*; compare this with (47) on page 194, as well as F4* below.

Now we require that the subgroup U of G have finite index:

(34*) $k := G : U < \infty.$

Denote by R a set of representatives of the cosets σU of G mod U. Given $\rho \in R$ and $\sigma \in G$, there is a unique ρ_σ in R such that $\sigma \rho U = \rho_\sigma U$; in other words,

(35*) $\sigma\rho = \rho_\sigma \sigma_\rho$ with unique $\rho_\sigma \in R$, $\sigma_\rho \in U$.

Given x in X^U and a coset $\nu = U\sigma$, the element x^ν is well defined via $x^\nu = x^\sigma$. We then set

(36*) $\text{cor}(x) := \prod_{\nu \in G/U} x^\nu = \prod_{\rho \in R} x^{\rho^{-1}}.$

(As ρ runs over R, $U\rho^{-1}$ runs over the set G/U of cosets ν.) Clearly $\text{cor}(x)$ lies in X^G. The homomorphism

(37*) $\text{cor}: X^U \to X^G$

is called *corestriction*. If G is *finite* and $x = N_U y$, with $y \in X$, we obviously have $\text{cor}(x) = N_G y$, so (37*) gives rise to a homomorphism

(38*) $\text{cor}: H^0(U, X) \to H^0(G, X).$

In view of (36*), the notation

(39*) $\text{cor} =: N_{G/U}$

is also appropriate. Incidentally, note the formula $N_{G/U} \circ N_U = N_G$, already used above. For $x \in X^G$ we have $\mathrm{cor}(x) = x^{G:U} = x^k$, so

$$(40^*) \qquad\qquad \mathrm{cor(res\ } x) = x^{G:U} = x^k$$

for every $x \in X^G$ and hence also, if G is finite, for every $x \in H^0(G, X)$.

Since the higher cohomology groups H^n arise in some sense from the construction of fixed modules, it is to be expected that there are natural homomorphisms

$$(41^*) \qquad\qquad \mathrm{cor} : H^n(U, X) \to H^n(G, X)$$

also for $n \geq 1$, and that equation (40^*) holds for $x \in H^n(G, X)$. This is indeed the case, and is a fact of great importance. To provide a proof for it here would take us too far afield, but we will handle the cases $n = 1$ and $n = 2$ by explicit computation. For a 1-cocycle $x = x(\sigma)$ of U in X, define $\mathrm{cor}(x)$ by

$$(42^*) \qquad\qquad \mathrm{cor}(x)(\sigma) = \prod_{\rho \in R} x(\sigma_\rho)^{\rho^{-1}},$$

where σ_ρ is defined by (35^*). For a 2-cocycle $x = x(\sigma, \tau)$ of U in X, define $\mathrm{cor}(x)$ by

$$(43^*) \qquad\qquad \mathrm{cor}(x)(\sigma, \tau) = \prod_{\rho \in R} x\big((\sigma\tau)_\rho \tau_\rho^{-1}, \tau_\rho\big)^{\rho^{-1}}.$$

In analogy with (35^*) we have $\tau\rho = \rho_\tau \tau_\rho$. Multiplication by σ yields $\sigma\tau\rho = \sigma\rho_\tau \tau_\rho = (\rho_\tau)_\sigma \sigma_{\rho_\tau} \tau_\rho$, hence

$$(44^*) \qquad\qquad (\rho_\tau)_\sigma = \rho_{\sigma\tau}, \quad (\sigma\tau)_\rho = \sigma_{\rho_\tau} \tau_\rho, \quad \sigma\rho_\tau = \rho_{\sigma\tau} \sigma_{\rho_\tau}.$$

With this one shows without too much trouble that (42^*) and (43^*) do indeed define cocycles, and do give rise to homomorphisms $\mathrm{cor} : H^1(U, X) \to H^1(G, X)$ and $\mathrm{cor} : H^2(U, X) \to H^2(G, X)$. One can also check by computation that these maps are independent of the choice of the set R of representatives.

F3*. *Let X be a G-module, and let U be a subgroup of finite Index in G. Then*

$$(45^*) \qquad\qquad \mathrm{cor(res\ } x) = x^{G:U}$$

on $H^1(G, X)$ and $H^2(G, X)$, as well as on X^G (or $H^0(G, X)$ for G finite). In particular, if G is finite, the elements of the cohomology groups $H^i(G, X)$, for $i = 0, 1, 2$, satisfy

$$(46^*) \qquad\qquad x^{G:1} = 1.$$

Proof. Take x in $C^1(G, X)$ or $C^2(G, X)$, and set

$$y = \mathrm{cor(res\ } x).$$

(a) Consider first $x \in C^1(G, X)$. For notational simplicity, define

$$a = \prod_{\rho} x(\rho^{-1}).$$

Now we know from (35*) that

$$x(\sigma_\rho)^{\rho^{-1}} x(\rho^{-1}) = x(\sigma_\rho \rho^{-1}) = x(\rho_\sigma^{-1}\sigma) = x(\rho_\sigma^{-1})^\sigma x(\sigma).$$

If we form the product over all ρ, we get

$$y(\sigma)a = a^\sigma x(\sigma)^k, \quad \text{so } y(\sigma) = x(\sigma)^k a^{\sigma-1}.$$

Thus y and x^k coincide up to a coboundary.

(b) Now suppose $x \in C^2(G, X)$. Again we introduce abbreviations:

$$a(\sigma) = \prod_{\rho} x(\sigma_\rho, \rho^{-1}), \quad b(\sigma) = \prod_{\rho} x(\rho_\sigma^{-1}, \sigma).$$

By (20*) we have

$$x((\sigma\tau)_\rho \tau_\rho^{-1}, \tau_\rho)^{\rho^{-1}} x((\sigma\tau)_\rho, \rho^{-1}) = x((\sigma\tau)_\rho \tau_\rho^{-1}, \tau_\rho \rho^{-1}) x(\tau_\rho, \rho^{-1})$$
$$= x(\sigma_{\rho_\tau}, \rho_\tau^{-1}\tau) x(\tau_\rho, \rho^{-1}).$$

Taking the product over all ρ yields

(47*) $$\qquad y(\sigma, \tau) a(\sigma\tau) = a(\tau) \prod_{\rho} x(\sigma_{\rho_\tau}, \rho_\tau^{-1}\tau).$$

Again from (20*) we get $x(\sigma_{\rho_\tau}, \rho_\tau^{-1}\tau)x(\rho_\tau^{-1}, \tau) = x(\sigma_{\rho_\tau}, \rho_\tau^{-1})^\tau x(\sigma_{\rho_\tau}\rho_\tau^{-1}, \tau) = x(\sigma_{\rho_\tau}, \rho_\tau^{-1})^\tau x(\rho_{\sigma\tau}^{-1}\sigma, \tau)$. Taking the product over all ρ and using (47*) we deduce

(48*) $$\qquad y(\sigma, \tau) a(\sigma\tau) = a(\tau) a(\sigma)^\tau b(\tau)^{-1} \prod_{\rho} x(\rho_{\sigma\tau}^{-1}\sigma, \tau).$$

Another application of (20*) yields

$$x(\rho_{\sigma\tau}^{-1}, \sigma)^\tau x(\rho_{\sigma\tau}^{-1}\sigma, \tau) = x(\rho_{\sigma\tau}^{-1}, \sigma\tau) x(\sigma, \tau).$$

Again taking the product over ρ we finally get, using (48*) and (44*),

$$y(\sigma, \tau) a(\sigma\tau) = a(\tau) a(\sigma)^\tau b(\tau)^{-1} b(\sigma\tau) x(\sigma, \tau)^k b(\sigma)^{-\tau},$$

and we see that y and x^k differ only by a coboundary. $\qquad\square$

Remark. If G is a *finite* group and U is a subgroup of G, there exist maps res and cor on H^i for every $i \in \mathbb{Z}$, and equalities (45*) and (46*) hold for all i; see Serre, *Local fields*, Chapter VII.

We also mention that formula (45*) can be applied to get a simple alternative proof for Theorem 3 (page 194).

The construction of $H^1(G, X)$ also works when X is a (not necessarily abelian) group on which the group G acts by automorphisms: We define the 1-cocycles of G in X via (13*) and take the equivalence classes defined by the *first* equation in (14*). But in this case $H^1(G, X)$ is generally not a group anymore.

As an example we take a finite Galois extension L/K with group G and make G act in the natural way on

$$X = \mathrm{GL}(n, L).$$

For $\sigma \in G$, denote by v_σ the self-map of L^n corresponding to σ. Then v_σ lies in $\mathrm{End}_K(L^n)^\times$, and

$$v_\sigma^{-1} x v_\sigma = x^\sigma \quad \text{for } x \in \mathrm{GL}(n, L).$$

Now let $\sigma \mapsto x_\sigma$ be any 1-cocycle of G in X, and consider the elements

$$u_\sigma := v_\sigma x_\sigma \quad \text{in } \mathrm{End}_K(L^n).$$

Any λ in L satisfies $\lambda u_\sigma = u_\sigma \lambda^\sigma$, and since $x_{\sigma\tau} = x_\sigma^\tau x_\tau$, we have $u_\sigma u_\tau = u_{\sigma\tau}$. Thus the subalgebra Γ of $\mathrm{End}_K(L^n)$ generated by the u_σ and the $\lambda \in L$ is the crossed product of L and G with cocycle $c_{\sigma,\tau} = 1$. Similarly, the v_σ and L generate a subalgebra Γ' of $\mathrm{End}_K(L^n)$, which is isomorphic to Γ. By the *Skolem–Noether Theorem* there exists $a \in \mathrm{End}_K(L^n)^\times$ such that $a\lambda a^{-1} = \lambda$ and $av_\sigma a^{-1} = u_\sigma$. It follows that $a \in \mathrm{GL}(n, L)$ and $x_\sigma = a^\sigma a^{-1}$. Thus we recover a famous result, which made an earlier appearance in problem §13.4 of volume I (and is also equivalent to the result in §8.21):

F4* (Hilbert's Theorem 90). *For every finite Galois extension L/K with group G we have*

$$(49^*) \qquad\qquad H^1(G, \mathrm{GL}(n, L)) = 1.$$

In particular, $H^1(G, L^\times) = 1$.

Remark. If G is a *profinite group*, as we have said, the G-module M is required to satisfy the continuity condition (4*), and in the definition of $H^n(G, M)$ we tacitly restrict consideration to *continuous* cocycles and coboundaries. (For the map (15*) continuity is automatically fulfilled.) The results in F1*–F3* then hold correspondingly (though in F2* we still need the restriction to G finite when $i = -1$ and $i = 0$). Also F4* remains valid for any (infinite) Galois extension L/K.

If K is a field, let K_s denote the separable closure of K in an algebraic closure C of K. Looking at the Galois group $G_K := G(K_s/K)$ as a profinite group, we get from the arguments in this chapter a *canonical isomorphism*

$$f_K : \mathrm{Br}\, K \to H^2(G_K, K_s^\times),$$

making the following diagram commutative for any finite Galois subextension L/K of K_s/K:

$$(50^*)$$

$$
\begin{array}{ccc}
\mathrm{Br}\, L/K & \longrightarrow & \mathrm{Br}\, K \\
{\scriptstyle f_{L/K}}\downarrow & & \downarrow{\scriptstyle f_K} \\
H^2(G(L/K), L^\times) & \overset{\mathrm{inf}}{\longrightarrow} & H^2(G_K, K_s^\times)
\end{array}
$$

Definition 4*. For every finite *separable* field extension E/K, the *algebra-theoretic corestriction* $\mathrm{cor}_{E/K} : \mathrm{Br}\, E \to \mathrm{Br}\, K$ is the unique homomorphism making the diagram

(51*)

$$
\begin{array}{ccc}
\mathrm{Br}\, E & \xrightarrow{\;\mathrm{cor}_{E/K}\;} & \mathrm{Br}\, K \\
f_E \downarrow \simeq & & \simeq \downarrow f_K \\
H^2(G_E, E_s^\times) & \xrightarrow{\;\mathrm{cor}\;} & H^2(G_K, K_s^\times)
\end{array}
$$

commute, where we can take $E_s = K_s$ since E/K is separable.

On the other hand, if E/K is *purely inseparable*, G_E can be identified with G_K. If we replace the lower horizontal arrow in (51*) by the homomorphism coming from the map $E_s^\times \to K_s^\times$ of raising to the p^i-th power, where $p^i = E : K$, we obtain an algebra-theoretic corestriction in the purely inseparable case all well.

Finally, for an *arbitrary finite* extension E/K, we define $\mathrm{cor}_{E/K}$ as

(52*)
$$
\mathrm{cor}_{E/K} = \mathrm{cor}_{F/K} \circ \mathrm{cor}_{E/F},
$$

where F denotes the separable closure of K in E. One establishes easily that the equality

(53*)
$$
(\mathrm{cor}_{E/K} \circ \mathrm{res}_{E/K})(x) = x^{E:K}
$$

still holds for all $x \in \mathrm{Br}\, K$. One also checks without difficulty that (52*) is valid for *any* intermediate field F of the finite extension E/K.

Our final result of this chapter is useful in a variety of contexts:

F5* (Projection formula). *Let K'/K be any finite field extension. For finite **cyclic** extensions L/K and L'/K', where $L' = LK'$, let σ and σ' be generators of $G(L/K)$ and $G(L'/K')$, respectively. Then, if σ' is chosen so that*

(54*)
$$
\sigma' = \sigma^r \quad \text{on } L, \quad \text{where } r = K' \cap L : K,
$$

we have for each nonzero $b \in K'$ the formula

(55*)
$$
\mathrm{cor}_{K'/K}[L', \sigma', b] = [L, \sigma, N_{K'/K} b].
$$

This can be proved by resorting—for K'/K separable—to the explicit description (43*) of the cohomological corestriction. The computation is notationally laborious and is left to the reader as an exercise.

(*Hint:* Consider first the case $K' \subseteq L$ and use formula (71) on page 199. The case where K'/K is purely inseparable is obvious, and the case where K'/K is separable and $K' \cap L = K$ is not difficult.)

Remark. Let G be a group and U a subgroup of finite index in G. With the notations of (35*), define a map Cor from G into the abelianization $U^{\mathrm{ab}} = U/U'$ of U (here U' is generated by the commutators in U) by setting

(56*)
$$
\mathrm{Cor}(\sigma) = \prod_\rho \sigma_\rho \bmod U'.
$$

It is easy to prove, using (44*), that Cor does not depend on the choice of R and is a *homomorphism*. It is called the *group-theoretic corestriction* from G to U. This notion has long played an important role in group theory, and is sometimes known by its German name *Verlagerung*.

31

The Brauer Group of a Local Field

1. In this chapter we show how one can reach a full determination of the Brauer group of a local field K. We have already seen that for $K = \mathbb{R}, \mathbb{C}$ the Brauer group has two and one elements, respectively. By contrast, when K is nonarchimedean the answer turns out to be richer, and yet still manageable enough: Br K is isomorphic to the group of all roots of unity in $\mathbb{C}$! More precisely, one can produce a well determined isomorphism inv_K from Br K onto the group $\mathbb{Q}/\mathbb{Z}$, taking each $[A]$ in Br K to its so-called *Hasse invariant*. We will show further that the Hasse invariant behaves as simply as one could wish under change of base, and we will deduce hence that whether an extension of K is a splitting field of $[A]$ depends solely on its degree over K.

We will assume throughout that

$$K \text{ is a local field with normalized valuation } w_K.$$

Recall what this means: The valuation w_K is such that

  (a) K is complete with respect to w_K,

  (b) the valuation group of K satisfies $w_K(K^\times) = \mathbb{Z}$, and

  (c) the residue field $\overline{K}$ of K with respect to w_K is finite.

To investigate Br K, it is convenient to generalize the notion of absolute values from fields do division algebras. This can be done forthrightly. Since we will be dealing only with nonarchimedean values, we can stick to the framework of *valuations*:

Definition. A *valuation on a division algebra* D is a map $v : D^\times \to \mathbb{R}$ with the following properties:

  (i) $v(ab) = v(a) + v(b)$.

  (ii) $v(a + b) \geq \min(v(a), v(b))$.

Remarks. (a) A homomorphism $v : D^\times \to \mathbb{R}$ is a valuation of D if and only if it has the property

  (ii') $v(x + 1) \geq \min(v(x), 0)$.

(b) Let v be a valuation on D. If we set $v(0) = \infty$, conditions (i) and (ii) still hold for all a, b in D, with the appropriate interpretation.

Thus, by choosing any real constant $0 < c < 1$ and setting

(1)
$$|x| = c^{v(x)},$$

we obtain a *nonarchimedean absolute value* on D, that is,

$$|x| = 0 \iff x = 0, \qquad |xy| = |x|\,|y|, \qquad |x + y| \le \max(|x|, |y|).$$

Just as in the case of fields, this makes D into a metric space, and we can talk about convergence, Cauchy sequences, and so on, all with respect to v, since the choice of c in (1) makes no difference.

Building on Theorem 4 of Chapter 23, about the unique extensibility of complete absolute values, we obtain without difficulty:

F1. *Let D be a division algebra with center K and suppose $D : K < \infty$. The normalized valuation w_K can be uniquely extended to a valuation v of D, and D is complete with respect to v. If we set $N = N_{D/K}$ and $D : K = n^2$ we have*

(2)
$$v(x) = \frac{1}{n^2} w_K(Nx).$$

Proof. We define v through (2) and show that it satisfies properties (i) and (ii$'$) of a valuation. Property (i) is clear since N is multiplicative. Now take a maximal subfield L of D. For $z \in L$ we have $Nz = N_{D/K}z = (N_{L/K}z)^{D:L}$; since $D : L = n$, this implies

(3)
$$v(z) = \frac{1}{n} w_K(N_{L/K}z),$$

that is, on L the map v defined by (2) coincides with the unique extension of w_K to a valuation on L. But since any element x of D lies in *some* maximal subfield L, this proves (ii$'$), because on L the map v does amount to a valuation, as we've seen. It also follows that v is unique. Finally, D is *complete* with respect to v because D has finite dimension over K (see F10 in Chapter 23). $\qquad\square$

Remark. It is clear from the proof of F1 that the result in fact holds for *any complete valuation w_K on a field K*. The next theorem, too, does not require the full force of the assumptions made about w_K: instead of condition (c) postulated at the start of the chapter, it suffices that $\overline{K}$ be *perfect*.

Theorem 1. *Take $[D] \in \operatorname{Br} K$, where D is a division algebra. Among the maximal subfields of D there exists L such that L/K is unramified. Thus every central-simple K-algebra A possesses unramified splitting fields over K of degree $s(A)$.*

Proof. We start just as in the proof of F21 in Chapter 29, with "separable" in lieu of "unramified": Take a maximal element L in the set of subfields of D containing K

that are unramified over K. It is enough to show that L coincides with its centralizer C in D. We know that C is a division algebra in any case, and that L is the center of C. By the maximality of L, the desired equality $C = L$ follows from the next result. $\square$

Lemma 1. *Take* $[D] \in \mathrm{Br}\, K$, *where* D *is a division algebra. If every subfield* $F \supsetneqq K$ *of* D *is ramified over* K, *then* $D = K$.

Proof. Let x be any element of D such that $v(x) \geq 0$, and consider the subfield $E = K[x]$ of D. If $\overline{E} \neq \overline{K}$, there exists by Theorem 3(iv) in Chapter 24 an intermediate field F of E/K such that F/K is unramified of degree $\overline{E} : \overline{K} \neq 1$, in contradiction with the assumption. (Note that $\overline{E}/\overline{K}$ is separable because $\overline{K}$ is perfect.) Hence

(4) $$\overline{E} = \overline{K}.$$

Now let Π be a prime element of D, meaning that $v(\Pi) > 0$ is minimal in $v(D^{\times}) \cap \{u \in \mathbb{R} : u > 0\}$. Because of (4), there exists $a \in K$ such $v(x - a) > 0$. Thus x has a representation

$$x = a + x_1 \Pi, \quad \text{with } v(x_1) \geq 0.$$

Applying the same reasoning to x_1 instead of x, we write $x_1 = a_1 + x_2 \Pi$ with $v(x_2) \geq 0$, that is,

$$x = a + a_1 \Pi + x_2 \Pi^2,$$

and continue in this way to define inductively a sequence (a_n) of elements *in the valuation ring of* K such that

(5) $$x = a_0 + a_1 \Pi + a_2 \Pi^2 + \cdots + a_{n-1} \Pi^{n-1} + x_n \Pi^n,$$

where $x_n \in D$ with $v(x_n) \geq 0$. Each partial sum

(6) $$\sum_{i=0}^{n} a_i \Pi^i$$

lies in the subfield $K[\Pi]$ of D. But $K[\Pi]$ is *complete* with respect to v, so the partial sums (6) converge in $K[\Pi]$; and because of (5), the limit can only be x itself. Hence $x \in K[\Pi]$. Now, x was any element of D with $v(x) \geq 0$, and any $y \in D$ can be multiplied by some positive power Π^i of Π so that $v(y\Pi^i) \geq 0$; therefore any $y \in D$ also belongs to $K[\Pi]$. Thus $D = K[\Pi]$ is commutative, and therefore $D = K$. $\square$

As an immediate consequence of Theorem 1 we have:

F2. *The Brauer group of* K *satisfies*

(7) $$\mathrm{Br}\, K = \bigcup_{\substack{L/K \\ \text{unramified}}} \mathrm{Br}(L/K),$$

that is, it is the union of the relative Brauer groups $\mathrm{Br}(L/K)$ *of (finite) unramified extensions* L/K. (Here the L are regarded as subfields of a fixed algebraic closure of K, as usual.)

Since unramified extensions of local fields are *cyclic* (by Theorem 4(iii) in Chapter 24), Theorem 1 implies that any $[A]$ in $\mathrm{Br}\, K$ has a splitting field L for which L/K is cyclic of degree $L : K = s(A)$. Even more is true:

F3. *Every central-simple algebra A over a (nonarchimedean) local field K is cyclic.*

Proof. Set $A : K = n^2$, $s = s(A)$, $r = l(A)$. Then

$$(8) \qquad\qquad\qquad n = rs,$$

by (36) in Chapter 29. For every natural number m let K_m be the unramified extension of degree m over K (in a fixed algebraic closure of K; see Theorem 4 in Chapter 24). By Theorem 1, K_s is a splitting field of A. But from (8) we know that K_s is a subfield of K_n; thus

$$L := K_n,$$

too, is a splitting field of A. Hence A is similar to a cyclic crossed product Γ with maximal subfield L (Theorem 4 in Chapter 30). By looking at dimensions we see that A is in fact isomorphic to Γ, so A is itself a cyclic algebra. $\qquad\square$

2. The introduction of valuations on division algebras has been useful so far only in providing the quickest path to the prominent result in Theorem 1. Here we would like to take a side trip and go over some observations that, although not strictly necessary to the later development of the subject, are of interest in themselves.

Assumptions and notation. Let a field K be given, together with a *complete* and *discrete* valuation w_K; suppose also that $w_K(K^\times) = \mathbb{Z}$, which entails no loss of generality. Let D be a central division algebra over K, of finite dimension

$$(9) \qquad\qquad\qquad D : K = n^2.$$

Denote by v the unique extension of w_K to a valuation on D (see remark after F1). Let $A = \{x \in D \mid v(x) \geq 0\}$ be the *valuation ring* of v, and choose a *prime element* Π of v. Clearly,

$$(10) \qquad\qquad\qquad \bar{D} := A/\Pi A$$

is a *division algebra*, and is also an algebra over $\bar{K} = R/\pi$, the residue field of K. We will soon see (Lemma 2) that

$$(11) \qquad\qquad\qquad f := \bar{D} : \bar{K}$$

is finite, so after setting

$$(12) \qquad\qquad f_0^2 = \bar{D} : Z(\bar{D}), \quad c = Z(\bar{D}) : \bar{K}$$

we get the equation

(13) $$f = f_0^2 c.$$

Finally, we denote by $e = v(D^\times) : v(K^\times)$ the *ramification index* of v with respect to w_K.

Lemma 2. *In the situation just described, A is finitely generated as an R-module.*

Proof. There exists a K-basis $\alpha_1, \ldots, \alpha_m$ of D consisting only of elements of A. Any $x \in A$ has a unique representation

$$x = x_1 \alpha_1 + \cdots + x_m \alpha_m, \quad \text{with } x_i \in K.$$

Multiply from the right by α_j and apply the reduced trace map $\mathrm{Tr} := \mathrm{Tr}_{D/K}^0$ (see Chapter 29, Definition 11) to obtain the equations

$$\mathrm{Tr}(x\alpha_j) = \sum_{i=1}^m x_i \, \mathrm{Tr}(\alpha_i \alpha_j) \quad \text{for } 1 \le j \le m.$$

Now, for every $y \in A$ we have $\mathrm{Tr}(y) \in R$; see for example (98) in Chapter 29. Since Tr is nondegenerate (Chapter 29, F26), we have $\delta := \det(\mathrm{Tr}(\alpha_i \alpha_j)_{i,j}) \ne 0$, and we can apply *Cramer's rule* to conclude that all the x_i lie in $\delta^{-1} R$. Thus

(14) $$A \subseteq \delta^{-1}(R\alpha_1 + \cdots + R\alpha_m)$$

is a submodule of a finitely generated R-module, and hence is itself finitely generated. □

Since the R-module on the right-hand side of (14) is a free R-module, one concludes that A is likewise a free R-module. Comparing dimensions we see that A has an R-basis with $m = n^2$ elements.

F4. *Let the setup be as above. If $\alpha_1, \ldots, \alpha_f$ is a set of representatives of a $\overline{K}$-basis of $\overline{D}$, the elements*

(15) $$\alpha_i \Pi^j, \quad \text{where } 1 \le i \le f \text{ and } 0 \le j < e,$$

form an R-basis of A, and hence also a K-basis of D. In particular,

(16) $$n^2 = ef.$$

Proof. Let N be the R-submodule of A generated by the elements of the form (15). One checks easily that

$$A = N + \pi A.$$

Now, the R-module A is finitely generated by Lemma 2, so *Nakayama's Lemma* (F32 in Chapter 28) applies, with $\Re(R) = \pi R$, to yield $A = N$. Thus the elements (15) certainly span the R-module A. To show that a sum

$$\sum_{i,j} a_{ij} \alpha_i \Pi^j = \sum_j \left(\sum_i a_{ij} \alpha_i \right) \Pi^j, \quad \text{where } a_{ij} \in R,$$

can only vanish if all the a_{ij} vanish, work as in the proof of F1 in Chapter 24. □

Lemma 3. *In the preceding situation, assume further that $\overline{K}$ is perfect. Then*

$$(17) \qquad\qquad n = f_0 c.$$

Proof. We consider maximal subfields Λ of the division algebra $\overline{D}$. Every such Λ must contain the center $Z(\overline{D})$ of $\overline{D}$, and hence (by Chapter 29, Theorem 16) have dimension

$$(18) \qquad\qquad \Lambda : \overline{K} = f_0 c$$

over $\overline{K}$, in the notation of (12). Since $\overline{K}$ is perfect, Λ has the form $\Lambda = \overline{K}[\bar{x}]$, with $x \in A$. The subfield $K[x]$ of D then satisfies

$$\overline{K}[\bar{x}] : \overline{K} \leq \overline{K[x]} : \overline{K} \leq K[x] : K \leq n,$$

and we get from (18) the bound $f_0 c \leq n$. By Theorem 1 and its preceding remark, D contains a maximal subfield L that is unramified over K. Hence $n = L : K = \overline{L} : \overline{K} \leq f_0 c$, because $\overline{L}$ is contained in some appropriate Λ. □

F5. *In the preceding situation (with $\overline{K}$ perfect), we have*

$$(19) \qquad\qquad n = e f_0.$$

Proof. Immediate from (16), (13) and (17). □

F6. *Let K be a nonarchimedean local field and D a central division algebra over K, of dimension n^2. Then*

$$(20) \qquad\qquad e = f = n,$$

where e is the ramification index of D and $f = \overline{D} : \overline{K}$ its residue class degree.

Proof. $\overline{K}$ is perfect because it is finite. By *Wedderburn's Theorem* (Chapter 29, Theorem 21), $\overline{D}$ is commutative; that is, $f_0 = 1$ and $c = f$ in (12). The conclusion then says the same as (19) and (17). □

3. With Theorem 1 and its corollary, equation (7), we have taken the first step in the determination of the Brauer group of a local field K. There remains the calculation of $\mathrm{Br}(L/K)$ for L/K unramified. Every unramified extension L/K is *cyclic* (Theorem 4 in Chapter 24), so by Theorem 4 in the previous chapter we have

$$(21) \qquad\qquad \mathrm{Br}(L/K) \simeq K^{\times}/N_{L/K}(L^{\times}),$$

and our problem boils down to understanding the group $K^{\times}/N_{L/K}(L^{\times})$, called the *norm residue group* of L/K.

Let π be a *prime element* of K. The multiplicative group $K^{\times}$ is the direct product

$$(22) \qquad\qquad K^{\times} = \langle \pi \rangle \times U_K$$

of the infinite cyclic group $\langle \pi \rangle$ generated by π with the group U_K of *units* in K: Every $a \in K^\times$ has a unique representation $a = \pi^i x$, with $i \in \mathbb{Z}$ and $x \in U_K$ (so $w_K(x) = 0$); moreover, $i = w_K(a)$. Since L/K is *unramified*, π is a prime of L as well. Similarly to (22), we have

$$(23) \qquad\qquad L^\times = \langle \pi \rangle \times U_L,$$

where U_L is the group of units of L. If $N = N_{L/K}$ is the norm map of L/K, we have

$$(24) \qquad\qquad NL^\times = \langle \pi^n \rangle \times NU_L, \quad \text{where} \quad n = L : K.$$

The product in (24) really is *direct*, because $NU_L \subseteq U_K$. From (22) and (24) we get the following isomorphism for the norm residue group in (21):

$$(25) \qquad\qquad K^\times / NL^\times \simeq \langle \pi \rangle / \langle \pi^n \rangle \times U_K / NU_L.$$

The first factor on the right is a cyclic group of order n. Regarding the second, we have (for unramified extensions) a simple yet fundamental characterization:

Theorem 2. *If L/K is an* **unramified** *extension of local fields, every unit in K is a norm of L:*

$$(26) \qquad\qquad U_K = NU_L.$$

Before we prove this key arithmetic fact, we observe that it can be reformulated as follows, thanks to (25):

Theorem 2′. *If L/K is an* **unramified** *extension of local fields, the norm residue group $K^\times / NL^\times$ is cyclic of order $n = L : K$, and is generated by $\pi NL^\times$, where π is any prime of K.*

Proof of Theorem 2. By Theorem 3(v) in Chapter 24, there is an isomorphism $\sigma \mapsto \bar{\sigma}$ from the Galois group $G = G(L/K)$ to that of the extension $\bar{L}/\bar{K}$ of residue fields, given by

$$(27) \qquad \bar{\sigma}(\bar{y}) = \overline{\sigma(y)} \quad \text{for any } y \in L \text{ such that } w_L(y) \geq 0.$$

Accordingly, the norms $N = N_{L/K}$ and $\bar{N} = N_{\bar{L}/\bar{K}}$ and the traces $\mathrm{Tr} = \mathrm{Tr}_{L/K}$ and $\overline{\mathrm{Tr}} = \mathrm{Tr}_{\bar{L}/\bar{K}}$ satisfy

$$(28) \qquad\qquad \bar{N}(\bar{y}) = \overline{N(y)}, \quad \overline{\mathrm{Tr}}(\bar{y}) = \overline{\mathrm{Tr}(y)}.$$

We know that the norm is surjective for an extension of finite fields (see (74) in Chapter 30), and that the trace is surjective for any finite separable extension (see F6 in Chapter 13, vol. I). Thus

$$(29) \qquad\qquad \bar{N} \text{ and } \overline{\mathrm{Tr}} \text{ are surjective.}$$

Using (29) and (28) for the norm first, we conclude that for any $x \in U_K$ there exists $y \in U_L$ such that

$$(30) \qquad\qquad x \equiv Ny \bmod \pi,$$

where π is a prime of K (hence of L). We will now approximate an arbitrary $x \in U_K$ by norms, starting with (30). To do this we use the fact that for every y in the valuation ring R_L of L — and every $n \in \mathbb{N}$ — we have a congruence

$$(31) \qquad\qquad N(1 + \pi^n y) \equiv 1 + \mathrm{Tr}(\pi^n y) \bmod \pi^{n+1}.$$

To show this we write $N(1 + \pi^n y) = \prod_{\sigma \in G}(1 + \pi^n y)^\sigma = \prod_{\sigma \in G}(1 + \pi^n y^\sigma) \equiv 1 + \sum_{\sigma \in G} \pi^n y^\sigma \bmod \pi^{2n}$, which, as a result of the equality $\mathrm{Tr}(\pi^n y) = \pi^n \mathrm{Tr}(y)$, implies that the congruence in (31) is actually true $\bmod \pi^{2n}$.

To construct for a given $x \in U_K$ a sequence (y_n) of approximants in U_L with

$$(32) \qquad\qquad x \equiv N(y_n) \bmod \pi^n, \quad y_n \equiv y_{n-1} \bmod \pi^{n-1},$$

we proceed by induction. The basis of the induction is (30). Now suppose (32) holds for some $n \geq 1$. Since all the elements involved are units, we have $x/Ny_n \equiv 1 \bmod \pi^n$, so there exists z in the valuation ring R_K of K such that

$$(33) \qquad\qquad x = N(y_n)(1 + \pi^n z).$$

Applying (29) and (28) to the trace, we find $y \in R_L$ such that $z \equiv \mathrm{Tr}(y) \bmod \pi$. Using (31), we then get

$$1 + \pi^n z \equiv 1 + \pi^n \mathrm{Tr}(y) \equiv N(1 + \pi^n y) \bmod \pi^{n+1},$$

so equation (33) leads to

$$x \equiv N(y_n) N(1 + \pi^n y) \bmod \pi^{n+1}.$$

Since $N(y_n) N(1 + \pi^n y) = N(y_n(1 + \pi^n y))$, we can make (32) true with $n+1$ instead of n if we set $y_{n+1} := y_n(1 + \pi^n y) \equiv y_n \bmod \pi^n$.

The result is now clear: from the second congruence in (32) we know that the sequence (y_n) converges in L; its limit y lies in U_L, because this set is closed in L. From the first congruence in (32) we obtain $x = N(y)$ by passing to the limit, because N is continuous. (The continuity of N follows for instance from (56) in Chapter 23 or F1 in Chapter 27.) $\square$

Remark. We emphasize that for an *unramified* extension L/K of local fields, the element $\pi NL^\times$ is a *canonical* generator of the cyclic group $K^\times/NL^\times$ (compare Theorem 2′). Indeed, the prime π of K was chosen arbitrarily, but since two such primes differ only by multiplication by a unit, and every unit is a norm (Theorem 2), the element $\pi NL^\times$ is independent of the choice of π.

As a result of Theorem 4 in Chapter 30, we can recast Theorem 2′ in algebra-theoretic terms:

Theorem 3. *Let the situation be as in Theorem 2', and*

let $\varphi_{L/K}$ be the Frobenius automorphism of L/K

(see Theorem 4(iii) in Chapter 24). *Then every $\alpha \in \mathrm{Br}(L/K)$ is represented by a cyclic algebra of the form*

$$(34) \qquad (L, \varphi_{L/K}, \pi^k),$$

and k is uniquely determined modulo $n = L : K$.

From Theorem 2 we can gain yet another result of some significance (although tangential to our goal of finding the Brauer group of a local field).

F7. *If K is a nonarchimedean local field and A is a central-simple K-algebra, the reduced norm $N^0 = N_{A/K}^0$ is surjective.*

Proof. By (83) in Chapter 29, it suffices to consider the case of a division algebra $A = D$. By Theorem 1, such an algebra admits a maximal subfield L unramified over K. Since N^0 agrees with $N_{L/K}$ on L, by (98) in Chapter 29, we conclude from Theorem 2 that all units in K lie in the image N^0. There remains to show that the image of N^0 contains a prime of K. Let Π be a prime of D, and take the subfield $K(\Pi)$ of D. Then the ramification index e of D/K satisfies $e \leq K(\Pi) : K \leq n$, with $n = \sqrt{D : K}$. But at the same time, $e = n$ (see F6); therefore $L := K(\Pi)$ is a maximal subfield of D. This shows that $N^0(\Pi) = N_{L/K}(\Pi)$. But $N_{L/K}(\Pi)$ is a prime of K, because $w_K(N_{L/K}(\Pi)) = nv(\Pi) = 1$. $\qquad\square$

4. With F2 and Theorem 3 the Brauer group of a local field can be regarded as being known in principle. Through careful analysis and the use of the right notions, however, the result can be summarized in a more suitable, and wholly satisfying, way. A key piece of information toward this end is that the Galois group $G(L/K)$ of an unramified extension L/K is not only cyclic, but has a *canonical* generator, the *Frobenius automorphism* $\varphi_{L/K}$.

In light of F2 and Theorem 2' we first look for a model group that has exactly one cyclic subgroup of order n, for all $n \in \mathbb{N}$, and that equals the union of all these subgroups. The group $W(\mathbb{C})$ of all roots of unity comes to mind with these properties, but it turns out to be more convenient to use instead the additive group

$$(35) \qquad \mathbb{Q}/\mathbb{Z}.$$

If two rational numbers a and b belong to the same coset of $\mathbb{Q}/\mathbb{Z}$ — in other words, if $a - b$ is an integer — we generally write

$$(36) \qquad a \equiv b \bmod 1,$$

rather than $a \equiv b \bmod \mathbb{Z}$. The coset $a + \mathbb{Z}$ of $a \in \mathbb{Q}$ is denoted by

$$(37) \qquad a \bmod 1.$$

Since $e^{2\pi i \mathbb{Z}} = 1$, the map $a \mapsto e^{2\pi i a}$ is a homomorphism from $\mathbb{Q}/\mathbb{Z}$ into $\mathbb{C}^{\times}$, and an isomorphism onto its image $W(\mathbb{C})$, so $\mathbb{Q}/\mathbb{Z}$ does have the required properties. For every $n \in \mathbb{N}$, the set

$$\text{(38)} \qquad \langle \tfrac{1}{n} \bmod 1 \rangle = \tfrac{1}{n}\mathbb{Z}/\mathbb{Z}$$

is a subgroup of order n in $\mathbb{Q}/\mathbb{Z}$ — the only one — and it is *cyclic*. If k, k' are integers, we have by definition

$$\text{(39)} \qquad \frac{k}{n} \equiv \frac{k'}{n} \bmod 1 \iff k \equiv k' \bmod n.$$

Building on Theorem 3 and F2 we then have:

Definition and proposition. Let L/K be an *unramified* extension of local fields of degree n. According to Theorem 3, every $\alpha \in \mathrm{Br}(L/K)$ is represented by a cyclic algebra of the form

$$\text{(40)} \qquad (L, \varphi_{L/K}, \pi^k),$$

where $k \bmod n$ is uniquely determined. Setting

$$\text{(41)} \qquad \mathrm{inv}_{L/K}(\alpha) = \frac{k}{n} \bmod 1$$

we obtain a homomorphism

$$\text{(42)} \qquad \mathrm{inv}_{L/K} : \mathrm{Br}(L/K) \to \mathbb{Q}/\mathbb{Z},$$

which maps $\mathrm{Br}(L/K)$ isomorphically onto the cyclic subgroup of order n of $\mathbb{Q}/\mathbb{Z}$ (which is to say, $\tfrac{1}{n}\mathbb{Z}/\mathbb{Z}$). The map $\mathrm{inv}_{L/K}$ can also be described as follows: Let

$$\text{(43)} \qquad \kappa : \mathrm{Br}(L/K) \to K^{\times}/NL^{\times}$$

be the canonical isomorphism given by Theorem 4 of Chapter 30 (canonical, that is, with reference to $\tau = \varphi_{L/K}$). Then, if $\alpha \in \mathrm{Br}(L/K)$ maps to $a \bmod NL^{\times}$ under κ, we have

$$\text{(44)} \qquad \mathrm{inv}_{L/K}(\alpha) = \frac{w_K(a)}{L:K} \bmod 1.$$

Now define a homomorphism

$$\text{(45)} \qquad \mathrm{inv}_K : \mathrm{Br}\, K \to \mathbb{Q}/\mathbb{Z}$$

on the full Brauer group as follows: Given $\alpha \in \mathrm{Br}\, K$, take (using F2) an *unramified* extension L/K (in a fixed algebraic closure of K) such that $\alpha \in \mathrm{Br}(L/K)$, and set

$$\text{(46)} \qquad \mathrm{inv}_K(\alpha) = \mathrm{inv}_{L/K}(\alpha).$$

Then $\mathrm{inv}_K(\alpha)$ is *well defined*. We call it the *Hasse invariant* of α (or of A, where $[A] = \alpha$).

Proof. Based on Theorem 3, we immediately deduce from (39) the well definedness of $\mathrm{inv}_{L/K}$ and the fact that it maps $\mathrm{Br}(L/K)$ isomorphically onto its image $\frac{1}{n}\mathbb{Z}/\mathbb{Z}$.

For $a \in K$, set $k = w_K(a)$, so that $a = \pi^k u$ for some unit u in K. By Theorem 2, we then have $a \equiv \pi^k \bmod NL^\times$; this shows (44).

There remains to show that $\mathrm{inv}_K(\alpha)$ is well defined by (46). Let L'/K be a second unramified extension with $\alpha \in \mathrm{Br}(L'/K)$. The composite LL'/K is also unramified (see for instance Theorem 4 in Chapter 24), so we can assume without loss of generality that $L \subseteq L'$. Suppose (44) holds, that is, assume α is represented in terms of the splitting field L by the crossed cyclic product

$$(47) \qquad\qquad (L, \varphi_{L/K}, a).$$

The Frobenius automorphism $\varphi_{L'/K}$ of L'/K induces on L/K the Frobenius $\varphi_{L/K}$ of L/K, so α is represented in terms of the splitting field L' by

$$(48) \qquad\qquad (L', \varphi_{L'/K}, a^{L':L}).$$

(see (87) in Chapter 30). But then we have, by definition,

$$(49) \qquad\qquad \mathrm{inv}_{L'/K}(\alpha) = \frac{w_K(a^{L':L})}{L':K} \bmod 1.$$

Since $w_K(a^{L':L}) = (L':L)w_K(a)$ and $L':K = (L':L)(L:K)$, a comparison of equation (49) with (44) yields the equality $\mathrm{inv}_{L'/K}(\alpha) = \mathrm{inv}_{L/K}(\alpha)$. $\square$

In the next, very important, theorem we will see that not only does the Hasse invariant map inv_K yield a isomorphism

$$(50) \qquad\qquad \mathrm{Br}\, K \simeq \mathbb{Q}/\mathbb{Z},$$

but it also helps describe, for any finite extension E/K, the restriction map $\mathrm{res}_{E/K}$ —the homomorphism $\mathbb{Q}/\mathbb{Z} \to \mathbb{Q}/\mathbb{Z}$ induced by $\mathrm{res}_{E/K}$ is just multiplication by the degree $E:K$ of the extension.

Theorem 4. *Let K be a (nonarchimedean) local field. The map*

$$(51) \qquad\qquad \mathrm{inv}_K : \mathrm{Br}\, K \to \mathbb{Q}/\mathbb{Z}$$

is an isomorphism. If E is a finite extension of K, every $\alpha \in \mathrm{Br}\, K$ satisfies

$$(52) \qquad\qquad \mathrm{inv}_E(\mathrm{res}_{E/K}(\alpha)) = (E:K)\,\mathrm{inv}_K(\alpha),$$

that is, the diagram

$$
(53) \qquad
\begin{array}{ccc}
\mathrm{Br}\, K & \xrightarrow{\ \mathrm{res}_{E/K}\ } & \mathrm{Br}\, E \\
{\scriptstyle \mathrm{inv}_K}\downarrow & & \downarrow{\scriptstyle \mathrm{inv}_E} \\
\mathbb{Q}/\mathbb{Z} & \xrightarrow{\ E:K\ } & \mathbb{Q}/\mathbb{Z}
\end{array}
$$

commutes, where the bottom arrow is the multiplication map by the degree $E:K$ of the extension E/K.

Proof. (a) inv_K is injective because its restriction (42) to each of the subgroups $\mathrm{Br}(L/K)$ that make up $\mathrm{Br}\,K$ is injective. To show the surjectivity of inv_K, take $n \in \mathbb{N}$ and consider an unramified extension L/K of degree n (Chapter 24, Theorem 4). Using (42) we see that inv_K maps $\mathrm{Br}(L/K)$ to $\langle \frac{1}{n} \bmod 1 \rangle$ in $\mathbb{Q}/\mathbb{Z}$. This suffices, since the elements of this form obviously generate $\mathbb{Q}/\mathbb{Z}$.

(b) For the proof of (52) we start by invoking Theorem 3(iv) in Chapter 24 (existence of a largest unramified subextension) to reduce the problem to two cases:

(i) *E/K is unramified*;   (ii) *E/K is purely ramified.*

Indeed, if L is a largest unramified extension of the finite extension E/K, the extension E/L is purely ramified, that is, it has ramification index $e(E/L) = E:L$, which is tantamount to $\bar{E} = \bar{L}$, by Theorem 1 of Chapter 24; and if (52) holds for L/K and E/L, it also holds for E/K, since $\mathrm{res}_{E/K} = \mathrm{res}_{E/L} \circ \mathrm{res}_{L/K}$ and $E:K = (E:L)(L:K)$.

(c) We take on first the case of *E/K unramified*. Given $\alpha \in \mathrm{Br}\,K$, take an unramified L/K such that $\alpha \in \mathrm{Br}(L/K)$, and assume without loss of generality that $E \subseteq L$. Let α be represented by

$$(54) \qquad \Gamma = (L, \varphi_{L/K}, \pi^k).$$

Then, by (89) in Chapter 30, $\mathrm{res}_{E/K}(\alpha) = [\Gamma \otimes_K E]$ is represented by

$$(55) \qquad (L, \varphi_{L/K}^{E:K}, \pi^k) = (L, \varphi_{L/E}, \pi^k);$$

here we have used the functorial property

$$(56) \qquad \varphi_{L/E} = \varphi_{L/K}^{E:K}$$

of the Frobenius, which is an easy consequence of its definition. By construction,

$$(57) \qquad \mathrm{inv}_E(\mathrm{res}_{E/K}(\alpha)) = \frac{k}{L:E} \bmod 1,$$

since π is a prime element E as well (E/K being unramified). At the same time,

$$(58) \qquad \mathrm{inv}_K(\alpha) = \frac{k}{L:K} \bmod 1,$$

and the desired formula (52) follows since $L:K = (E:K)(L:E)$.

(d) Now suppose we are in the case of a *purely ramified* extension E/K. As before, choose $\alpha \in \mathrm{Br}(L/K)$, where L/K is unramified. Denote by $F = EL$ the composite of E and L (in an algebraic closure as usual). Consider the diagram of fields

(59)

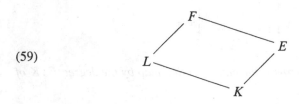

It is easy to see that F/L is purely ramified of degree $E:K$, and F/E is unramified of degree $L:K$. Using (88) of Chapter 30 we see that $\operatorname{res}_{E/K}(\alpha)$ is represented by

$$(F, \varphi_{F/E}, \pi^k)$$

if α is represented by (54) as above. Then, by definition, $\operatorname{res}_{E/K}(\alpha)$ has Hasse invariant

(60) $$\operatorname{inv}_E(\operatorname{res}_{E/K}(\alpha)) = \frac{w_E(\pi^k)}{F:E} \bmod 1.$$

(Here we have applied (44) to the extension F/E.) But $w_E(\pi^k) = k w_E(\pi) = k(E:K)$, since E/K is purely ramified. Given that $F:E = L:K$, equality (60) then says that

$$\operatorname{inv}_E(\operatorname{res}_{E/K}(\alpha)) = (E:K) \frac{k}{L:K} \bmod 1.$$

Now a comparison with (58) yields the conclusion (52). $\qquad\square$

Theorem 5. *Let E/K be a finite extension of local fields. For $[A] \in \operatorname{Br} K$, there is equivalence between*:

(i) *The exponent $e(A)$ divides $E:K$.*

(ii) *E is a splitting field of A.*

(Thus, whether E is a splitting field of A depends only on the degree of E over K!) As a consequence, *the exponent and the Schur index of A coincide*: $s(A) = e(A)$. Finally, *every central-simple algebra A over a (nonarchimedean) local field K is a cyclic algebra.*

Proof. Set $\alpha = [A]$. From the definition of $e(A) = e(\alpha)$ we see that (i) is equivalent to $\alpha^{E:K} = 1$, and then again, since inv_K is an isomorphism, to

(61) $$(E:K) \operatorname{inv}_K(\alpha) = 0.$$

Now formula (52) from Theorem 4 comes into play, for it says that (61) amounts to $\operatorname{inv}_E(\operatorname{res}_{E/K}(\alpha)) = 0$, and hence to $\operatorname{res}_{E/K}(\alpha) = 1$. This last equation says exactly that E is a splitting field of $\alpha = [A]$, so the equivalence of (i) and (ii) is proved. Of course, the implication (ii) $\Rightarrow$ (i) is good for any field K (see once more Theorem 19 in Chapter 29 and F3 in Chapter 30).

From Theorem 4 in Chapter 24 (for instance), we know there exists an extension E/K of degree $e(A)$. The first part of our theorem then says that E is a least-degree splitting field of A, so we have $s(A) = E:K = e(A)$, by Theorem 19 in Chapter 29.

To verify the last statement of the theorem, let $n = \sqrt{A:K}$ be the reduced degree of A. The Schur index $s(A)$ divides n, by (36) in Chapter 29. Hence, if L/K is the unramified degree-n extension, L is a splitting field of A. By Theorem 3, therefore, A is similar to a cyclic algebra of the form (34). A dimension comparison now shows that the two algebras are isomorphic, concluding the proof. $\qquad\square$

Theorem 6. *Let D be a division algebra of Schur index n over a local field K. For any extension E/K of degree n, the field E is isomorphic over K to a maximal subfield of D.*

Proof. By Theorem 5, any such E is a splitting field of D, and hence K-isomorphic to a maximal subfield of D (Chapter 29, Theorem 19).   □

Theorem 6 is somewhat surprising. It shows in particular that different maximal subfields of D by no means have to be isomorphic.

Theorem 7. *Let n be a natural number and let L/K be an unramified extension of degree n. If E/K is any extension of degree n we have*

$$(62) \qquad\qquad \mathrm{Br}(E/K) = \mathrm{Br}(L/K).$$

*If in addition E/K is Galois, therefore, the second cohomology group $H^2(G, E^\times)$ of the Galois group G of E/K with coefficients in E is cyclic of order n; it has a canonical generator $\gamma_{E/K}$, called the **canonical class** of E/K. If we identify $H^2(G, E^\times)$ with $\mathrm{Br}(E/K)$, the element $\gamma_{E/K}$ is uniquely characterized by the condition*

$$(63) \qquad\qquad \mathrm{inv}_K(\gamma_{E/K}) = \frac{1}{n} \bmod 1.$$

Proof. The first statement follows immediately from Theorem 5, which says that the elements of $\mathrm{Br}\,K$ that split over E are exactly the same that split over L.

If E/K is Galois, we have

$$(64) \qquad\qquad H^2(G, E^\times) \simeq \mathrm{Br}(E/K) = \mathrm{Br}(L/K),$$

the isomorphism being canonical. But as we saw above (and also as Theorem 4 implies, in almost automatic fashion), the invariant map takes $\mathrm{Br}(E/K) = \mathrm{Br}(L/K)$ isomorphically onto the cyclic subgroup $\frac{1}{n}\mathbb{Z}/\mathbb{Z}$ of $\mathbb{Q}/\mathbb{Z}$, which has order n. The element of $\mathrm{Br}(E/K)$ with invariant $\frac{1}{n} \bmod 1$ is a canonical generator of the cyclic group $\mathrm{Br}(E/K)$ and thus gives rise to a canonical generator of $H^2(G, E^\times)$, in view of (64). (In fact, the isomorphism (43) says that the canonical generator of $\mathrm{Br}(E/K) = \mathrm{Br}(L/K)$ corresponds precisely to the canonical generator $\pi N_{L/K}L^\times$ of $K^\times/N_{L/K}L^\times$ under the canonical isomorphism $\mathrm{Br}(L/K) \simeq K^\times/N_{L/K}L^\times$; see the remark before Theorem 3.)   □

Now if E/K is also assumed to be *cyclic*, Theorem 7 implies, together with Theorem 4 of Chapter 30, a result of special importance to the arithmetic of local fields, one that opens the way to *local class field theory*, the subject of Chapter 32:

Theorem 8. *If E/K is a **cyclic** extension of local fields, of degree n, the group $H^0(G, E^\times) = K^\times/N_{E/K}E^\times$ is cyclic of order n.*

Theorem 9. *For any extension E/K of nonarchimedean local fields, the map $\mathrm{res}_{E/K}: \mathrm{Br}\,K \to \mathrm{Br}\,E$ is **surjective**.*

Proof. This statement, too, follows from the all-important Theorem 4, specifically from the commutative diagram (53).   □

Again in light of Theorem 4, the next result suggests itself as a counterpoint to Theorem 9:

Theorem 10. *Let E/K be a extension of local fields. There is a unique homomorphism* Br $E \to$ Br K *such that the diagram*

(65)

$$
\begin{array}{ccc}
\mathrm{Br}\, E & \longrightarrow & \mathrm{Br}\, K \\
\Big\downarrow{\scriptstyle \mathrm{inv}_E} & & \Big\downarrow{\scriptstyle \mathrm{inv}_K} \\
\mathbb{Q}/\mathbb{Z} & \xrightarrow{\ \mathrm{id}\ } & \mathbb{Q}/\mathbb{Z}
\end{array}
$$

*commutes, namely, the algebra-theoretic **corestriction*** $\mathrm{cor}_{E/K}$ *(see Definition 4* on page 221). Thus* $\mathrm{cor}_{E/K}$ *is a isomorphism in this case.*

Proof. The existence and uniqueness of the desired map are clear. There remains to show that $\mathrm{cor}_{E/K}$ preserves the Hasse invariant. Take an arbitrary $\beta \in$ Br E. By Theorem 9 there exists $\alpha \in$ Br K such that $\beta = \mathrm{res}_{E/K}(\alpha)$. Using (52) from Theorem 4 together with (53*) from the last chapter (page 221) it follows that $\mathrm{inv}_K(\mathrm{cor}(\beta)) = \mathrm{inv}_K(\mathrm{cor} \circ \mathrm{res}(\alpha)) = \mathrm{inv}_K(\alpha^{E:K}) = [E : K]\mathrm{inv}_K(\alpha) = \mathrm{inv}_E(\mathrm{res}(\alpha)) = \mathrm{inv}_E(\beta)$. Thus, as desired,

(66) $$\mathrm{inv}_K(\mathrm{cor}_{E/K}(\beta)) = \mathrm{inv}_E(\beta).$$ □

Let L/K be a Galois extension of local fields with Galois group G. If L/K is *unramified*, then G is cyclic, and since $H^2(G, U_L) \simeq H^0(G, U_L)$ it then follows from Theorem 2 that $H^2(G, U_L) = 1$. What can one say about $H^2(G, U_L)$ if L/K is *not* assumed to be unramified?

Theorem 11. *Let L/K be a Galois extension of local fields with Galois group G. $H^2(G, U_L)$ is taken by the natural map $H^2(G, U_L) \to H^2(G, L^\times)$ isomorphically onto the subgroup of the cyclic group $H^2(G, L^\times)$ having order e, the ramification index of L/K. In particular, $H^2(G, U_L)$ is cyclic of order $e = e(L/K)$.*

Proof. We start from the exact sequence of G-modules

(67) $$1 \longrightarrow U_L \longrightarrow L^\times \xrightarrow{\ w_L\ } \mathbb{Z} \longrightarrow 0.$$

Passing to the exact sequence of cohomology groups (see F2* in Chapter 30, page 216) we obtain the exact sequence

(68) $$H^1(G, \mathbb{Z}) \longrightarrow H^2(G, U_L) \longrightarrow H^2(G, L^\times) \xrightarrow{\ w_L\ } H^2(G, \mathbb{Z}).$$

But $H^1(G, \mathbb{Z}) = \mathrm{Hom}(G, \mathbb{Z}) = 1$ (since G is finite), so $H^2(G, U_L) \to H^2(G, L^\times)$ is indeed *injective*. By Theorem 7, the group $H^2(G, L^\times)$ is cyclic of order $n = L:K$, and is generated by the canonical class $\gamma_{L/K}$. Because (68) is exact, we will be done if we can prove that

(69) $\qquad\qquad w_L(\gamma_{L/K})$ has order $f = n/e$,

where f is the residue class degree of L/K. Let M/K be the *unramified* extension of same degree n as L/K, and L_0/K the largest unramified subextension of L/K. We have the diagram of fields

(70)

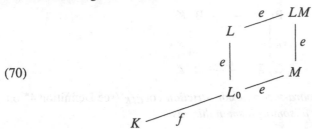

in which M/L_0 and LM/L are *unramified* of degree e, whereas L/L_0 and LM/M are *purely ramified* of degree e. Introducing the abbreviation

(71) $\qquad\qquad H^2(E/K) = H^2(G(E/K), E^\times)$

for Galois extensions E/K of finite degree, we obtain from (70) the commutative diagram

(72)

$$
\begin{array}{ccc}
H^2(L/K) & \xrightarrow{\ w_L\ } & H^2(G(L/K), \mathbb{Z}) \\
\downarrow & & \downarrow \\
H^2(LM/K) & \xrightarrow{\ w_{LM}\ } & H^2(G(LM/K), \mathbb{Z}) \\
\uparrow & & \uparrow \\
H^2(M/K) & \xrightarrow{\ e w_M\ } & H^2(G(M/K), \mathbb{Z})
\end{array}
$$

Here the vertical arrows stand for the appropriate inflation maps, which are all *injective* (see (33*) in the previous chapter). By their definition, and because of (70), the canonical classes satisfy

$$\inf(\gamma_{L/K}) = \gamma_{LM/K}^e = \inf(\gamma_{M/K}).$$

Because of (72) one then gets

(73) $\qquad\qquad \mathrm{ord}(w_L(\gamma_{L/K})) = \mathrm{ord}(e\, w_M(\gamma_{M/K})).$

But M/K is unramified; therefore $H^2(G(M/K), U_M) = H^0(G(M/K), U_M) = 1$ and hence $\mathrm{ord}(w_M(\gamma_{M/K})) = M : K = n = ef$. Now (69) follows in view of (73).   □

32

Local Class Field Theory

1. According to local class field theory, whose rudiments we will lay out in this chapter, every *abelian* extension L/K of local fields implies an associated *canonical* isomorphism

$$(1) \qquad\qquad K^\times / N_{L/K} L^\times \simeq G(L/K)$$

between its norm residue group and its Galois group. This momentous fact admits of a remarkably simple demonstration based on the algebra-theoretic results of the previous chapter. This proof, though very satisfying in several regards and dating back to *H. Hasse* and *E. Noether*, is apparently lesser known than it deserves to be.

To formulate the argument as clearly and concisely as possible, we will adopt a bit of formalism. At first we let L/K be *any finite Galois* field extension, with Galois group $G = G(L/K)$. We consider the set

$$(2) \qquad\qquad G^* = \mathrm{Hom}(G, \mathbb{Q}/\mathbb{Z})$$

of all homomorphisms $\chi : G \to \mathbb{Q}/\mathbb{Z}$ from G into the (additive) group $\mathbb{Q}/\mathbb{Z}$. This set has a natural (abelian) group structure; its elements χ are called the (*linear*) *characters* of G, and G^* is the *character group* of G. (Since G is finite, this is in agreement with the definition on page 149 of vol. I.) Any character $\chi \in G^*$ maps the *commutator subgroup* G' of G to the identity (see Problem §15.2 in vol. I), so χ can also be regarded as a character of the *abelianization*

$$(3) \qquad\qquad G^{\mathrm{ab}} := G/G',$$

that is, the *maximal abelian* quotient group of G. Conversely, any character of G^{ab} can be interpreted as a character of G. Thus we can identify G^* with $(G^{\mathrm{ab}})^*$, and will do so as a rule in the future. Incidentally, we can also view (3) as the Galois group of the *largest abelian subextension* L^{ab}/K of L/K:

$$(4) \qquad\qquad G(L/K)^{\mathrm{ab}} = G(L^{\mathrm{ab}}/K).$$

Now let $\chi \in G^*$ be given, and denote by L_χ the intermediate field of L/K corresponding to the subgroup $\ker \chi$ of G. Since $\ker \chi$ is a normal subgroup of G, the extension L_χ/K is Galois. We then have

(5) $$G(L_\chi/K) \simeq G/\ker \chi \simeq \operatorname{im} \chi,$$

with canonical isomorphisms. Consequently, L_χ/K is *cyclic* of degree equal to the order of χ.

In the sequel the term *local field* will tacitly refer only to *nonarchimedean* ones. But with a few obvious exceptions, the results about to be stated can be completed by folding in the (trivial) archimedean case with little effort—not an insignificant matter when it comes to the consideration of *global fields*.

Definition 1. Let L/K be a *Galois* extension of *local fields*. Define a map

(6)
$$K^\times \times G(L/K)^* \to \mathbb{Q}/\mathbb{Z}$$
$$(a, \chi) \quad \mapsto \langle a, \chi \rangle$$

by setting

(7) $$\langle a, \chi \rangle = \operatorname{inv}_K(L_\chi, \sigma_\chi, a).$$

Here σ_χ denotes the unique generator of $G(L_\chi/K) \simeq G/\ker \chi$ such that

(8) $$\chi(\sigma_\chi) = \frac{1}{n_\chi} \bmod 1, \quad \text{where } n_\chi = L_\chi : K.$$

F1. *The map* (6) *is bilinear; it is called the* **canonical pairing** *of* L/K.

Proof. Linearity in the first variable is clear; see (68) in Chapter 30. As for the second variable, first a preliminary remark: Set $G = G(L/K)$. For any $\chi \in G^*$ define a function $\tilde\chi : G \to \mathbb{Q}$ by assigning to $\mu \in G$ the value $\tilde\chi(\mu) = i/n$ if μ gives rise on L_χ to the automorphism σ_χ^i, where $0 \le i < n = n_\chi$. Then $\chi(\mu) = \tilde\chi(\mu) \bmod 1$, and one verifies easily that the inflation of the *uniformized* cocycle on G corresponding to the cyclic algebra (L_χ, σ_χ, a) is given by

(9) $$c(\mu, \nu) = a^{\tilde\chi(\mu) + \tilde\chi(\nu) - \tilde\chi(\mu\nu)}$$

—see (62) in Chapter 30. Now, suppose $\chi_1, \chi_2, \chi_3 \in G^*$ satisfy $\chi_1 + \chi_2 = \chi_3$, and let $c_1(\mu, \nu), c_2(\mu, \nu), c_3(\mu, \nu)$ be the corresponding cocycles in the sense just described. Then a straightforward computation shows that $c_3(\mu, \nu)$ only differs from the product $c_1(\mu, \nu)c_2(\mu, \nu)$ by a *coboundary*. □

Because of F1, the map (6) yields naturally a homomorphism $K^\times \to G(L/K)^{**}$; see F4 in Chapter 14 (vol. I). Since $G(L/K)^*$ equals $G(L/K)^{\mathrm{ab}*}$ and $G(L/K)^{\mathrm{ab}**}$ is canonically isomorphic to $G(L/K)^{\mathrm{ab}}$, we obtain a canonical homomorphism

(10) $$K^\times \to G(L/K)^{\mathrm{ab}},$$

which we denote by $(\cdot, L/K)$. This is called the *norm residue symbol* of L/K. For any $a \in K^\times$, by definition, $(a, L/K)$ is the element of $G(L/K)^{ab}$ characterized by

$$(11) \qquad \chi(a, L/K) = \langle a, \chi \rangle = \operatorname{inv}_K(L_\chi, \sigma_\chi, a) \quad \text{for every } \chi \in G^*.$$

Because of (4) we can view $(a, L/K)$ as an *automorphism of the largest abelian subextension L^{ab}/K of L/K.*

The name "norm residue symbol" for (10) can be justified as follows:

F2. *For every Galois extension L/K of local fields, $N_{L/K}L^\times$ lies in the kernel of $(\cdot, L/K)$. Hence $(\cdot, L/K)$ gives rise to a homomorphism (of the same name)*

$$(12) \qquad (\cdot, L/K) : K^\times / N_{L/K}L^\times \to G(L/K)^{ab}.$$

Proof. Take $a \in N_{L/K}L^\times$ and $\chi \in G(L/K)^*$. Since a is a norm of L_χ/K as well as of L, the right-hand side of (11) vanishes, that is, $(a, L/K) = 1$. $\qquad \square$

F3. *If L/K is a* **cyclic** *extension of local fields and σ is a generator of $G(L/K)$, we have*

$$(13) \qquad (a, L/K) = \sigma^{[L:K]\operatorname{inv}_K(L,\sigma,a)}$$

for every $a \in K^\times$.

Proof. Set $n = L : K$. Then $\operatorname{inv}_K(L, \sigma, a) = \frac{m}{n} \bmod 1$, with m well defined mod n, and the right-hand side of (13) will be σ^m. Now take the character χ of $G(L/K)$ such that $\chi(\sigma) = \frac{1}{n} \bmod 1$. Since $G(L/K)^* = \langle \chi \rangle$ and $L_\chi = L$, the assertion follows from the equality $\chi(a, L/K) = \operatorname{inv}_K(L, \sigma, a) = \chi(\sigma^m)$. $\qquad \square$

F4. *If L/K is* **unramified** *and has Frobenius automorphism $\varphi_{L/K}$, the norm residue symbol is given by the explicit formula*

$$(14) \qquad (a, L/K) = \varphi_{L/K}^{w_K(a)},$$

where $a \in K^\times$. Thus the norm residue symbol of an unramified extension L/K is distinguished by the property that every prime π of K maps to the Frobenius automorphism of L/K.

Proof. Since $\operatorname{inv}_K(L, \varphi_{L/K}, a) = \dfrac{w_K(a)}{L : K} \bmod 1$, the result follows from (13) by taking $\sigma = \varphi_{L/K}$. $\qquad \square$

F5. *If L/K is a Galois subextension of a Galois extension $\tilde{L}/K$ of local fields, there holds for every $a \in K^\times$ the compatibility relation*

$$(15) \qquad (a, L/K) = (a, \tilde{L}/K)^L,$$

where $\sigma \mapsto \sigma^L$ denotes the canonical map $G(\tilde{L}/K)^{ab} \to G(L/K)^{ab}$.

Proof. To each character χ of $G(L/K) = G(\tilde{L}/K)/G(\tilde{L}/L)$ we associate the character $\tilde{\chi} = \inf(\chi)$ of $G(\tilde{L}/K)$, where $\tilde{\chi}(\sigma) = \chi(\sigma^L)$. We have $\ker \chi = \ker \tilde{\chi}/G(\tilde{L}/L)$, and hence $\tilde{L}_{\tilde{\chi}} = L_\chi$, with $\sigma_{\tilde{\chi}} = \sigma_\chi$. Thus

$$\chi(a, L/K) = \mathrm{inv}_K(L_\chi, \sigma_\chi, a) = \mathrm{inv}_K(\tilde{L}_{\tilde{\chi}}, \sigma_{\tilde{\chi}}, a) = \tilde{\chi}(a, \tilde{L}/K) = \chi((a, \tilde{L}/K)^L),$$

as desired.   □

A deeper, functorial property of the norm residue symbol emerges in this sharpening of F5:

F6. *Let L/K and L'/K' be Galois extensions of local fields, where $K \subseteq K'$ and $L \subseteq L'$. For any $b \in K'^\times$ we have*

$$(16) \qquad (b, L'/K')^L = (N_{K'/K}b, L/K),$$

where $\sigma \mapsto \sigma^L$ denotes the canonical map $G(L'/K')^{\mathrm{ab}} \to G(L/K)^{\mathrm{ab}}$.

Proof. (a) In view of F5 we may as well assume that $L' = LK'$ and so view $G(L'/K')$ as a subgroup of $G(L/K)$. Denote by $\chi \mapsto \chi' = \mathrm{res}(\chi)$ the corresponding restriction map $G(L/K)^* \to G(L'/K')^*$. Taking (11) into account, our task boils down to proving that

$$(17) \qquad \langle b, \chi' \rangle_{K'} = \langle N_{K'/K}b, \chi \rangle_K$$

for every $\chi \in G(L/K)^*$, where for clarity the pairing brackets are subscripted with the base field they refer too. By definition, (17) is tantamount to the equality

$$(18) \qquad \mathrm{inv}_{K'}(L'_{\chi'}, \sigma_{\chi'}, b) = \mathrm{inv}_K(L_\chi, \sigma_\chi, N_{K'/K}b)$$

of the corresponding Hasse invariants, and it is easy to derive (18) by invoking the *projection formula* (55*) of Chapter 30 (page 221), because the corestriction $\mathrm{cor}_{K'/K}$ preserves Hasse invariants (Theorem 10 in Chapter 31). But we would like to justify (17) on arithmetic grounds, as follows.

(b) Let f be the residue class degree of K'/K. Take elements $\varphi \in G(L/K)$ and $\varphi' \in G(L'/K')$ inducing on the *largest unramified subextensions* L_0/K of L/K and L'_0/K' of L'/K' the corresponding *Frobenius automorphisms*. If the residue field $\bar{K}$ of K has q elements, the residue field of K' has q^f elements. Hence

$$(19) \qquad \varphi' = \varphi^f \quad \text{on } L_0.$$

To prove (17) we first consider the case that L_χ/K is *unramified*. Since $L'_{\chi'} = L_\chi K'$, the extension $L'_{\chi'}/K'$ is also unramified, and it follows from (19) that $\chi'(\varphi') = \chi(\varphi^f) = f\chi(\varphi)$. We now apply F4 (and F5), obtaining $\langle b, \chi' \rangle_{K'} = \chi'(b, L'/K') = \chi'(b, L'_{\chi'}/K') = \chi'(\varphi'^{w_{K'}(b)}) = w_{K'}(b)\chi'(\varphi') = w_{K'}(b)f\chi(\varphi)$; that is,

$$\langle b, \chi' \rangle_{K'} = f w_{K'}(b)\chi(\varphi).$$

Correspondingly, we get

$$(20) \qquad \langle N_{K'/K}b, \chi \rangle_K = \chi(N_{K'/K}b, L_\chi/K) = \chi(\varphi^{w_K(N_{K'/K}b)})$$
$$= w_K(N_{K'/K}b)\chi(\varphi) = f w_{K'}(b)\chi(\varphi),$$

proving (17) for L_χ/K unramified.

(c) Since both sides of (17) behave multiplicatively in b and the multiplicative group of K' is generated by prime elements, we can assume henceforth that $w_{K'}(b) = 1$. Using F5 we can also choose L large enough that the largest unramified subextension L_0 of L/K has a given degree. Now, by considering the characters of the cyclic group $G(L_0/K)$ as characters of $G(L/K)$, we can choose $\psi \in G(L/K)^*$ such that L_ψ/K is *unramified* and

$$(21) \qquad f\psi(\varphi) = \langle b, \chi' \rangle_{K'}.$$

From (20) and the assumption $w_{K'}(b) = 1$, we see that the left-hand side of (21) also equals

$$\langle b, \psi' \rangle_{K'} = \langle N_{K'/K} b, \psi \rangle_K.$$

Subtracting this last equation from (17) and using the additivity of the pairing, we see that the problem has been reduced to the case where the left-hand side of (17) vanishes. In this case — and setting $L_\chi = L$ without loss of generality, which means that $L'_{\chi'} = LK' = L'$ — we see that $b = N_{L'/K'} x$ for some x in L'. It follows that $N_{K'/K} b = N_{K'/K}(N_{L'/K'} x) = N_{L'/K} x = N_{L/K}(N_{L'/L} x)$, and the right-hand side of (17) vanishes as required. $\qquad\square$

F7. *Let L/K be a Galois extension of local fields and let ρ be an isomorphism from L onto some field $L' = \rho L$. This makes $L'/K' = \rho L/\rho K$ canonically into a Galois extension of local fields, whose norm residue symbol satisfies*

$$(22) \qquad (\rho x, \rho L/\rho K) = \rho(x, L/K)\rho^{-1}.$$

We eschew the fuss of a detailed proof and content ourselves with the thought that the conclusion is unavoidable, since only *natural* isomorphisms appear anywhere in the construction.

Theorem 1 (Local reciprocity law). *For any Galois extension L/K of local fields, the norm residue symbol $(\cdot, L/K) : K^\times \to G(L/K)^{\mathrm{ab}}$ of L/K is surjective, and its kernel coincides with the subgroup $N_{L/K} L^\times$ of $K^\times$. Thus, regarded as a map*

$$(23) \qquad (\cdot, L/K) : K^\times/N_{L/K} L^\times \to G(L/K)^{\mathrm{ab}},$$

the symbol is an isomorphism. In particular, when L/K is abelian, we obtain a canonical isomorphism $G(L/K) \simeq K^\times/N_{L/K} L^\times$.

Proof. Since $G(L/K)^* = (G(L/K)^{\mathrm{ab}})^*$, what we need to show is that the canonical pairing

$$(24) \qquad K^\times/N_{L/K} L^\times \times G(L/K)^* \to \mathbb{Q}/\mathbb{Z}$$

of L/K is *nondegenerate* in each of the two variables (see F4 in Chapter 14, vol. I). For $\chi \neq 0$ we have $L_\chi : K > 1$, so there exists $a \in K^\times$ that is not a norm of L_χ/K (see Theorem 8 in Chapter 31). It follows that $\langle a, \chi \rangle = \mathrm{inv}_K(L, \sigma_\chi, a) \neq 0$, and the nondegeneracy in the second variable is proved.

To show the nondegeneracy of the pairing (24) in the *first* variable, we suppose that for some $a \in K^\times$ we have

$$(25) \qquad\qquad \langle a, \chi \rangle = 0 \quad \text{for all } \chi \in G(L/K)^*.$$

Let E/K be any cyclic subextension of L/K. Clearly $E = L_\chi$ for some χ, so the condition $\langle a, \chi \rangle = 0$ says that a is a norm of E/K. So a *is a norm of any cyclic subextension of L/K*, and we must show that it is a norm of L/K itself. We apply induction on the degree n of L/K. The case $n = 1$ being obvious, take $n > 1$. Since $G(L/K)$ is *solvable* (Theorem 5 in Chapter 25), there is a cyclic subextension E/K of L/K such that $E : K > 1$. We know there is $b \in E$ such that $a = N_{E/K} b$, and we wish to find $b' \in N_{E/K}^{-1}(a)$ such that $(b', L/E) = 1$. By induction, b' will be a norm of L/E, and hence a will be a norm of L/K.

If L/K is *abelian*, we can simply take $b' = b$, because F6 yields

$$(b, L/E) = (N_{E/K} b, L/K) = (a, L/K) = 1.$$

Thus, *the local reciprocity law is proved for L/K abelian.*

Otherwise we look at the largest abelian subextension L_0/E of L/E. Since $G(L_0/E) = G(L/E)^{\text{ab}}$, the condition $(b', L/E) = 1$ is equivalent to $(b', L_0/E) = 1$. After renaming our fields, then, we can carry out the proof under the assumption that L/E is *abelian*.

Let $[G, A]$ be the subgroup of G generated by all the commutators $\rho \tau \rho^{-1} \tau^{-1}$, where $\rho \in G = G(L/K)$ and $\tau \in A = G(L/E)$, and let F be the corresponding fixed field. Being a central extension of a cyclic group, $G(F/K) = G/[G, A]$ is *abelian*. Then F6 implies that

$$(b, L/E)^F = (N_{E/K} b, F/K) = (a, F/K) = 1,$$

that is, $(b, L/E)$ lies in $G(L/F)$. Thus $(b, L/E)$ is a product of elements of the form $\rho(x, L/E)\rho^{-1}(x, L/E)^{-1}$, where $\rho \in G$ and $x \in E^\times$. By (22), such an element can also be written $(\rho x, L/E)(x, L/E)^{-1} = (\rho(x)x^{-1}, L/E)$. Thus, modulo $N_{L/E} L^\times$ our b is a product of elements of the form $\rho(x)x^{-1}$. Since all such lie in the kernel of $N_{E/K}$, an appropriate modification to b leads to $b' \in N_{L/E} E^\times$ satisfying $N_{E/K}(b') = a$. $\qquad\qquad\qquad\qquad\qquad\qquad\qquad\qquad\qquad\qquad\qquad\qquad \Box$

Remark. In the situation of Theorem 1, let E be an intermediate field of L/K. By the local reciprocity law there exists a unique homomorphism f making the diagram

$$(26)$$

$$\begin{array}{ccc}
K^\times / N_{L/K} L^\times & \xrightarrow{\ (\cdot, L/K)\ } & G(L/K)^{\text{ab}} \\
\downarrow & & \downarrow{\scriptstyle f} \\
E^\times / N_{L/E} L^\times & \xrightarrow{\ (\cdot, L/E)\ } & G(L/E)^{\text{ab}}
\end{array}$$

commute.

It would be pity to leave out that f is the *group-theoretic corestriction* Cor from $G = G(L/K)$ to $U = G(L/E)$ (see last remark in Chapter 30). This can be seen as follows: Let ψ be a character of U. Using the explicit corestriction formulas from the appendix to Chapter 30, one shows that, in general,

$$(27) \qquad \mathrm{cor}_{E/K}[L_\psi, \sigma_\psi, a] = [L_\chi, \sigma_\chi, a],$$

for every $a \in K^\times$, with $\chi := \mathrm{cor}(\psi) = \psi \circ \mathrm{Cor}$. Applying this to local fields and taking into account that $\mathrm{cor}_{E/K}$ preserves Hasse invariants, we conclude that $\psi(a, L/E) = \chi(a, L/K)$ and (since $\chi = \psi \circ \mathrm{Cor}$) that

$$(28) \qquad (a, L/E) = \mathrm{Cor}(a, L/K) \quad \text{for all } a \in K^\times. \qquad \square$$

Here is a first noteworthy consequence of the *local reciprocity law* (together with the property of the norm residue symbol expressed in F6):

F8. *Let E/K be an arbitrary extension of local fields, and let $F = E^{\mathrm{ab}}$ be the largest abelian subextension of E/K. Then*

$$(29) \qquad N_{E/K} E^\times = N_{F/K} F^\times.$$

Hence, by Theorem 1, the degree $K^\times : N_{E/K} E^\times$ is finite.

Proof. If E/K is *Galois*, the claim follows immediately from Theorem 1 (and F5), thanks to the canonical identification $G(E/K)^{\mathrm{ab}} = G(E^{\mathrm{ab}}/K)$.

If E/K is *separable*, consider the normal closure L/K of E/K. For a in $N_{F/K} F^\times$ we have $1 = (a, F/K) = (a, L^{\mathrm{ab}}/K)^F$, hence $(a, L^{\mathrm{ab}}/K) \in G(L^{\mathrm{ab}}/F)$. Since $F = E \cap L^{\mathrm{ab}}$ and (23) is surjective, this shows that there exists $b \in E^\times$ satisfying $(a, L^{\mathrm{ab}}/K) = (b, EL^{\mathrm{ab}}/E)^{L^{\mathrm{ab}}} = (N_{E/K}b, L^{\mathrm{ab}}/K)$; for the last equality we have used F6. Because (23) is injective, a only differs from $N_{E/K}b$ by a factor in $N_{L/K} L^\times \subseteq N_{E/K} E^\times$. Hence $N_{F/K} F^\times \subseteq N_{E/K} E^\times$, which proves (29).

Finally, to prove (29) in general, there remains to consider the case of a *purely inseparable* extension E/K of degree $p = \mathrm{char}\, K$. Here what has to be shown is that $K^\times = N_{E/K} E^\times$. By F2 in Chapter 25, the field E is of the form $E = \mathbb{F}_q((X))$. Hence $E^p = \mathbb{F}_q((X^p)) \subseteq K$, and by comparing degrees we get $E^p = K$. Therefore $N_{E/K} E^\times = K^\times$ as needed. $\qquad \square$

F9. *Let m be a natural number and K a local field containing a primitive m-th root of unity. Set $L = K(\sqrt[m]{K^\times})$, so L/K is the largest Kummer extension of K of exponent m. Then this extension is finite and*

$$(30) \qquad N_{L/K} L^\times = K^{\times m},$$

so the norm residue symbol gives a canonical isomorphism between $K^\times/K^{\times m}$ and $G(L/K)$.

Proof. From the assumption we know char K does not divide m, and Theorem 7 in Chapter 25 implies $K^\times/K^{\times m}$ is finite. We now apply *Kummer theory* (Theorem 4 of Chapter 14, vol. I), to derive the finiteness of $L:K$ and a canonical isomorphism

$$(31) \qquad\qquad K^\times/K^{\times m} \simeq G(L/K)^*.$$

At the same time, the *local reciprocity law* gives a *local reciprocity law reciprocity law!local* canonical isomorphism

$$(32) \qquad\qquad K^\times/N_{L/K}L^\times \simeq G(L/K).$$

Since $G(L/K)$ is of exponent m, we have $K^{\times m} \subseteq N_{L/K}L^\times$, which implies (30) in view of (31) and (32). $\qquad\qquad\qquad\qquad\qquad\qquad\qquad\qquad\qquad\qquad\qquad\square$

Definition 2. Let K be a local field. Denote by K_a/K the *largest abelian extension* of K (in a fixed algebraic closure of K) and define the *norm residue symbol*

$$(33) \qquad\qquad (\cdot, K): K^\times \to G(K_a/K),$$

by decreeing (for $a \in K^\times$) that $(a, K)^L = (a, L/K)$, where L/K is any finite subextension of K_a/K. Then (a, K) is well defined because of F5. If L/K is a finite subextension of K_a/K, denote by

$$(34) \qquad\qquad \mathcal{N}_L = N_{L/K}L^\times$$

the norm group of L/K. By definition, the *kernel* of (33) equals

$$(35) \qquad\qquad \mathcal{D}_K := \bigcap_L \mathcal{N}_L.$$

Using F8, we can also express $\mathcal{D}_K$ as the intersection of the norm groups of *all finite extensions* E/K. (Compare F11 further down.)

A subgroup H of $K^\times$ is called a *norm group* if it equals $N_{E/K}E^\times$ for some finite extension E/K. As mentioned, there must then be some finite *abelian* extension L/K such that $H = \mathcal{N}_L$; such an L is called the *class field* of H ("the" because it is easily proved to be unique; see Theorem 2 on page 249).

F10. *Any norm group of a local field K is* **closed** *in $K^\times$. It is also* **open***, because it has finite index in $K^\times$ (by F8). Thus the norm residue symbol (33) is* **continuous***.*

Every subgroup H of $K^\times$ containing a norm group is itself a norm group of K.

Proof. (a) Let E/K be finite and set $N = N_{E/K}$. Given a sequence (y_n) in E, suppose the images Ny_n converge toward $x \in K^\times$. Because the integers $w_K(Ny_n)$ converge to $w_K(x)$, we can assume that $w_K(Ny_n) = w_K(x)$ for all n. By replacing y_n with $y_n y_1^{-1}$, we reduce to the case that x, and so also each y_n, lies in U_E. But U_E is *compact*, so the y_n have an accumulation point y in U_E. Since N is continuous, $x = Ny$. Hence $NE^\times$ is closed in $K^\times$. As for openness, just note that a closed subgroup H of finite index in a topological group is the complement of the union of the (finitely many) cosets other than H, all of which are closed.

(b) Suppose a subgroup H of K contains the norm group of a finite extension E/K, which we assume without loss of generality to be *abelian*. The image $(H, E/K)$ under the norm residue symbol of E/K has the form $G(E/L)$, for some intermediate field L of E/K. Since $N_{E/K}E^\times \subseteq H$, an element $a \in K^\times$ lies in H if and only if $(a, E/K) \in G(E/L)$. Thus H consists of all $a \in K^\times$ such that $1 = (a, E/K)^L = (a, L/K)$. It follows that $H = N_{L/K}L^\times$.   □

We now undertake to prove, in a sort of converse to F10, that for every open subgroup H of finite index in $K^\times$ there is also an abelian extension L/K such that $H = N_{L/K}L^\times$. This is very easy to justify when char $K = 0$, but we wish to prove it also for the case char $K > 0$, which turns out a bit more obstinate.

Lemma 1. *If $m \in \mathbb{N}$ is not divisible by* char K, *then $K^{\times m}$ is a norm group of the local field K.*

Proof. By the assumption, we can take the extension K' of K obtained by adjoining a primitive m-th root of unity of K. By F9, $K'^{\times m}$ is the norm group of an abelian extension E/K'. Then

$$N_{E/K}E^\times = N_{K'/K}(N_{E/K'}E^\times) = N_{K'/K}(K'^{\times m}) \subseteq K^{\times m}.$$

Hence $K^{\times m}$ contains a norm group of K, and thus, by the last statement in F10, is itself a norm group.   □

Lemma 2. *If K is a local field of characteristic $p > 0$ and $x \in K^\times$ is a norm of every cyclic extension of K of degree p, then x is a p-th power in K.*

Proof. Let C be an algebraic closure of K and define $\wp : C \to C$ by $\wp(x) = x^p - x$. For $a \in K$, let α be an element of C such that $\wp(\alpha) = a$. (If $\wp(\alpha') = a$ as well, then $\alpha' - \alpha$ lies in the prime field $\mathbb{F}_p$ of K.) Now, either $\alpha \in K$, or $K(\alpha)/K$ is a cyclic extension of degree p. Conversely, every cyclic extension of degree p is obtained in this way; see Theorem 3 in Chapter 14 (vol. I). We now define a map $(\cdot, \cdot] : K^\times \times K \to \mathbb{F}_p$ by setting

$$(36) \qquad\qquad (b, a] = (b, K(\alpha)/K)\alpha - \alpha \quad \in \mathbb{F}_p,$$

where $\alpha \in C$ is required to satisfy $\wp(\alpha) = a$. One verifies easily that $(\cdot, \cdot]$ is well defined and *bilinear*. Moreover, $(b, a]$ vanishes if and only if b is a norm of $K(\alpha)/K$ (where $\wp(\alpha) = a$). Thus

$$(37) \qquad\qquad (a, a] = 0 \quad \text{for every } a \in K^\times,$$

since for $\alpha \notin K$ we have $N_{K(\alpha)/K}\alpha = (-1)^p(-a) = a$.

To actually prove Lemma 2 we consider the set M of $x \in K^\times$ such that $(x, a] = 0$ for all $a \in K^\times$ (or equivalently, such that $(x, ax] = 0$ for all $a \in K^\times$). From (37) we have

$$(x, ax] + (a, ax] = (ax, ax] = 0,$$

so M is the set of $x \in K^\times$ such that

$$(38) \qquad\qquad (a, ax] = 0 \quad \text{for every } a \in K^\times.$$

But then it is apparent that $M \cup \{0\}$ is an *additive* subgroup of K. On the other hand, M is a subgroup of $K^\times$ containing $K^{\times p}$, as follows from the first description of it. Taking it all together, then, we see that $M \cup \{0\}$ is an *intermediate field* of K/K^p. Since $K = \mathbb{F}_q((X))$ we have $K : K^p = p$, so $M \cup \{0\}$ coincides with either K^p or K. If the second case is excluded, Lemma 2 is proven.

So assume for a contradiction that $M \cup \{0\} = K$. Then the pairing $(\cdot, \cdot]$ is zero overall, so the equation $X^p - X = a$ is solvable already in K for *every* $a \in K$. But this is unpossible, because there exists an unramified extension L/K of degree p. (Alternative justification: If π is a prime element of K, one can show easily by applying w_K that there is no $\alpha \in K$ such that $\alpha^p - \alpha = \pi^{-1}$.) $\qquad\square$

F11. *For every nonarchimedean local field K,*

$$\mathfrak{D}_K = 1.$$

Proof. (a) By Lemma 1, we have $\mathfrak{D}_K \subseteq K^{\times m}$ for every $m \in \mathbb{N}$ such that char K does not divide m (and so for *all* m if char $K = 0$).

(b) It follows easily from F6 in Chapter 25 that $\bigcap_m K^{\times m} = 1$. By part (a), this implies $\mathfrak{D}_K = 1$ when char $K = 0$.

(c) Now suppose char $K = p > 0$. First we show that

$$(39) \qquad\qquad \mathfrak{D}_K \subseteq N_{F/K}\mathfrak{D}_F \quad \text{for any finite } F/K.$$

Take $x \in \mathfrak{D}_K$. For a given finite extension L/F, then, there exists $z \in L^\times$ such that $x = N_{L/K}z = N_{F/K}(N_{L/F}z)$; that is, $X_L := N_{F/K}^{-1}(x) \cap N_{L/F}L^\times$ is nonempty. Clearly $N_{F/K}^{-1}(x)$ is *compact*, and hence, because of F10, so is every X_L. The intersection of finitely many X_L is nonempty, so by compactness their intersection over *all* L is also nonempty. For an element y in the intersection we have $N_{F/K}y = x$, and moreover $y \in \mathfrak{D}_F$.

(d) Take $x \in \mathfrak{D}_K$. By Lemma 2 we have $x = y^p$ for some $y \in K^\times$. We claim that y also lies in $\mathfrak{D}_K$, so

$$(40) \qquad\qquad \mathfrak{D}_K = \mathfrak{D}_K^p.$$

Indeed, for every finite F/K (39) implies first that $x = N_{F/K}z$ for some $z \in \mathfrak{D}_F$. Again by Lemma 2 we have $z = t^p$ for some $t \in F^\times$. Now $y^p = x = N_{F/K}z = (N_{F/K}t)^p$ implies that $y = N_{F/K}t$. Thus y is a norm for every finite F/K; that is, $y \in \mathfrak{D}_K$.

(e) From (40) we see that $\mathfrak{D}_K \subseteq K^{\times p^n}$ for all n. This and part (a) of the proof imply that $\mathfrak{D}_K \subseteq K^{\times m}$ for *all* $m \in \mathbb{N}$. It follows that $\mathfrak{D}_K = 1$; see first item of part (b). $\qquad\square$

F12. *Let K_u/K be the largest unramified extension of the local field K, that is, the union of all finite unramified extensions of K (in a fixed algebraic closure of K). We can regard K_u/K as a subextension of K_a/K. The norm residue symbol (33) of K induces by restriction to U_K a topological isomorphism*

$$(41) \qquad\qquad U_K \to G(K_a/K_u).$$

Proof. For every finite unramified E/K and every $x \in U_K$ we have $(x, K)^E = (x, E/K) = 1$; see F4. Thus the map (33) does take U_K into $G(K_a/K_u)$. Let α be an element of K_a kept invariant under all (x, K) such that $x \in U_K$. Set $L = K(\alpha)$ and let $L_0 = L \cap K_u$ be the largest unramified subextension of L/K. For every unit y of L_0, the element $x = N_{L_0/K} y$ is a unit of K, so

$$(y, L/L_0) = (x, L/K) = 1.$$

Now, L/L_0 is *purely ramified*, so applying N_{L/L_0} to a prime element of L yields a prime of L_0. Thus $(y, L/L_0)$ equals 1 for *any* $y \in L_0^\times$, so $L = L_0$. Hence $\alpha \in K_u$. Altogether we have shown that the image of (41) is *dense* in $G(K_a/K_u)$. But $(\cdot, K)$ is *continuous* by F10, and U_K is compact, so (41) is surjective. Injectivity follows from F11. Having compact domain, the bijective continuous map (41) is *open* as well. $\qquad\square$

Theorem 2 (Local existence theorem). *Let K be a local field. The map*

$$(42) \qquad\qquad L \mapsto \mathcal{N}_L = N_{L/K} K^\times$$

induces a one-to-one correspondence between finite abelian extensions L/K of K (in a fixed algebraic closure of K) and open subgroups H of finite index in $K^\times$. Moreover,

$$L \subseteq L' \iff \mathcal{N}_{L'} \subseteq \mathcal{N}_L.$$

If char $K = 0$, every subgroup H of finite index in $K^\times$ is open.

Proof. (a) By F10, each $\mathcal{N}_L$ is open and has finite index in $K^\times$.

(b) Let L/K and L'/K be finite subextensions of K_a/K. The local reciprocity law immediately implies that

$$(43) \qquad\qquad \mathcal{N}_{LL'} = \mathcal{N}_L \cap \mathcal{N}_{L'}.$$

In particular, $\mathcal{N}_{L'} \subseteq \mathcal{N}_L$ is equivalent to $\mathcal{N}_{LL'} = \mathcal{N}_{L'}$. This last equality is, again by the reciprocity law, equivalent to $LL' = L'$, and so to $L \subseteq L'$.

(c) Let H be a subgroup of $K^\times$ of index $m < \infty$. Then $K^{\times m} \subseteq H$. If char K does not divide m, Lemma 1 says that $K^{\times m}$ is a norm group, and hence so is H (see F10). This takes care of the characteristic-zero case.

(d) Now suppose char $K = p > 0$. Set $H_0 = H \cap U_K$. Since H is open, so is H_0. Thus, by Galois theory, the image of H_0 under the *isomorphism* (41) has the form $G(K_a/F)$, with $F : K_u < \infty$. Since $F \subseteq K_a$, there is a finite subextension L/K of

K_a/K such that $F = K_u L$. For any $a \in U_K \cap N_{L/K} L^\times$, then, (a, K) is trivial both on K_u and on L, and hence $(a, K) \in G(K_a/F)$. It follows that

$$(44) \qquad\qquad U_K \cap N_{L/K} L^\times \subseteq H_0.$$

Introduce the abbreviation $\mathcal{N} = N_{L/K} L^\times$. By (44) we have

$$(45) \qquad\qquad \mathcal{N} \cap (\mathcal{N} \cap H) U_K \subseteq H,$$

and it suffices to prove that $(\mathcal{N} \cap H) U_K$ is a norm group, by (43) and F10. Clearly $\mathcal{N} \cap H$ has finite index in $K^\times$, since $\mathcal{N}$ and H do. Thus there exists $f \in \mathbb{N}$ such that $\pi^f \in \mathcal{N} \cap H$, where π is a prime of K. Hence $\langle \pi^f \rangle U_K$ is contained in $(\mathcal{N} \cap H) U_K$, which reduces the problem to showing that $\langle \pi^f \rangle U_K$ is a norm group. But this is obvious: $\langle \pi^f \rangle U_K$ is the norm group of the unramified extension of degree f (see F4). $\qquad\qquad\qquad\square$

F13. *Every norm group $\mathcal{N}$ of a local field K with prime element π contains a norm group of the form $\langle \pi^f \rangle \times U_K^{(m)}$.*

Proof. Since $\mathcal{N}$ is open in $K^\times$, there exists m such that $U_K^{(m)} \subseteq \mathcal{N}$. Since $K^\times : \mathcal{N}$ is finite, there is f such that $\pi^f \in \mathcal{N}$. Hence $\mathcal{N}$ contains a subgroup of the specified form. Now, by Theorem 2, such a subgroup must be a norm group: more precisely, we can write it as the intersection of two norm groups of a more special form, namely

$$(46) \qquad \langle \pi^f \rangle \times U_K^{(m)} = (\langle \pi^f \rangle \times U_K) \cap (\langle \pi \rangle \times U_K^{(m)}),$$

where $\langle \pi^f \rangle \times U_K$ is the norm group of the *unramified extension* $K(\zeta_{q^f - 1})/K$ of degree f, and $\langle \pi \rangle \times U_K^{(m)}$ is the norm group of a *certain purely ramified extension* $K_{\pi,m}/K$, of degree $(q-1)q^{m-1}$. $\qquad\qquad\square$

A closer look at these extensions $K_{\pi,m}/K$ would be our natural next project, but we must deny ourselves the pleasure; we refer instead to J. Neukirch, *Class field theory*, p. 63. But for the case $K = \mathbb{Q}_p$ we show:

F14. *The subgroup $\langle p \rangle \times U_{\mathbb{Q}_p}^{(m)}$ of $\mathbb{Q}_p^\times$ is the norm group of $\mathbb{Q}_p(\zeta_{p^m})/\mathbb{Q}_p$.*

Proof. Set $K = \mathbb{Q}_p$, $U = U_K$ and $L = \mathbb{Q}_p(\zeta)$, for some primitive p^m-th root of unity ζ. As stated in the example at the end of Chapter 24, the extension L/K is purely ramified of degree $n = p^{m-1}(p-1)$, and p equals $N_{L/K}(1 - \zeta)$. Thus what we need to do is prove that

$$(47) \qquad\qquad U^{(m)} \subseteq \mathcal{N}_L,$$

for once this is done, we have $H := \langle p \rangle \times U^{(m)} \subseteq \mathcal{N}_L$, and since $K^\times = \langle p \rangle \times W_{p-1} \times U^{(1)}$ (for $p \neq 2$) or $K^\times = \langle p \rangle \times \langle -1 \rangle \times U^{(2)}$ (for $p = 2$), the group $K^\times/H$ has order $(p-1)p^{m-1} = n = K^\times : \mathcal{N}_L$.

To prove (47) we first consider the case $p \neq 2$. It is trivial that $U^{(1)n} \subseteq \mathcal{N}_L$. By the results in Chapter 25 (in particular F9), we have

$$U^{(1)n} = U^{(1)(p-1)p^{m-1}} = U^{(1)p^{m-1}} = U^{(m)}.$$

For the case $p = 2$ (and assuming $m \geq 2$, as we may) we first obtain

$$U^{(2)2^{m-2}} = U^{(m)}, \quad \text{and so} \quad U^{(2)n} = U^{(m+1)}.$$

Since $U^{(2)} = 5^{\mathbb{Z}_2}$ (see F11 in Chapter 25), this leads to $5^{2^{m-2}} \in U^{(m)} \setminus U^{(m+1)}$. If we can show that $5^{2^{m-2}}$ is a norm of L/K, we will be done, since $U^{(m)} : U^{(m+1)} = 2$. And indeed,

$$N_{L/K}(2+i) = N_{\mathbb{Q}_p(i)/\mathbb{Q}_p}(2+i)^{2^{m-2}} = 5^{2^{m-2}}. \qquad \square$$

Theorem 3. *Any finite **abelian** extension $E/\mathbb{Q}_p$ is contained in some extension of the form $\mathbb{Q}_p(\zeta)/\mathbb{Q}_p$, where ζ is a root of unity. In other words: The largest abelian extension of $\mathbb{Q}_p$ is obtained from $\mathbb{Q}_p$ by adjunction of all roots of unity.*

Proof. In view of Theorem 2 and F13, the field E is contained in the class field L of some group of the form (46). By F14, however, L is the composite of $\mathbb{Q}_p(\zeta_{p^f - 1})$ with $\mathbb{Q}_p(\zeta_{p^m})$; that is, $L = \mathbb{Q}_p(\zeta)$ for some root of unity ζ of order $(p^f - 1)p^m$. $\square$

Remark. Theorem 3 is a local version of the famous *Kronecker–Weber Theorem* (see p. 259 in vol. I), which states the same thing for $\mathbb{Q}$ instead of $\mathbb{Q}_p$. This theorem finds its natural place today in *global class field theory* (see F. Lorenz, *Algebraische Zahlentheorie*, p. 273). But it should be mentioned that its proof can be reduced to the local result in Theorem 3, using only relatively modest means from algebraic number theory; see for instance J. Neukirch, *Class field theory*, p. 46.

33

Semisimple Representations of Finite Groups

1. Definition 1. (a) Let A be a K-algebra and V a K-vector space. A *representation* of A in V is a homomorphism of K-algebras

$$T : A \to \mathrm{End}_K(V)$$

from A into the K-algebra $\mathrm{End}_K(V)$. If $V : K < \infty$, we say that T has *finite degree* $V : K$. The function associating to each $a \in A$ the trace of $T(a)$ is called the *character* of T, and is denoted by χ_V, χ_T or simply χ. Thus

$$\chi(a) = \mathrm{Tr}\ T(a) \quad \text{for } a \in A.$$

If K has characteristic zero, $\chi(1)$ is always the degree of T.

(b) A representation T of A in V bestows on V a natural A-*module* structure:

$$av = T(a)v.$$

This is called the *representation module* of T. Conversely, if A is a K-algebra and V is an A-module, the map $a \mapsto a_V$ taking each $a \in A$ to a_V (left multiplication by a in V) is a representation of A in V. *In this sense representations of a K-algebra A and A-modules amount to the same thing.* However, it is convenient to use the two languages — representation-theoretic and module-theoretic — side by side and pass from one to the other as needed. Thus, a representation of A in V is called *irreducible* if V is a simple (irreducible) A-module. A representation is *fully reducible* or *semisimple* when the corresponding module has the same property (Chapter 28). Two representations T, T' of A in V, V' are called *equivalent* or *isomorphic* if the A-modules V, V' are isomorphic, that is, if there is a K-isomorphism $\varphi : V \to V'$ such that

(1) $$T'(a) = \varphi \circ T(a) \circ \varphi^{-1} \quad \text{for every } a \in A.$$

The representation $T = T_{\mathrm{reg}}$ of A corresponding to the A-module A_l is called the *regular representation* of A. If $A : K < \infty$, the character $\chi = \chi_{\mathrm{reg}} = \chi_A$ is the *regular character* of A. (Thus, in the notation of vol. I, p. 134 we have $\chi_{\mathrm{reg}} = \mathrm{Tr}_{A/K}$.) And not surprisingly, the character of an irreducible representation of A is called an *irreducible character* of A.

(c) Let V, W be A-modules, and T, S the respective representations. We say that S is a *subrepresentation* of T if W is a sub-A-module of V. If the A-module V is a direct sum $V = V_1 \oplus \cdots \oplus V_m$ of A-modules V_i and $T, T_1, \ldots, T_m$ are the respective representations, we call T the *(direct) sum* of the representations $T_1, \ldots, T_m$, and we write $T = T_1 + \cdots + T_m$. If $V_i : K < \infty$ for all i, the corresponding characters satisfy

$$(2) \qquad \chi_{V_1 \oplus \cdots \oplus V_m} = \chi_{V_1} + \cdots + \chi_{V_m}.$$

Finally, let V and N be A-modules, with respective representations T and S. If S is *irreducible* and T is *semisimple*, we say that S *occurs r times* in T if the *isogenous component* of type N of V has length r (in symbols, $V : N = r$; see F19 in Chapter 28).

Remark. The endomorphism algebra of the K-vector space K^n can be canonically identified with the matrix algebra $M_n(K)$. A representation T of a K-algebra A in K^n then amounts to a K-algebra homomorphism

$$T : A \to M_n(K),$$

and vice versa. We often talk of *matrix representations* of degree n of a K-algebra A. Of course, any representation of degree n of A is equivalent to a matrix representation of A of the same degree.

Definition 2. Let V be a K-vector space. A *representation of a group G in V* is a homomorphism

$$T : G \to \mathrm{GL}(V) \subset \mathrm{End}_K(V)$$

of G in the automorphism group of the K-vector space V.

We wish to show that the notion of a group representation is subsumed naturally under the notion of representation introduced in Definition 1. To do this, we consider the *group algebra KG* of the group G over the field K (see Section 6.2 in vol. I). Every element a in KG has the form

$$(3) \qquad a = \sum_{g \in G} a_g g,$$

with uniquely defined coefficients $a_g \in K$, equal to 0 for almost all g. Thus the elements $g \in G$ form a K-basis of KG, and *as a K-vector space*, KG is isomorphic to the K-vector space $K^{(G)}$ of all functions $f : G \to K$ that vanish at almost all $g \in G$. Multiplication in the K-algebra KG extends G's group operation $(g, h) \mapsto gh$ distributively:

$$\left(\sum_g a_g g \right) \left(\sum_g b_g g \right) = \sum_{g,h} a_g b_h gh.$$

If we convert the element on the right-hand side back to canonical form (3), we obtain

$$(4) \qquad \left(\sum_g a_g g \right) \left(\sum_g b_g g \right) = \sum_g \left(\sum_t a_{gt^{-1}} b_t \right) g.$$

Thus in the model $K^{(G)}$ of KG multiplication is given by *convolution*:

$$(a * b)(g) = \sum_t a(gt^{-1})b(t).$$

F1. *Every representation T of a group G in a K-vector space V defines, by linear extension, a representation — still denoted by T — of the group algebra KG in V:*

$$T\left(\sum_g a_g g\right) := \sum_g a_g T(g).$$

Conversely, if T is a representation of the K-algebra KG in V, restriction to G yields a representation of the group G in V, also denoted by T. If V is finite-dimensional, we often consider also the character of the representation T of KG just as a function $\chi : G \to K$. This does not lose any information, because

$$\chi\left(\sum_g a_g g\right) = \sum_g a_g \chi(g).$$

Definition 3. Representations of a group G in vector spaces over a given field K are called *K-representations* of G, and their characters are called *K-characters* of G.

From the preceding discussion, we see that *if K is a field, the K-representations of G are in one-to-one correspondence with the representations of the group algebra KG, and so also with KG-modules.*

The group algebra KG always admits the *trivial representation*, corresponding to the trivial KG-module K, itself characterized by the action $gx = x$ for all $g \in G$ and $x \in K$. (Thus for any element a of KG written in the form (3) we have $ax = \left(\sum_g a_g\right)x$ for all $x \in K$.) The corresponding K-representation of G is the trivial homomorphism $G \to K^\times = \mathrm{GL}(1, K)$. The character of the trivial representation of KG is called the *trivial character* of KG, and clearly it maps every element $g \in G$ to the number 1. This character is of course irreducible.

We turn to an important feature of *group* representations: Let V_1, V_2 be KG-modules, with respective representations T_1, T_2. We obtain a representation $T = T_1 \otimes T_2$ of G in $V_1 \otimes_K V_2$ by associating to each $g \in G$ the endomorphism $T(g) := T_1(g) \otimes T_2(g)$ of $V_1 \otimes_K V_2$ (see item (iii) in §6.11). As a KG-module, $V_1 \otimes_K V_2$ is then given by

$$(5) \qquad\qquad g(x_1 \otimes x_2) = gx_1 \otimes gx_2.$$

The corresponding characters, regarded as functions on G, satisfy

$$(6) \qquad\qquad \chi_{V_1 \otimes_K V_2} = \chi_{V_1} \chi_{V_2}.$$

The following simple fact will turn out to be significant:

F2. *Let G be a finite group of order n. The character $\rho = \chi_{KG}$ of the regular representation of KG is given by*

$$(7) \qquad\qquad \rho(g) = \begin{cases} n \ \textit{for} \ g = 1, \\ 0 \ \textit{for} \ g \neq 1, \end{cases}$$

where n is viewed as an element of K (in particular, it may equal 0).

Proof. Let $g_1, g_2, \ldots, g_n$ be the elements of G, and denote by T the regular representation of KG. Then $T(g)g_i = gg_i$. For $g \neq 1$ we have $gg_i = g_j \neq g_i$, hence $\operatorname{Tr} T(g) = 0$. On the other hand, $\rho(1) = n$ is the degree of T. $\qquad\qquad \square$

Now let's investigate the *center* $Z(KG)$ of the group algebra KG. An element a of the form (3) lies in $Z(KG)$ if and only if $ax = xa$ for every $x \in G$. This equality is equivalent to $x^{-1}ax = a$, or by comparison of coefficients, to $a_{xgx^{-1}} = a_g$ for all $g, x \in G$; in other words, the function $f : g \mapsto a_g$ must be *constant on each conjugacy class*

$$[g] = \{xgx^{-1} \mid x \in G\}$$

of G. This motivates the following terminology: for $g \in G$, the *class sum* of $[g]$ is the element

$$(8) \qquad\qquad k_g = k_{[g]} = \sum_{t \in [g]} t$$

of KG, where we assume the conjugacy class $[g]$ to be finite. We therefore have:

F3. *If $c_1, c_2, \ldots, c_h$ are the distinct conjugacy classes of a finite group G, the class sums $k_{c_1}, k_{c_2}, \ldots, k_{c_h}$ form a K-basis of the center $Z(KG)$ of KG.*

Definition 4. A function $f : G \to K$ in $K^{(G)}$ is called a *class function* or *central function* of G with values in K if f is constant on each conjugacy class of G; that is, if $f(xgx^{-1}) = f(g)$ for all $x, g \in G$. The set of such functions, with the operations of pointwise addition and multiplication of function values, forms a K-algebra, which we denote by

$$(9) \qquad\qquad Z_K(G) = Z_K G.$$

Its unit element is the *trivial character* of KG, which we denote by 1_G.

As K-vector spaces, the subalgebra $Z_K G$ of $K^{(G)}$ and the center of KG can be identified with each other, but multiplication in the two algebras is fundamentally different: in $Z_K G$ it is pointwise multiplication and in $Z(KG)$ it is convolution— see (4).

Every K-character χ of G belongs to $Z_K G$, so it is a class function:

$$(10) \qquad\qquad \chi(xgx^{-1}) = \chi(g) \quad \text{for all } x, g \in G.$$

Indeed, if T is a representation with character χ, then

$$\operatorname{Tr} T(xgx^{-1}) = \operatorname{Tr} T(x)T(g)T(x)^{-1} = \operatorname{Tr} T(g).$$

Incidentally, the product $\chi_1\chi_2$ in $Z_K G$ of two K-characters χ_1 and χ_2 of G is also a K-character of G; see (6). The set

(11) $X_K(G)$

of all $\mathbb{Z}$-linear combinations of K-characters χ of G is therefore a subring (a $\mathbb{Z}$-subalgebra) of $Z_K G$. We call $X_K(G)$ the *K-character ring* of G. The elements of $X_K(G)$ are called *generalized K-characters* of G; they obviously coincide with differences $\chi - \chi'$ of *proper characters* χ, χ' (characters of K-representations of G).

F4. *Let χ be a character of a group G (that is, a character of a representation T of finite degree d of G in a vector space over a field K). Suppose $g \in G$ satisfies $g^m = 1$. Then $\chi(g)$ is a d-fold sum of m-th roots of unity (in an algebraic closure C of K).*

Proof. Let $\varepsilon_1, \ldots, \varepsilon_d$ be the roots of the characteristic polynomial of the endomorphism $T(g)$ in C, listed with multiplicity. Then $\chi(g) = \varepsilon_1 + \cdots + \varepsilon_d$. Since $T(g)^m = T(g^m) = 1 = \mathrm{id}_V$, we have $\varepsilon_i^m = 1$ for every i.   □

As an exercise, the reader can prove that the if χ is a $\mathbb{C}$-character of a finite group G, then

(12) $\chi(g^{-1}) = \overline{\chi(g)}$ for all $g \in G$.

Before we proceed with the study of group algebras, we formulate some basic rules of behavior of representations under base change, for algebras in general:

Definition 5. Let $T : A \to \mathrm{End}_K(V)$ be a representation of the K-algebra A. For every field extension E/K, we obtain by *base change* a representation

$$T^E : A_E = A \otimes_K E \to \mathrm{End}_K(V) \otimes_K E = \mathrm{End}_E(V \otimes_K E).$$

The representation module $V_E = V \otimes_K E$ of T^E is characterized by $(a \otimes \alpha)(v \otimes \beta) = av \otimes \alpha\beta$. If $T : A \to M_n(K)$ is a matrix representation, we have $T^E(a) = T(a)$ for $a \in A$, where we view A as a subset of $A \otimes E$ and $M_n(K)$ as a subset of $M_n(E)$. The characters χ, χ^E of T, T^E satisfy $\chi^E(a) = \chi(a)$ for $a \in A$; for this reason we write simply χ instead of χ^E.

The representation T and the A-module V are called *absolutely irreducible* if the A_E-module V_E is irreducible for *every* extension E/K. For instance, any degree-1 representation is absolutely irreducible (this is trivial). We say that a character χ of a K-algebra A is *absolutely irreducible* if χ belongs to an absolutely irreducible representation of A.

F5. *A finite-degree representation T of a K-algebra A in V is absolutely irreducible if and only if $T(A) = \mathrm{End}_K(V)$. If T is semisimple, both properties are equivalent to the equality $\mathrm{End}_A(V) = K$. If T is irreducible and K is algebraically closed, T is absolutely irreducible.*

Proof. If $T(A) = \text{End}_K(V)$, then $T^E(A_E) = \text{End}_E(V_E)$ — and therefore V_E is irreducible — for every E/K. If, on the contrary, $T(A)$ and $\text{End}_K(V)$ are distinct, we have $T^E(A_E) \neq \text{End}_E(V_E)$ for all E/K; but then V_E is reducible if we take for E an algebraic closure of K, as can be seen by applying F11 from Chapter 29 to E and V_E.

The equality $T(A) = \text{End}_K(V)$ always implies $\text{End}_A(V) = K$. Conversely, suppose $\text{End}_A(V) = K$. If V is a semisimple A-module this implies that the map $T : A \to \text{End}_K(V)$ is surjective; see for instance F20 in Chapter 28. $\qquad\square$

Definition 6. An extension E of K is called a *splitting field* of a K-algebra A if every irreducible representation of finite degree of the E-algebra $A_E = A \otimes_K E$ is absolutely irreducible. (This agrees with Definition 8 in Chapter 29 if A is a central-simple K-algebra.) If K is a splitting field of the group algebra KG of a group G over K, we say that K is a *splitting field of the group* G.

F6. *Let A be a **commutative** K-algebra. Every absolutely irreducible finite-degree representation of A has degree 1. Thus the following two statements are equivalent*:

 (i) *K is a splitting field of A.*

 (ii) *Every irreducible finite-degree K-representation of A has degree 1.*

*Since the group algebra of an abelian group is commutative, each of the preceding conclusions still holds verbatim if A denotes an **abelian group**.*

Proof. This is clear by F5. $\qquad\square$

Definition 7. Let KG be the group algebra of a *finite group* G over a field K. A special role attaches to the element

$$N_G := \sum_{g \in G} g$$

of KG. If V is a KG-module, N_G will also denote the multiplication of elements of V by the element $N_G \in KG$:

$$N_G x = \sum_{g \in G} g x \quad \text{for } x \in V.$$

The map $N_G : V \to V$ is called the *norm map*. The submodule

$$V^G = \{ x \in V \mid g x = x \text{ for every } g \in G \}$$

of elements in V invariant under G is called the *fixed module of V under G*. The representation of G corresponding to V^G is the $(V^G : K)$-fold sum of the trivial representation of G, so its character is $(V^G : K) 1_G$. The element $N_G x$ lies in V^G for every $x \in V$. This is the primary reason N_G is of interest.

If V and W are KG-modules, so is $\text{Hom}_K(V, W)$, in a natural way: For each linear map $F : V \to W$, define gF for $g \in G$ by

$$(gF)(x) = gF(g^{-1}x).$$

It follows immediately from the definition that

(13) $$\mathrm{Hom}_K(V, W)^G = \mathrm{Hom}_{KG}(V, W),$$

that is, the KG-homomorphisms $V \to W$ are exactly the G-invariant elements of the KG-module $\mathrm{Hom}_K(V, W)$.

F7. *Let G be a finite group of order n, which we assume is not divisible by the characteristic of K. For every exact sequence*

$$0 \to X \to Y \xrightarrow{\ p\ } Z \to 0$$

of KG-modules, the corresponding sequence of fixed modules,

$$0 \to X^G \to Y^G \to Z^G \to 0,$$

is exact as well.

Proof. Only the surjectivity of $Y^G \to Z^G$ needs to be checked. Take $z \in Z^G$. There exists $y \in Y$ such that $z = p(y)$. Since $z \in Z^G$ and $n \in K^\times$, we have

$$z = \frac{1}{n} N_G z = \frac{1}{n} N_G p(y) = p\left(\frac{1}{n} N_G y\right) \in p(Y^G). \qquad \square$$

F8. *Let V, W be KG-modules of finite dimension over K, with characters χ_V, χ_W.*

  (a) *The character of G associated to the KG-module $V^* = \mathrm{Hom}_K(V, K)$ is $\chi_{V^*}(g) = \chi_V(g^{-1})$.*

  (b) *The character of G associated to the KG-module $M = \mathrm{Hom}_K(V, W)$ is $\chi_M(g) = \chi_V(g^{-1})\chi_W(g)$. (Note that the canonical isomorphism $V^* \otimes W \simeq \mathrm{Hom}_K(V, W)$ of K-vector spaces is an isomorphism of KG-modules.)*

The demonstration of these simple but important facts is left to the reader, as is that of the next result.

F9. *Let G be a finite group of order n and K a field whose characteristic does not divide n. Let χ be the character of the KG-module M. Then the dimension of the fixed module M^G satisfies*

(14) $$M^G : K = \frac{1}{n} \sum_{g \in G} \chi(g).$$

(Here both sides are elements of K, so the equation only determines the dimension of $M^G : K$ modulo char K.)

2. We now come to a seminal result of representation theory:

F10. *Let G be a finite group of order n, which we assume prime to the characteristic of K. Then every submodule W of a KG-module V is a direct summand of V.*

Proof. The exact sequence $0 \to W \to V \to V/W \to 0$ of KG-modules gives rise naturally to a sequence

$$0 \to \mathrm{Hom}_K(V/W, W) \to \mathrm{Hom}_K(V, W) \to \mathrm{Hom}_K(W, W) \to 0$$

of KG-modules, which is clearly exact. By F7, then, the corresponding sequence of fixed modules is exact. In view of (13), this proves the *surjectivity* of

$$\mathrm{Hom}_{KG}(V, W) \to \mathrm{Hom}_{KG}(W, W).$$

In particular, the identity map $\mathrm{id} : W \to W$ has an associated KG-homomorphism $f : V \to W$, satisfying $fx = x$ for all $x \in W$. Set $W' = \ker f$. Clearly, $W \cap W' = 0$, and for every $y \in V$ the difference $y - fy$ lies in W', that is, $y = (y - fy) + fy \in W' + W$. Hence $V = W \oplus W'$ is the direct sum of the KG-modules W and W'. □

Theorem 1 (Maschke). *If G is a finite group, its group algebra KG over a field K is semisimple if and only if* char K *does not divide the order of G.*

Proof. The *sufficiency* of the condition for the semisimplicity of KG follows from F10 (because of F10 in Chapter 28).

For the converse, suppose that char $K = p > 0$ and that $n = G : 1$ is a multiple of p. The nonzero element $x = N_G$ of KG satisfies $xg = x = gx$ for every $g \in G$ (see Definition 7). Thus x lies in the center $Z(KG)$ of KG. Moreover,

$$x^2 = xx = x \sum_{g \in G} g = \sum_{g \in G} xg = nx = 0.$$

Thus $Z(KG)$ contains a *nilpotent* element $\neq 0$, so the algebra KG cannot be semisimple (by F5 in Chapter 29, for instance). □

Remark. Let G be a finite group of order n. The fact that KG is *semisimple* when $n \in K^\times$ is so important that we wish to justify it in another way. Suppose that KG is not semisimple, so the *radical* of KG contains a *nonzero* element a, by F29 in Chapter 28. By F38 in the same chapter, then, ax is *nilpotent* for every $x \in KG$, so ax has trace 0 in the regular representation, that is, $\chi_{KG}(ax) = 0$ for every $x \in KG$. But this contradicts F2, unless $n = 0$ in K.

Now that we know that the group algebra KG of a group of order n is semisimple if $n \in K^\times$, the results of Chapters 28 and 29 can provide valuable information about KG in this case. Our task is to mine out this information, or at least some of it. Although we will limit ourselves to the semisimple case, we should at least mention the importance of *modular representation theory*, which studies especially group algebras over *finite fields of arbitrary characteristic*. For modular theory, incidentally, results on *semisimple representations* are an important tool: see J.-P. Serre, *Representations linéaires des groupes finis*.

We start by compiling a catalog of properties of representations of semisimple algebras that arise from the material in Chapters 28 and 29:

F11 (Synopsis). Let A be a *semisimple K-algebra*.

(a) *Every representation of A is fully reducible, so every A-module is a direct sum of irreducible submodules. These irreducible summands are uniquely determined up to isomorphism and order.* (See Chapter 28, F13 and F19.)

(b) *Every simple A-module is isomorphic to a simple submodule of A_l (which is to say, a minimal left ideal of A).* In other words, *every irreducible representation of A is a subrepresentation of the regular representation of A.* (Chapter 28, F12.)

(c) Consider the decomposition of A into simple components:

$$(15) \qquad\qquad A = A_1 \times A_2 \times \cdots \times A_k$$

(Chapter 29, Theorem 1). As A-modules, the A_i coincide with the distinct isogenous components of A (Chapter 28, F18). Hence: *There are exactly k isomorphism classes of simple A-modules. Thus, A has exactly k (inequivalent) irreducible representations.*

Next, let V_i, for $1 \le i \le k$, be irreducible A-modules, each isomorphic to a submodule of the A-module A_i. Then $V_1, V_2, \ldots, V_k$ represent all the isomorphism classes of simple A-modules, and no two of them are isomorphic; for brevity we say they form a *system of representatives of simple A-modules.* Let $T_1, T_2, \ldots, T_k$ be the representations corresponding to the V_i. *Every irreducible representation of A is equivalent to one and only one of the representations* $T_1, T_2, \ldots, T_k$. In other words: *Every irreducible A-module is isomorphic to exactly one of* $V_1, V_2, \ldots, V_k$.

Thus the irreducible representations of A correspond in the sense above, up to equivalence, to the simple components A_i of A. Incidentally, as an A_i-module, V_i is (up to isomorphism) the only *simple* one (Chapter 29, F2). Note also that for $i \ne j$ the product $A_j A_i$ is zero, so $A_j V_i$ is zero as well.

(d) For every $1 \le i \le k$, let $D_i = \mathrm{End}_A(V_i) = \mathrm{End}_{A_i}(V_i)$ be the algebra of endomorphisms of the simple A-module V_i. By Theorem 3 in Chapter 29, the natural map $A_i \to \mathrm{End}_{D_i}(V_i)$ is an isomorphism; thus $T_i : A \to A_i \to \mathrm{End}_{D_i}(V_i) \to \mathrm{End}_K(V_i)$ has image $\mathrm{End}_{D_i}(V_i)$. The D_i-vector space V_i has finite dimension n_i coinciding with the length of A_i. Thus $A_i \simeq M_{n_i}(D_i^0)$. From (15) we then have

$$(16) \qquad\qquad A \simeq M_{n_1}(D_1^0) \times M_{n_2}(D_2^0) \times \cdots \times M_{n_k}(D_k^0).$$

(e) Since A_i has length n_i, *the regular representation of A contains the irreducible representation T_i exactly n_i times.* Denote by K_i the center of A_i. By (16),

$$(17) \qquad\qquad ZA = K_1 \times K_2 \times \cdots \times K_k$$

is the center of A. The K_i are extensions of K. It follows from the definitions that *the representation T_i of A maps K_i isomorphically onto the center of $D_i = \mathrm{End}_A(V_i) \subseteq \mathrm{End}_K(V_i)$.*

(f) For $1 \le i \le k$, let e_i be the unity of the simple algebra A_i in (15). The elements $e_1, e_2, \ldots, e_k$, which are uniquely determined by A (apart from their order), are called the *primitive central idempotents of A.* They satisfy

$$(18) \qquad\qquad 1 = e_1 + e_2 + \cdots + e_k \quad \text{and} \quad e_i e_j = \delta_{ij} e_i.$$

These elements suffice to recover the direct product decomposition of A, because $A_i = e_i A = A e_i$. In terms of the representations T_i of A in V_i we have

$$(19) \qquad T_i e_i = \mathrm{id}_{V_i} \quad \text{and} \quad T_i e_j = 0 \quad \text{for } j \neq i.$$

(g) We now assume in addition that the semisimple K-algebra A has *finite dimension* $n = A : K$. Then *every irreducible representation of A has finite degree. If $d_i = V_i : K$ denotes the degree of T_i, we have*

$$(20) \qquad d_i = n_i (D_i : K) = n_i \, s_i^2 (K_i : K),$$

where s_i is the Schur index of D_i. We also have

$$(21) \qquad n = \sum_{i=1}^{k} n_i^2 \, s_i^2 (K_i : K).$$

The first assertion follows from (b), the second and third from (d).

(h) For $1 \leq i \leq k$, let χ_i be the character of T_i. Then

$$(22) \qquad \chi_i(e_i) = V_i : K = d_i \quad \text{and} \quad \chi_j(e_i) = 0 \quad \text{for } j \neq i.$$

By part (e), the *regular character* $\rho = \chi_A$ is given by

$$(23) \qquad \rho = \sum_{i=1}^{k} n_i \chi_i.$$

If χ is any irreducible character of A, we have $\chi = \chi_i$ for some $1 \leq i \leq k$, again by (c).

(i) Now assume for the time being that K has characteristic 0. In view of (a) and (c), it follows easily using (22) that *two representations (of finite degree) of A having the same character are equivalent*; in other words, *two A-modules V and W satisfying $\chi_V = \chi_W$ are equivalent*. Up to equivalence, then, in the characteristic zero case a (finite-dimensional) representation of a semisimple finite-dimensional K-algebra A is fully determined by its character!

All the statements in F11 hold, in particular, for the group algebra $A = KG$ of a finite group G of order n over a field K such that $n \in K^\times$.

Remarks. (1) For a *semisimple* group algebra $A = KG$, let A_1 (say) be the simple component belonging to the trivial representation of KG, so $V_1 = K$ and hence $D_1 = K \simeq A_1$ and $n_1 = 1$. By F11(e), therefore, *the trivial representation of G appears exactly once in the regular representation of KG*. The unique one-dimensional trivial submodule of KG is in fact KN_G, with N_G as in Definition 7.

(2) For any group G we have $(KG)^0 \simeq KG$, since the self-map $g \mapsto g^{-1}$ of G defines, when extended by linearity, an isomorphism t from KG onto $(KG)^0$. But note that if A_i is a component of a semisimple group algebra $A = KG$, it is not always the case that $A_i^0 \simeq A_i$; only that there is an involutive permutation τ of $\{1, 2, \ldots, k\}$ such that $A_i^0 \simeq A_i$ for $\tau(i) = i$ and $A_i^0 \simeq A_j \not\simeq A_i$ for $\tau(i) = j \neq i$. For $A = KG$, we obtain from (16) the isomorphism

$$(16') \qquad A \simeq M_{n_1}(D_1) \times M_{n_2}(D_2) \times \cdots \times M_{n_k}(D_k).$$

F12. *Let A be a semisimple K-algebra. With notation as in F11, the following statements are equivalent if $A : K < \infty$:*

(i) *K is a splitting field of A.*

(ii) *For every $1 \leq i \leq k$, the representation T_i is absolutely irreducible.*

(iii) *For every $1 \leq i \leq k$ we have $D_i = K_i = K$ (with the usual identifications); that is, the simple components A_i of A are central-simple K-algebras that split over K.*

(iv) *$A \simeq M_{n_1}(K) \times M_{n_2}(K) \times \cdots \times M_{n_k}(K)$.*

If $n = A : K$ is the dimension of A over K, which we have assumed finite, these four statements are also equivalent to each of the following:

(v) $\sum_{i=1}^{k} (V_i : K)^2 = n;$ (vi) $\sum_{i=1}^{k} n_i^2 = n;$ (vii) *$d_i = n_i$ for $1 \leq i \leq k$.*

Proof. Since $V_1, V_2, \ldots, V_k$ form a system of representatives of simple A-modules, the result follows via F5 from parts (d) and (g) of F11. □

Specializing F12 to group algebras, we obtain for any finite group G a characterization of the splitting fields K whose characteristic does not divide the order of G.

F13. *If K is a splitting field of a finite group G of order $n \in K^\times$, every extension L of K is a splitting field of G.*

Proof. From F12(iv) we get $LG \simeq KG \otimes_K L \simeq M_{n_1}(L) \times \cdots \times M_{n_k}(L)$. □

F14. *If A is a semisimple K-algebra of dimension n over K, there is equivalence between:*

(i) *A is commutative and K is a splitting field of A.*

(ii) *Every irreducible K-representation of A has degree 1.*

(iii) *A has exactly n inequivalent irreducible K-representations.*

Since a group algebra KG is commutative if and only if G is commutative, this result remains true if A is replaced by a group G of order $n \in K^\times$.

Proof. Suppose (ii) holds. By (20) we have $n_i = 1$ and $D_i = K_i = K$ for all i. By F12, therefore, K is a splitting field of A, and A is isomorphic to $K \times K \times \cdots \times K$, and hence commutative. This shows (i) and (iii).

Now assume (iii). By (21) and (20), all the d_i equal 1, which shows (ii).

Suppose (i) is satisfied. Then all the n_i in F12(iv) equal 1, which implies that $n = k$. That shows (iii).   □

Theorem 2. *Let G be a finite group of order n, and K a field with $n \in K^\times$. If G has, up to equivalence, precisely $k = k(G, K)$ irreducible K-representations, and if h is the number of conjugacy classes of G, then $k \leq h$. The following conditions are equivalent:*

(i) $k = h$;   (ii) $Z_K(G) : K = k$;

(iii) $Z(KG) : K = k$;   (iv) $Z(KG) \simeq K \times K \times \cdots \times K$.

If K is a splitting field of G, these four conditions always hold.

Proof. By F3, the center $Z(KG)$ of KG has dimension h over K. Then (17) implies that $k \leq h$, as well as the equivalence of (i)–(iv). If K is a splitting field of G, it follows from F12 that condition (iv) hold.   □

Note that condition (iv) amounts to K being a splitting field of the commutative algebra $Z(KG)$.

Theorem 3. *Let $A = KG$ be the group algebra of a finite group G of order n over a field such that $n \in K^\times$. For each simple component A_i of KG (see the notations of F11), the unit element e_i can be expressed in terms of the irreducible character χ_i and the invariant n_i of A_i by the formula*

$$(24) \qquad e_i = \frac{n_i}{n} \sum_{g \in G} \chi_i(g^{-1})g.$$

In particular, $n_i \in K^\times$. Observe that if K is a splitting field of G, the invariant n_i is the degree of the corresponding absolutely irreducible representation.

Proof. Write $e_i = \sum_{x \in G} a_x x$, with $a_x \in K$. For each $g \in G$ we have $g^{-1}e_i = \sum_x a_x g^{-1}x$. Applying the regular character ρ we obtain, by (7),

$$\rho(g^{-1}e_i) = \sum_x a_x \rho(g^{-1}x) = a_g \rho(1) = n a_g.$$

On the other hand, (23) says that $\rho(g^{-1}e_i) = \sum_j n_j \chi_j(g^{-1}e_i) = n_i \chi_i(g^{-1}e_i) = n_i \chi_i(g^{-1})$, because $g^{-1}e_i \in A_i$ and $\chi_j(A_i) = 0$ for $j \neq i$ (see item (c) in F11). Hence $n a_g = n_i \chi_i(g^{-1})$, which proves (24).   □

Theorem 4 (Orthogonality relations). *With the same assumptions and notation as in Theorem 3 we have, for each $g \in G$,*

$$(25) \qquad \frac{1}{n} \sum_{x \in G} \chi_i(xg)\chi_j(x^{-1}) = \begin{cases} \dfrac{1}{n_i}\chi_i(g) \text{ for } i = j, \\ 0 \qquad \text{ for } i \neq j. \end{cases}$$

Proof. We know that $e_i e_j = \delta_{ij} e_i$ for all i, j. Multiplying together two instances of (24) yields, thanks to (4),

$$e_i e_j = \frac{n_i}{n} \frac{n_j}{n} \sum_g \left(\sum_x \chi_i(xg^{-1})\chi_j(x^{-1}) \right) g.$$

The validity of (25) follows by comparing coefficients.   □

F15. *Let G be a finite group of order $n \in K^\times$. Let m be the exponent of G (the least common multiple of the orders of elements of G). If K contains a primitive m-th root of unity ζ_m, then K is a splitting field of the center $Z(KG)$ of KG, so the equivalent conditions is Theorem 2 are satisfied.*

(In actuality K is a splitting field of G as well, but for char $K = 0$ we will only be able to prove this at the end of the chapter.)

Proof. Let L be an algebraic closure of K. Then L is certainly a splitting field of G (see F5). Let $e_1, \ldots, e_h$ be the primitive central idempotents of LG. By Theorem 2, h is the *class number* of G. Now apply Theorem 3 to LG and use F4, to conclude that the assumption $\zeta_m \in K$ implies $e_i \in KG$. Hence KG contains the (central) subalgebra $Ke_1 \times \cdots \times Ke_h$. Since $Z(KG) : K = h$, therefore, we have $Z(KG) = Ke_1 \times \cdots \times Ke_h \simeq K \times \cdots \times K$.   □

F16. *Let G be a finite group of order n. If K is a field of characteristic $p > 0$ not dividing n, the equivalent conditions of Theorem 2 suffice to imply that K is a splitting field of G.*

Proof. Since the coefficients of the primitive central idempotents e_i of KG are algebraic over the prime field $\mathbb{F}_p$ of K — this by virtue of (24) and F4 — they all lie in a subfield F of K of finite degree over $\mathbb{F}_p$. All the e_i lie in FG. By assumption, there are h of them, where h is the class number of G. As before, it follows that $Z(FG) = Fe_1 \times \cdots \times Fe_h \simeq F \times \cdots \times F$. But F is a *finite field*, so Wedderburn's theorem (Chapter 29, Theorem 21) and F12 together say that F is a splitting field of G. Hence the extension K of F is also a splitting field of G (see F13).   □

Theorem 5. *Let G be a finite group of exponent m. Every field of characteristic $p > 0$ containing a primitive m-th root of unity is a splitting field of G.*

Proof. Clearly p does not divide m, much less the order n of G. Now use F15 and F16.   □

Now we examine more closely the remarkable formula in Theorem 4. For $g = 1$ we obtain

(26)
$$\frac{1}{n} \sum_{x \in G} \chi_i(x)\chi_j(x^{-1}) = \frac{d_i}{n_i}\delta_{ij}.$$

This motivates the following:

Definition 8. Let G be a finite group of order n and K a field such that $n \in K^\times$. If $f, g \in Z_K G$ are class functions, set

$$(27) \qquad \langle f, g \rangle = \frac{1}{n} \sum_{x \in G} f(x) g(x^{-1}).$$

The map $\langle , \rangle : Z_K G \times Z_K G \to K$ is obviously a *symmetric bilinear form*. By (26) the characters $\chi_1, \chi_2, \ldots, \chi_k$ are all orthogonal with respect to it.

F17. *Let the assumptions and notation be as in Definition* 8.

(a) *The symmetric bilinear form* $\langle , \rangle$ *on* $Z_K G$ *is nondegenerate.*

(b) *The characters* $\chi_1, \ldots, \chi_k$ *form a K-basis of* $Z_K G$ *if and only if* $k = h$.

(c) *If K is a splitting field of G then* $\chi_1, \ldots, \chi_k$ *form an orthonormal basis of* $Z_K G$.

(d) *If* $\langle \chi_i, \chi_i \rangle = 1$ *for* $1 \le i \le k$ *(that is, if the* $\chi_1, \ldots, \chi_k$ *are orthonormal) and* char $K = 0$, *then K is a splitting field of G.*

Proof. (a) Take a nonzero $f \in Z_K G$. There exists $y \in G$ such that $f(y^{-1}) \neq 0$. If k_y is the class sum of y, seen as an element of $Z_K G$, we have $\langle k_y, f \rangle = \frac{1}{n} c f(y^{-1})$, where c is the number of elements in the conjugacy class of g. But c divides n, so $\langle k_y, f \rangle \neq 0$.

(b) Suppose $k = h$. By Theorem 2 and (20), the degrees d_i of the χ_i equal $d_i = n_i s_i^2$, so we get from (26) the relations

$$(28) \qquad \langle \chi_i, \chi_j \rangle = s_i^2 \delta_{ij}.$$

This implies the linear independence of the χ_i (recall from F16 that in nonzero characteristic the s_i equal 1).

(c) If K is a splitting field of G, all the s_i equal 1.

(d) Together with (26), the assumption implies that $d_i = n_i$ for all i. By F12, then, K is a splitting field of G. $\qquad \square$

F18. *Let G be a finite group of order n and K a field such that $n \in K^\times$. If χ_V, χ_W are characters of G corresponding to the KG-modules V and W, we have*

$$(29) \qquad \langle \chi_V, \chi_W \rangle = \dim_K \operatorname{Hom}_{KG}(V, W).$$

Proof. We consider the KG-module $M := \operatorname{Hom}_K(V, W)$, with character χ_M. By (14) we have

$$\frac{1}{n} \sum_{x \in G} \chi_M(x) = \dim_K M^G.$$

This yields the claim, because $M^G = \operatorname{Hom}_{KG}(V, W)$ and $\chi_M(x) = \chi_V(x^{-1}) \chi_W(x)$ for all $x \in G$: see (13) and F8. $\qquad \square$

Remark 1. Applying (29) to irreducible modules V, W, we obtain a new proof of the orthogonality relations (26) (compare also item (d) and (20) in F11). Naturally one can work the other way around and prove (29) from (26), since both sides of (29) behave bilinearly in V and W.

Remark 2. Let G and K be as in F18 and let $\chi_1, \ldots, \chi_k$ be all the irreducible K-characters of G. In view of (26), the χ_i form a $\mathbb{Z}$-basis of the *character ring* $X_K(G)$ of G if none of the degrees d_i is divisible by char K (and in particular if char $K = 0$).

F19 (Dual orthogonality relations). *Let K be a splitting field of a finite group G of order n, where $n \in K^\times$. If $\chi_1, \ldots, \chi_h$ are the irreducible K-characters of G, we have for $a, b \in G$*

$$(30) \qquad \sum_{i=1}^{h} \chi_i(a^{-1})\chi_i(b) = \begin{cases} n_G(a) & \text{if } [a] = [b], \\ 0 & \text{if } [a] \neq [b]. \end{cases}$$

Here $n_G(a)$ is the order of the centralizer $Z_G(a)$ of a in G.

Proof. Let $c = n/n_G(a)$ be the number of elements in the conjugacy class $[a]$. Since $\chi_1, \ldots, \chi_h$ form an orthonormal basis of $Z_K G$, the class sum k_a can be written as

$$(31) \qquad k_a = \sum_{i=1}^{h} \langle k_a, \chi_i \rangle \chi_i = \sum_{i=1}^{h} \frac{c}{n} \chi_i(a^{-1})\chi_i.$$

Now substitute an arbitrary $b \in G$ to obtain (30). $\square$

We end this section by taking a closer look at representations of *abelian groups*:

F20. Let G be an *abelian* group of order n and exponent m. Let K be a field such that $n \in K^\times$, C an *algebraic closure* of K, ε a primitive m-th root of unity and G^* the set of homomorphisms $\lambda : G \to C^\times$. Since every irreducible C-representation of G has dimension one (F14), the elements of G^* are precisely the irreducible C-characters of G (which in turn are C-representations of G). Since $k = h = n$, the set G^* has precisely n elements; moreover G^* is an abelian group with respect to multiplication of characters. The elements of G^* only have values in the cyclic subgroup $\langle \varepsilon \rangle$ of $C^\times$. We call G^* the *character group* of G (see Section 14.2 in vol. I). For $\lambda \in G^*$, let

$$(32) \qquad K(\lambda) = K(\lambda G) \subseteq K(\varepsilon) \subseteq C$$

be the intermediate field of C/K obtained from K by adjoining all the $\lambda(g)$, for $g \in G$. The map T defined by

$$(33) \qquad T(x) = \lambda(x)_{K(\lambda)} \quad \text{for } x \in G,$$

where the right-hand side is multiplication by $\lambda(x)$, is obviously a K-representation of G; we also denote it by T_λ. The corresponding character $\chi = \chi_\lambda$ satisfies $\chi(x) = \mathrm{Tr}\,\lambda(x)_{K(\lambda)} = S_{K(\lambda)/K}(\lambda(x))$, that is,

$$(34) \qquad \chi(x) = \sum_\sigma \lambda(x)^\sigma = \sum_\sigma \lambda^\sigma(x),$$

where σ runs over the Galois group of the Galois extension $K(\lambda)/K$ and λ^σ is defined by $\lambda^\sigma(x) = \lambda(x)^\sigma$. Thus, if we set

$$(35) \qquad \mathrm{Sp}_K(\lambda) = \sum_\sigma \lambda^\sigma,$$

we have the following results:

(a) *For every* $\lambda \in G^*$, $\mathrm{Sp}_K(\lambda)$ *is the character of an irreducible K-representation of the abelian group G.* The representation (33) really is irreducible, because for every nonzero $\alpha \in K(\lambda)$ we have $T(KG)\alpha = K(\lambda)\alpha = K(\lambda)$.

Now let $KG = K_1 \times \cdots \times K_k$ be the decomposition of KG into simple components. Since KG is commutative, all the K_i are *fields*. Let $e_1, \ldots, e_k$ be their unit elements, and set

$$\lambda_i(x) = xe_i \in K_i \quad \text{for } x \in G.$$

Further, define $T_i(x) = \lambda_i(x)_{K_i}$ for $x \in G$; then $T_1, \ldots, T_k$ form a system of representatives of the irreducible K-representations of G, by F11(c). Moreover $K_i = K(\lambda_i(G))$. But each K_i is isomorphic over K to a subfield of $K(\varepsilon) \subseteq C$, so:

(b) *Every irreducible K-character of an abelian group G is of the form* $\mathrm{Sp}_K(\lambda)$, *for* $\lambda \in G^*$.

3. We set up the conventions for all of this section:

Assumptions and notation. *Let G be a finite group of order n. Let K be a field of* **characteristic zero** *and T an* **absolutely irreducible representation** *of G in a K-vector space V, with character χ and degree $d = \chi(1)$. Set*

$$(36) \qquad Z = Z(G), \quad \text{the center of } G.$$

Since T is absolutely irreducible, $\mathrm{End}_{KG}(V) = K\,\mathrm{id}_V$ (see F5, for example). Thus the restriction of T to the center $Z(KG)$ of the group algebra of KG has the form

$$(37) \qquad T(x) = \lambda(x)\,\mathrm{id}_V \quad \text{for all } x \in Z(KG),$$

for some homomorphism $\lambda : Z(KG) \to K$. Hence the class sum k_g of an elements $g \in G$ satisfies

$$T(k_g) = \lambda(k_g)\,\mathrm{id}_V,$$

which shows that $\chi(k_g) = d\,\lambda(k_g)$. Letting C be the conjugacy class of g in G, we therefore obtain

$$(38) \qquad \lambda(k_C) = \frac{|C|}{d}\chi(C),$$

where $|C|$ is the cardinality of C and $\chi(C)$ is the common value of $\chi(y)$ for $y \in C$.

F21. *Let the assumptions be as just stated. For every conjugacy class C of G, the element $(|C|/d)\chi(C)$ of K is integral over $\mathbb{Z}$.*

Proof. The product $k_{C_1} k_{C_2}$ of two class sums is obviously a linear combination of class sums k_C of G with *integer* coefficients. All these linear combinations form a $\mathbb{Z}$-subalgebra A of the $\mathbb{Z}$-algebra $Z(KG)$. Since A is finitely generated as a $\mathbb{Z}$-module, all the elements of A are *integral* over $\mathbb{Z}$ (see §28.8 on p. 323). In particular, each k_C is integral over $\mathbb{Z}$, and so is its homomorphic image $\lambda(k_C)$. Now a glance at (38) justifies the conclusion. □

Theorem 6. *Under the preceding assumptions, the degree d of χ divides the order n of the group G.*

Proof. For every $g \in G$, the value $\chi(g)$ is integral over $\mathbb{Z}$ (see F4 again). Because of F21, this implies that

$$(39) \qquad \sum_C \frac{|C|}{d} \chi(C)\chi(C^{-1}) \quad \text{is integral over } \mathbb{Z},$$

where the sum is over all the conjugacy classes C of G; note also that, if C is the conjugacy class of g, then $C^{-1} = \{y^{-1} \mid y \in C\}$ is the conjugacy class of g^{-1}. But the expression in (39) is none other than n/d, by (29). Being *integral* over $\mathbb{Z}$ and a rational number, n/d must in fact be an integer. □

A stronger result than Theorem 6 holds:

Theorem 6′. *Under the same assumptions, the degree d of χ divides the index $G : Z$ of the center Z of G.*

Proof. For $g \in G$ and $z \in Z$ we have $T(zg) = T(z)T(g) = \lambda(z)T(g)$, by (37); it follows that

$$(40) \qquad \chi(zg) = \lambda(z)\chi(g).$$

As the reader can easily check, the representation T can be assumed to be *faithful*, meaning that its kernel is trivial. Now use, as above, the relation

$$(41) \qquad \frac{n}{d} = \frac{1}{d} \sum_C |C|\chi(C)\chi(C^{-1}).$$

If C is the conjugacy class of g, the conjugacy class of zg is zC, for any $z \in Z$. By the faithfulness assumption, if $\chi(C)$ is nonzero, zC and C can only coincide if $z = 1$. Thus, if $C_1, \ldots, C_r$ form a complete set of representatives for the operation of Z on the conjugacy classes C of G, the right-hand side of (41) can be written as

$$(Z : 1) \sum_{i=1}^{r} \frac{|C_i|}{d} \chi(C_i)\chi(C_i^{-1}).$$

But then F21 says that the rational number

$$\frac{n}{d} \frac{1}{Z:1}$$

is integral over $\mathbb{Z}$, that is, an integer. Thus d divides $G:Z$.   □

F22. *Let χ be the character of a representation T of a finite group G in a $\mathbb{C}$-vector space V of dimension $d = \chi(1)$.*

(a) *For every $x \in G$, the absolute value of $\chi(x)$ satisfies $|\chi(x)| \leq d$, and equality obtains if and only if $T(x)$ is a homothety $\varepsilon \, \mathrm{id}_V$. Thus*

$$\ker T = \{x \in G \mid \chi(x) = d\}.$$

(b) *Let $g \in G$. If $\chi(g)/d$ is integral over $\mathbb{Z}$, then $|\chi(g)| = d$ or $|\chi(g)| = 0$.*

Proof. (a) We have $\chi(x) = \varepsilon_1 + \varepsilon_2 + \cdots + \varepsilon_d$, where the ε_i are roots of unity in $\mathbb{C}$. By the triangle inequality this implies $|\chi(x)| \leq d$, and equality only holds if $\varepsilon_1 = \varepsilon_2 = \cdots = \varepsilon_d$, that is, if $T(x) = \varepsilon \, \mathrm{id}_V$; for one can assume to begin with that $T(x)$ is a diagonal matrix, since *the restriction of T to the cyclic group $\langle x \rangle$ is semisimple* (see F6).

(b) Suppose $\chi(g) \neq 0$. Like $\chi(g)$, all the conjugates of $\chi(g)$ over $\mathbb{Q}$ are d-fold sums of roots of unity, so all have absolute value at most d. Thus all the conjugates $\alpha_1, \ldots, \alpha_r$ of the number $\alpha := \chi(g)/d$ have absolute value at most 1. But we know that if α is *integral* over $\mathbb{Z}$, the product $a_0 := \alpha_1 \alpha_2 \ldots \alpha_r$ is a nonzero integer. Hence $1 \leq |a_0| \leq |\alpha_1||\alpha_2| \ldots |\alpha_r| \leq 1$, that is, $|\alpha_i| = 1$ for all i. In particular, $|\chi(g)| = d$.   □

Lemma. *Let χ be an irreducible $\mathbb{C}$-character of a finite group G, and C a conjugacy class such that $\chi(C) \neq 0$. If $|C|$ is relatively prime to $d = \chi(1)$, then $|\chi(g)| = d$ for all $g \in C$.*

Proof. By assumption there exist $a, b \in \mathbb{Z}$ such that $a|C| + bd = 1$. Multiplication by $\chi(C)/d$ gives

$$a \frac{|C|}{d} \chi(C) + b \chi(C) = \frac{\chi(C)}{d}.$$

From F21 we see that the left-hand side is *integral* over $\mathbb{Z}$; the same being true of the other side, the claim follows using F22(b).   □

Theorem 7 (Burnside). *If a finite group G contains a conjugacy class with p^m elements, where p is prime and $m > 0$, then G has a nontrivial normal subgroup.*

Proof. Let C be such a conjugacy class, and let $\chi_1 = 1_G$, χ_2, ..., χ_h be all the irreducible $\mathbb{C}$-characters of G. By (30) we have

$$1 + \chi_2(1)\chi_2(C) + \cdots + \chi_h(1)\chi_h(C) = 0.$$

Dividing by p we conclude that for some $i \geq 2$ the quotient $\chi_i(1)\chi_i(C)/p$ is *not integral* over $\mathbb{Z}$ (otherwise $1/p$ would be integral); we fix such an i. The theorem's assumption implies that $|C|$ is relatively prime to $d_i = \chi_i(1)$; moreover, $\chi_i(C) \neq 0$. The Lemma then says that

(42) $$|\chi_i(C)| = d_i.$$

Let T_i be a representation with character χ_i. If T_i is not faithful, $\ker T_i$ is a nontrivial normal subgroup of G, and we're done. If T_i is faithful, F22 implies that the set

$$N = \{g \in G \mid |\chi_i(g)| = d_i\}$$

is the *center* of G. By (42), N contains C, and so is distinct from $\{1\}$. The case $N = G$ is excluded because $|C| > 1$. Hence G has a nontrivial normal subgroup. $\square$

This is a remarkable *group-theoretical application of representation theory*. It has a famous consequence:

Theorem 8 (Burnside's Theorem). *Any group G of order $p^a q^b$, where p and q are prime, is solvable.*

Proof. We can assume that G is neither abelian nor a p-group. By induction on the order of G, it suffices to show that G has a nontrivial normal subgroup N, since the solvability of N and G/N implies that of G.

Let Q be a Sylow q-group of G. Being a nontrivial q-group, Q contains an element $g \neq 1$ that commutes with everything in Q; thus the centralizer of g in G contains Q. This means the number of elements of the conjugacy class C of g in G divides $G : Q$, and so has the form p^m. If $m = 0$, then g is central in G and generates a normal subgroup; if $m > 0$, the claim follows from Theorem 7. $\square$

4. *In this section, G is a group and H is a subgroup of G.*

Any KG-module V is (by restriction) also a KH-module; when necessary we denote this KH-module by

$$\mathrm{res}_H(V).$$

Let T is the corresponding representation of G and — if $V : K$ is finite — let χ be its character. The representation T_H associated with $\mathrm{res}_H(V)$ is the restriction of T to KH, and its character is the restriction $\mathrm{res}_H(\chi)$ of χ to KH (or to H).

Definition 9. Let V be a KG-module, and assume that, as a K-vector space, it has a direct-sum decomposition

$$V = \bigoplus_{i \in I} W_i$$

into subspaces W_i that are *permuted transitively* by G (meaning that for any $g \in G$ and any W_i there exists W_j such that $g W_i = W_j$, and for any W_i and W_j there exists $g \in G$ such that $g W_i = W_j$). Now fix one of the summands, say W, and set

$$H = \{g \in G \mid g W = W\}.$$

Any family $(g_i)_{i\in I}$ such that $W_i = g_i W$ is a set of unique representatives for the left cosets gH of G mod H. Thus

$$(43) \qquad V = \bigoplus_{i\in I} g_i W = \bigoplus_{\rho\in G/H} \rho W.$$

In this situation we say that the KG-module V is induced by the KH-module W, and also, if $V:K < \infty$, that the character χ_V is induced by χ_W. Now one reads off from (43) the relation between the K-dimensions: $\dim V = (G:H)\dim W$. Moreover, if the KG-module V is irreducible, so is the KH-module W.

Definition 10. If W is *any* KH-module, we set

$$(44) \qquad \operatorname{ind}_H^G(W) := KG \otimes_{KH} W;$$

compare §28.9. (Note that KG is canonically a KH-*right module*.) Hence $\operatorname{ind}_H^G(W)$ has a natural KG-*module structure*, characterized by $g(x\otimes w) = gx \otimes w$ (which makes sense because $g(xh)\otimes w = (gx)h\otimes w = gx\otimes hw$).

F23. *A KG-module V is induced by a KH-module W if and only if the natural map*

$$(45) \qquad KG \otimes_{KH} W \to V$$

is an isomorphism (of KG-modules).

Proof. The map $(x, w) \mapsto xw$ from $KG \times W$ to V is *balanced* in the sense of §28.9, so it gives rise to a K-linear map (45), characterized by $x\otimes w \mapsto xw$. This map is clearly KG-linear. If $(g_i)_{i\in I}$ represent the cosets of H, we have

$$KG \otimes_{KH} W = \left(\bigoplus_i g_i KH\right) \otimes_{KH} W = \bigoplus_i g_i(KH \otimes_{KH} W).$$

Thus every element in the module (44) has a unique representation in the form $\sum_i g_i \otimes w_i$, with $w_i \in W$. Hence (45) is an isomorphism if and only if (43) holds. $\qquad\square$

We see in addition that the KG-module $KG \otimes_{KH} W$ is induced by the KH-module $W_1 = \{1\otimes w \mid w \in W\} \simeq W$. Together with the foregoing, this yields:

F24. *For every KH-module W there exists a KG-module V induced by W, and this module is unique up to isomorphism.*

Remark 1. Let W be a KH-module and V a KG-module. There is a natural isomorphism of K-vector spaces

$$(46) \qquad \operatorname{Hom}_{KH}(W, \operatorname{res}_H V) \simeq \operatorname{Hom}_{KG}(\operatorname{ind}_H^G W, V)$$

and a natural isomorphism of KG-modules

$$(47) \qquad \operatorname{ind}_H^G W \otimes_K V \simeq \operatorname{ind}_H^G(W \otimes_K \operatorname{res}_H V).$$

Indeed, let's assign to $f \in \mathrm{Hom}_{KG}(KG \otimes_{KH} W, V)$ the map

$$W \to \mathrm{Hom}_{KG}(KG, V),$$
$$w \mapsto (x \mapsto f(x \otimes w)).$$

Then there is an isomorphism between $\mathrm{Hom}_{KH}(W, \mathrm{Hom}_{KG}(KG, V))$ and the right-hand side of (46). Taken together with the isomorphism $\mathrm{Hom}_{KG}(KG, V) \simeq V$ defined by $s \mapsto s(1)$, this yields (46). For (47) we observe that a natural isomorphism $(KG \otimes_{KH} W) \otimes_K V \simeq KG \otimes_{KH} (W \otimes_K V)$ of KG-modules is obtained by the prescription $(g \otimes w) \otimes v \mapsto g \otimes (w \otimes g^{-1}v)$.

Remark 2. Let G be a finite group of order n, K a field such that $n \in K^{\times}$, W an *absolutely irreducible* KH-module and V an *absolutely irreducible* KG-module. By (46), V *appears in* $\mathrm{ind}_H^G W$ *with the same multiplicity as W in* $\mathrm{res}_H V$. This fact is known as *Frobenius reciprocity*.

F25. *Induction is transitive: If U is a subgroup of G and $H \subseteq U$, every KH-module W satisfies*

$$(48) \qquad \mathrm{ind}_H^G(W) = \mathrm{ind}_U^G\big(\mathrm{ind}_H^U(W)\big).$$

Proof. The isomorphism $KG \otimes_{KH} W \simeq KG \otimes_{KU}(KU \otimes_{KH} W)$ is an isomorphism of KG-modules. $\qquad \square$

Definition 11. Let G be finite and suppose the order of H is not divisible by char K. Define a map $\mathrm{ind}_H^G : Z_K H \to Z_K G$ on *class functions* by setting

$$(49) \qquad \mathrm{ind}_H^G(\psi)(x) = \frac{1}{|H|} \sum_{y \in G} \psi(y^{-1}xy),$$

with the stipulation that $\psi(y^{-1}xy)$ is 0 if $y^{-1}xy \notin H$.

The reason for this definition lies in the next result:

F26. *In the situation of Definition 11, if a KG-module V is induced by the KH-module W and $V : K < \infty$, then $\chi_V = \mathrm{ind}_H^G(\chi_W)$.*

Proof. As a K-vector space, V is a direct sum $\bigoplus g_i W$, as in (43). Choose bases for each subspace $g_i W$ and combine them together into a basis of V. For $x \in G$ we have

$$x g_i W = g_{i(x)} W,$$

for a unique index $i(x)$. This index equals i if and only if $g_i^{-1} x g_i \in H$. Since for $i(x) \neq i$ there is no contribution to the trace $\chi_V(x)$, we obtain

$$\chi_V(x) = \sum_{g_i^{-1} x g_i \in H} \mathrm{Tr}(x_{g_i W}),$$

where the sum is over all i such that $g_i^{-1} x g_i \in H$. Since g_i gives rise to a K-isomorphism $W \to g_i W$, we then have $\mathrm{Tr}(x_{g_i W}) = \mathrm{Tr}((g_i^{-1} x g_i)_W) = \chi_W(g_i^{-1} x g_i)$ for the i in question. The claim follows, because $(g_i h)^{-1} x (g_i h) = h^{-1}(g_i^{-1} x g_i) h$ for all $h \in H$ (and χ_W is a class function on H). $\qquad \square$

F27. *Let G be finite of order $n \in K^{\times}$. For every $\varphi \in Z_K G$ and $\psi \in Z_K H$ we have*

$$\langle \varphi, \operatorname{ind}_H^G(\psi) \rangle_G = \langle \operatorname{res}_H(\varphi), \psi \rangle_H, \tag{50}$$

$$\varphi \operatorname{ind}_H^G(\psi) = \operatorname{ind}_H^G(\operatorname{res}_H(\varphi)\,\psi). \tag{51}$$

As a consequence of (51), $\operatorname{ind}_H^G(X_K H)$ *is an ideal of the ring* $X_K(G)$.

Proof. The two equations can be verified by a direct computation starting from the definition in (49). They can also be proved conceptually: By enlarging K, one can assume that every class function on G and H is a K-linear combination of K-characters (see F17 and F15). Thus φ and ψ can be assumed to be K-characters. Now (50) follows from (46) via (29), and (51) follows from (47).   □

F28 (Clifford's Theorem). *Let V be an **irreducible** KG-module with $V : K < \infty$, and let N be a **normal subgroup** of G. Then V is semisimple as a KN-module, and we are in one of two cases:*

(a) *The KN-module V is isogenous.*

(b) *There is a subgroup H of G satisfying $N \subseteq H \neq G$ and such that V is induced by an irreducible KH-module W.*

Proof. Normality comes in as follows: If U is a KN-submodule of V, so is gU. If U is irreducible, so is gU; if U_1 and U_2 are isomorphic KN-submodules, so are gU_1 and gU_2.

Because it has finite dimension over K, the KN-module V admits an *irreducible* submodule U. Since V is irreducible as a KG-module, it is the sum of its submodules of the form gU. Hence, as a KN-module, V is *semisimple* (see F6 in Chapter 28). Now let W be an isogenous component of the KN-module V. Then gW is also one, for every $g \in G$. Since the KG-module V is *irreducible*, V is the sum of the components gW. But the KN-module V is the *direct* sum of its isogenous components. Thus, if we set $H = \{g \in G \mid gW = W\}$, the KN-module V is induced by the KH-module W. Moreover $N \subseteq H$. If $V = W$, we are in case (a); otherwise, we have $H \neq G$ and (b) is satisfied.   □

Theorem 9 (Ito's Theorem). *If char $K = 0$, the degree of an absolutely irreducible K-representation T of a finite group G divides the index of every abelian normal subgroup A of G.*[1]

Proof. Clearly we can assume that K is *algebraically closed*. We then proceed by induction on $\operatorname{ord}(G)$. If we're in case (b) of F28 with $N = A$, we can assume by induction that the dimension $\dim_K W$ divides $(H : A)$, so the degree in question, $\deg T = (G : H) \dim_K W$, divides $(G : A)$. Now suppose we're in case (a) instead. Since every irreducible K-representation of an abelian group A is one-dimensional, T is given by $T(a) = \lambda(a) \operatorname{id}_V$ for some character λ of A, and in particular $T(A)$ is contained in the center of $T(G)$. Therefore the degree of T is a divisor of $T(G) : T(A)$, by Theorem 6′, and so surely it divides $G : A$ as well.   □

---

[1] In fact the weaker assumption $\operatorname{ord}(G) \in K^{\times}$ suffices; see Huppert, *Endliche Gruppen*.

Definition 12. A finite group G is called *supersolvable* if there is a sequence $1 = N_0 \subseteq N_1 \subseteq \cdots \subseteq N_r = G$ of normal subgroups N_i *of G* such that all the quotients N_i/N_{i-1} are *cyclic*.

Remark. Like solvability (which requires merely that N_{i-1} be normal in N_i), supersolvability is inherited by subgroups and quotient groups. Every finite p-group is supersolvable, as is of course any finite abelian group. (Compare with F1 in Chapter 15 of vol. I.)

Theorem 10. *Let G be a finite group of order n and K an algebraically closed field such that $n \in K^\times$. If G is supersolvable, every irreducible K-representation of G is* **monomial**, *that is, induced by a representation of degree 1.*

Proof. We can assume the given $T : G \to \mathrm{GL}_K(V)$ is *faithful*. If G is abelian, T is already of degree 1; hence assume G is not abelian. Then G has an abelian normal subgroup A not contained in the center Z of G. (To see this consider the group $\bar{G} = G/Z \neq 1$. Since it too is supersolvable, it has a cyclic normal subgroup $\bar{A} \neq 1$. The inverse image A of $\bar{A}$ in G is a normal subgroup of G not contained in Z, and it is abelian because it is a cyclic extension of a central subgroup.) Since T is faithful, $T(A)$ is not in the center of $T(G)$, so the KA-module V cannot be isogenous. Hence we are in case (b) of F28 with $N = A$, and T is induced by an irreducible K-representation of a proper subgroup H of G. By induction on n this suffices to prove the theorem. $\square$

5. We saw in Theorem 10 that every irreducible $\mathbb{C}$-representation of a supersolvable group is monomial, that is, induced by a one-dimensional representation. This is not necessarily the case if the finite group G is not supersolvable. But as *R. Brauer* recognized, every $\mathbb{C}$-character of G is at least a $\mathbb{Z}$-linear combination of characters of monomial representations. This fact is very important in group theory and also plays a role in number theory, where the question originally arose.

Here is the set-up for the rest of this section:

Assumptions and notation. Let G be a finite group of order n, and m a common multiple of the orders of the elements of G (for instance, the order or the exponent of G). Let K be a field of characteristic zero, ε a primitive m-th root of unity in an algebraic closure C of K, and $R = \mathbb{Z}[\varepsilon]$ the subring of C generated by ε. Next, given a subgroup H of G,

let $X_K(H, R)$ be the ring of R-linear combinations of K-characters of H.

Finally, let p be a prime number and

$$(52) \qquad\qquad n = p^{w_p(n)} n_p$$

the decomposition of n, the group order, with respect to p. Note that passing from $\mathbb{Z}$ to $R = \mathbb{Z}[\varepsilon]$ in the definition of $X_K(H, R)$ is practicable, because the values of all characters of H lie in R. The ring $X_K(H, R)$ contains the character ring $X_K(H)$

of H as a subring; see (11). Now, the totality of irreducible K-characters of H, say $\chi_1, \ldots, \chi_k$, forms an R-basis of $X_K(H, R)$, because they form an *orthogonal basis* relative to the symmetric bilinear form $\langle , \rangle$ on $X_K(H, R) \subseteq Z_C H$. And obviously $\operatorname{ind}_H^G(X_K(H, R)) \subseteq X_K(G, R)$.

Theorem 11 (Artin's induction theorem). *For every K-character χ of G there are cyclic subgroups $A_1, \ldots, A_r$ of G and irreducible K-characters ξ_i of the A_i such that*

$$(53) \qquad n\chi = \sum_{i=1}^r a_i \operatorname{ind}_{A_i}^G(\xi_i),$$

for appropriate $a_i \in \mathbb{Z}$.

Proof. (a) For any cyclic subgroup A of G, consider the function $\gamma_A : A \to \mathbb{Z} \subseteq K$ defined by

$$(54) \qquad \gamma_A(x) = \begin{cases} |A| & \text{if } \langle x \rangle = A, \\ 0 & \text{otherwise.} \end{cases}$$

Set $\gamma_A^* = \operatorname{ind}_A^G(\gamma_A)$. By definition, for any $x \in G$ we have

$$\gamma_A^*(x) = \frac{1}{|A|} \sum_{y \in G} \gamma_A(y^{-1}xy) = |\{y \in G \mid \langle y^{-1}xy \rangle = A\}|.$$

Summing over all A yields $\sum_A \gamma_A^*(x) = |G| = n$, because every $y \in G$ determines a unique cyclic subgroup A with $A = \langle y^{-1}xy \rangle$. Thus we obtain

$$(55) \qquad n \, 1_G = \sum_A \operatorname{ind}_A^G(\gamma_A).$$

(b) Apply (55) to a *fixed* cyclic subgroup A of G, rather than G itself, to obtain

$$|A| \, 1_A = \sum_{A'} \operatorname{ind}_{A'}^A(\gamma_{A'}) = \gamma_A + \sum_{A' \neq A} \operatorname{ind}_{A'}^A(\gamma_{A'}),$$

where A' runs over the cyclic subgroups of A. By induction on $|A|$, this shows that $\gamma_A \in X_K(A)$ is a generalized K-character of A.

(c) Now just multiply (55) by χ to get

$$(56) \qquad n\chi = \sum_A \operatorname{ind}_A^G(\psi_A),$$

with $\psi_A = \operatorname{res}_A^G(\chi)\gamma_A \in X_K(A)$ by (51). Decomposing each ψ_A into irreducible K-characters of A yields the conclusion of the theorem in the form of (53); of course the A_i in that expression are not necessarily distinct. $\qquad\square$

Definition 13. For every σ in the Galois group $G(K(\varepsilon)/K)$, there is a unique associated $j = j_\sigma \in (\mathbb{Z}/m\mathbb{Z})^\times$ satisfying

(57) $$\sigma(\zeta) = \zeta^j \quad \text{for all } m\text{-th roots of unity } \zeta \text{ in } K(\varepsilon).$$

The assignment $\sigma \mapsto j_\sigma$ identifies $G(K(\varepsilon)/K)$ with a subgroup of $(\mathbb{Z}/m\mathbb{Z})^\times$. When necessary, we write the σ described by (57) as σ_j. For every $x \in G$ we can then set

(58) $$x^\sigma := x^j.$$

We say that $x, y \in G$ are *K-conjugates* if there exist $z \in G$ and $\sigma \in G(K(\varepsilon)/K)$ such that

(59) $$x^\sigma = z^{-1} y z.$$

We then write $x \overset{K}{\sim} y$. This clearly defines an equivalence relation on G, whose classes are the *K-classes* of G. If $\varepsilon \in K$, the K-classes are just the usual conjugacy classes. If $K = \mathbb{Q}$, two elements x, y are K-conjugate if and only if the cyclic groups $\langle x \rangle$ and $\langle y \rangle$ are conjugate, since $G(\mathbb{Q}(\varepsilon)/\mathbb{Q}) = (\mathbb{Z}/m\mathbb{Z})^\times$ (Gauss).

F29. *Any K-character χ is constant on any K-class of G.*

Proof. It suffices to show that $\chi(x^j) = \chi(x)$ for $\sigma = \sigma_j \in G(K(\varepsilon)/K)$ and for every $x \in G$. Let χ be the character from the K-representation T, and let $\varepsilon_1, \ldots, \varepsilon_d$, with $d = \chi(1)$, be the eigenvalues of $T(x)$. The ε_i are m-th roots of unity, and we can write $\chi(x^j) = \operatorname{Tr} T(x)^j = \sum_i \varepsilon_i^j = \sum_\sigma \varepsilon_i^\sigma = \chi(x)^\sigma = \chi(x)$, because $\chi(x) \in K$. $\square$

F30. *Let the function $f : G \to R$ be constant on the K-classes of G. Then*

(60) $$nf = \sum_A \operatorname{ind}_A^G(\psi_A) \quad \text{for some } \psi_A \in X_K(A, R),$$

where the sum is taken over all cyclic groups A of G. In particular, $nf \in X_K(G, R)$.

Proof. Multiplying (55) by f we obtain (60) with $\psi_A = \gamma_A \cdot \operatorname{res}_A(f)$, where we still have to show that $\psi_A \in X_K(A, R)$. Let $\psi_A = \sum_\lambda a_\lambda \lambda$ be the decomposition of ψ_A into the absolutely irreducible C-characters λ of A. Since the values of ψ_A lie in $(A : 1)R$, all the a_λ belong to R. Now, for any $\sigma = \sigma_j$ we have $\psi_A(x) = \psi_A(x^\sigma) = \sum_\lambda a_\lambda \lambda(x)^\sigma$. Thus conjugate λ's occur in ψ_A equally often; hence ψ_A is a linear combination of traces $\operatorname{Tr}_K(\lambda_i)$ of certain λ's, and with coefficients in R to boot. The conclusion now follows from F20(a). $\square$

F31. *On the cyclic group $A = \langle a \rangle$, consider the function $\xi_a : A \to \mathbb{Z}$ defined by*

(61) $$\xi_a(x) = \begin{cases} |A| & \text{if } x \overset{K}{\sim} a \text{ in } A, \\ 0 & \text{otherwise.} \end{cases}$$

Then ξ_a is an element of $X_K(A, R)$.

Proof. We have $\xi_a = (A : 1)f$ for some $f : A \to \mathbb{Z}$. The assertion then follows easily from F30, applied to the group A in place of G.   □

Recall that a group G is a *semidirect product* of two subgroups N and V, of which N is normal in G, if $G = NV$ and $N \cap V = 1$.

Definition 14. A subgroup H of G is called *K-elementary* (for the prime p) if H is the semidirect product of a cyclic group A with a Sylow p-group P of H, and for every $y \in P$ there exists $\sigma \in G(K(\varepsilon)/K)$ such that

$$(62) \qquad y^{-1}xy = x^\sigma \quad \text{for all } x \in A.$$

If it is always possible to choose $\sigma = 1$ in (62) — equivalently, if H is a *direct* product of A and P — we call H *elementary*. If $\varepsilon \in K$, every K-elementary subgroup is elementary.

F32. *If H is K-elementary, every K-representation of A can be extended to a K-representation of H.*

Proof. It suffices to show that every *irreducible* K-representation T of A on H can be extended. By F20 we can imagine that $T = T_\lambda$ is given in the form (33). Take $y \in P$. By assumption there is a corresponding $\sigma \in G(K(\varepsilon)/K)$, such that (62) holds for all $x \in A$. We define $T(y)$ as the restriction of σ^{-1} to the subfield $K(\lambda)$ of $K(\varepsilon)$; this is independent of the choice of σ. Clearly T is multiplicative on P, and one can check easily that

$$(63) \qquad T(y)^{-1}T(x)T(y) = T(x^\sigma) \quad \text{for } x \in A.$$

Thus, if we define $T(xy) = T(x)T(y)$, we obtain a K-representation T of H extending the given representation of A.   □

We say an element of G is *p-regular* if its order is relatively prime to p.

F33. *Let $a \in G$ be p-regular, and denote by $N_G^K(a)$ the set of $y \in G$ for which there is $\sigma \in G(K(\varepsilon)/K)$ such that equation (62) holds for all $x \in A = \langle a \rangle$. Let P be a Sylow p-group of $N_G^K(a)$, and set $H = AP$. (Obviously, H is a K-elementary subgroup of G.)*
   There exists $\psi \in X_K(H, R)$ such that $\varphi = \mathrm{ind}_H^G(\psi)$ satisfies these properties:

   (i) $\varphi(a) = N_G^K(a) : P \not\equiv 0 \bmod p$.

   (ii) $\varphi(b) = 0$ *for any p-regular element b of G that is not K-conjugate to a.*

Proof. We start with the function $\xi_a \in X_K(A, R)$ in (61). By F32, ξ_a can be extended to a function $\psi \in X_K(H, R)$. For any $b \in G$ we have

$$(64) \qquad \varphi(b) = \frac{1}{|H|} \sum_{y \in G} \psi(y^{-1}by).$$

Now, if b is p-regular, the same is true of its conjugate $y^{-1}by$. Thus the condition $y^{-1}by \in H$ implies $y^{-1}by \in A$ and hence $\psi(y^{-1}by) = \xi_a(y^{-1}by)$. Therefore, if b is not K-conjugate to a, it follows by the definition of ξ_a in (61) that $\varphi(b) = 0$.

Next apply (64) to $b = a$, to obtain

$$\varphi(a) = \frac{1}{|H|}|N_G^K(a)|\,|A| = N_G^K(a) : P \not\equiv 0 \bmod p. \qquad \square$$

Before we come to the main result of this section, we need some preparation:

Lemma. *For $x \in G$ let $x = x_r x_s$ be the decomposition of x into a p-regular component x_r and a p-component x_s. For every character χ of a representation T of G we have*

$$(65) \qquad\qquad \chi(x) \equiv \chi(x_r) \bmod \mathfrak{r},$$

were $\mathfrak{r}/pR$ is the radical of the $\mathbb{Z}/p\mathbb{Z}$-algebra R/pR.

Proof. Let $\varepsilon_1, \ldots, \varepsilon_d$, where $d = \chi(1)$, be the eigenvalues of $T(x)$. For every p-power q we have $\chi(x^q) = \sum \varepsilon_i^q \equiv (\sum \varepsilon_i)^q = \chi(x)^q \bmod pR$; analogously, $\chi(x_r^q) \equiv \chi(x_r)^q \bmod pR$. At the same time, for large enough p-powers q we have $x^q = x_r^q$, so we get

$$\chi(x)^q \equiv \chi(x_r)^q \bmod pR, \quad \text{that is,} \quad (\chi(x) - \chi(x_r))^q \equiv 0 \bmod pR.$$

But the radical of R/pR contains all nilpotents, so $\chi(x) - \chi(x_r) \in \mathfrak{r}$ as needed. $\square$

Theorem 12 (Brauer–Witt induction theorem). *Every K-character of G is an integer linear combination of induced K-characters of K-elementary subgroups of G. In other words, if E_K denotes the set of K-elementary subgroups H of G, the homomorphism*

$$(66) \qquad\qquad \mathrm{ind}: \bigoplus_{H \in E_K} X_K(H) \to X_K(G)$$

given by the induction maps $\mathrm{ind}_H^G : X_K(H) \to X_K(G)$ is surjective. (The domain of ind is the direct sum of the $X_K(H)$ taken as abelian groups, not, say, the coproduct of the rings $X_K(H)$.)

Proof. (a) Denote by $Y_K(G)$ the image of (66) and by $Y_K(G, R)$ the image of the analogous map with $X_K(H)$ replaced by $X_K(H, R)$, where $R = \mathbb{Z}[\varepsilon]$. Since the powers $1, \varepsilon, \varepsilon^2, \ldots, \varepsilon^{d-1}$, where $d = \varphi(m)$, form a $\mathbb{Z}$-basis of $\mathbb{Z}[\varepsilon]$, they form an $X_K(H)$-basis of $X_K(H, R)$ as well, and this for *any* subgroup H of G. Hence

$$(67) \qquad\qquad Y_K(G, R) \cap X_K(G) = Y_K(G).$$

Now, $Y_K(G)$ is an *ideal* of the ring $X_K(G)$ (see F27), so the theorem will be proved if we show that 1_G lies in $Y_K(G)$. By (67) this will be the case if $1_G \in Y_K(G, R)$.

(b) For a given prime p, let $a_1, a_2, \ldots, a_r$ be unique representatives of all the K-classes of p-regular elements of G. By F33 there exists for every $1 \le i \le r$ some $\varphi_i \in Y_K(G, R)$ such that $\varphi_i(a_i) \in \mathbb{N}$ is relatively prime to p and $\varphi_i(a_j) = 0$ whenever $j \ne i$. The sum $\varphi = \sum_i \varphi_i^{p-1}$ lies in $Y_K(G, R)$, and in view of F29 it satisfies $\varphi(x) \equiv 1 \bmod pR$ *for every p-regular* element x of G. Now use the preparatory lemma to deduce that

(68)
$$\varphi(x) \equiv 1 \bmod \mathfrak{r} \quad \text{for } all \; x \in G.$$

(c) Take $\alpha \in R$. If $\alpha \equiv 1 \bmod \mathfrak{r}^i$, clearly $\alpha^p \equiv 1 \bmod \mathfrak{r}^{i+1}$. So by replacing φ in (68) by an appropriate power of φ, we obtain for a predetermined e a function $\tilde{\varphi} \in Y_K(G, R)$ such that

(69)
$$\tilde{\varphi}(x) \equiv 1 \bmod p^e R \quad \text{for all } x \in G.$$

In particular this holds for $e = w_p(n)$; see (52). Thus the function $n_p(\tilde{\varphi} - 1_G)$ has values in nR; that is,

(70)
$$n_p(\tilde{\varphi} - 1_G) = nf$$

for some function $f : G \to R$, which is constant on K-classes of G because of F29. By F30, therefore, nf belongs to $Y_K(G, R)$, because cyclic subgroups are elementary, hence K-elementary, subgroups of G. Thus, like $\tilde{\varphi}$, the function in (70) belongs to $Y_K(G, R)$, and it follows that

(71)
$$n_p 1_G \in Y_K(G, R).$$

We derived (71) for an arbitrary prime p. But the greatest common divisor of the n_p is 1, so 1 is an integer linear combination of the n_p. Thus $1_G \in Y_K(G, R)$, which proves the result by part (a). $\square$

Theorem 13. *A class function $\varphi : G \to \mathbb{C}$ is a generalized K-character of G if and only if for every K-elementary subgroup H of G the restriction $\mathrm{res}_H(\varphi)$ is a generalized K-character.*

Proof. Only the "if" part needs to be proved. By Theorem 12 we have

$$1 = \sum_{H \in E_K} \mathrm{ind}_H^G(\psi_H) \quad \text{with } \psi_H \in X_K(H).$$

Multiplication by φ yields $\varphi = \sum_H \mathrm{ind}_H^G(\Theta_H)$, with $\Theta_H = \mathrm{res}_H(\varphi)\psi_H$. Thus φ lies in $X_K(G)$ whenever $\mathrm{res}_H(\varphi)$ lies in $X_K(H)$ for all $H \in E_K$. $\square$

Theorem 14 (Brauer). *Every $\mathbb{C}$-character of G is an integer linear combination of characters of **monomial** $\mathbb{C}$-representations.*

Proof. This too follows from Theorem 12, using Theorem 10 and the obvious fact that an elementary subgroup $H = A \times P$ is supersolvable. $\square$

Definition 15. A C-representation T of G (or its representation module, or its character χ) is called *realizable over K* if there exists a K-representation T_0 of G such that T is equivalent to T_0^C.

If V_0 is the representation module for T_0, there must be an isomorphism of CG-modules $V \simeq V_0 \otimes_K C$. As a function on G, χ is also the character for V_0. In particular, the KG-module V_0 is unique up to isomorphism. Further, if V is irreducible, V_0 is an absolutely irreducible KG-module.

F34. *K is a splitting field of G if and only if every C-representation of G is realizable over K.*

Proof. (a) Let $\chi_1, \ldots, \chi_k$ be the irreducible K-characters of G. If K is a splitting field of G, then $k = h$, and $\chi_1, \ldots, \chi_k$ form an *orthonormal basis* of $Z_K G$ with respect to $\langle , \rangle$. Thus every C-character χ of G can be written in the form $\chi = \sum_i \langle \chi, \chi_i \rangle \chi_i$. But since χ and χ_i are characters of C-representations of G, the numbers $\langle \chi, \chi_i \rangle$ must be nonnegative integers. Hence χ is a K-character, that is, χ is realizable over K.

(b) Now $\chi_1, \ldots, \chi_h$ will denote the irreducible C-characters of G. If they are all realizable over K, they form an orthonormal set of K-characters of G. Since $\langle \chi_i, \chi_i \rangle = 1$, each χ_i is an *irreducible* K-character. Hence $\chi_1, \ldots, \chi_h$ are all the distinct irreducible K-characters of G (see Theorem 2). Because they form an orthonormal set, K is a splitting field of G (see F17).   □

Theorem 15 ("Schur's conjecture"). *Let m be the exponent of G. If K contains a primitive m-th root of unity, every C-representation is realizable over K, that is, K is a splitting field of G (see F34).*

Proof. Let χ be any C-character. By Theorem 14,

$$(72) \qquad \chi = \sum_i a_i \operatorname{ind}_{H_i}^G (\lambda_i) \quad \text{with } a_i \in \mathbb{Z},$$

where the λ_i are C-characters of degree 1 on certain subgroups H_i of G. Since the values of the λ_i are m-th roots of unity, they all lie in K by assumption. Thus λ_i is a K-character of H_i, for every i. From (72) it follows that $\chi \in X_K(G)$; that is, $\chi = \sum_j b_j \chi_j$ is an *integer* linear combination of the irreducible K-characters $\chi_1, \ldots, \chi_k$ of G, with coefficients b_j given by $\langle \chi, \chi_j \rangle = b_j \langle \chi_j, \chi_j \rangle$. Since χ and the χ_j are proper characters, we must have $\langle \chi, \chi_j \rangle \geq 0$ and so also $b_j \geq 0$. Therefore χ is a K-character, and so realizable over K.   □

34

The Schur Group of a Field

1. In this chapter we address the question of which simple algebras can occur as simple components of group algebras KG of finite groups over fields K of characteristic 0.

Notation. Let G be a finite group, K a field of characteristic 0, C an algebraic closure of K, χ an irreducible C-character of G. We denote by $K(\chi)$ the subfield of C obtained by adjoining to K all values $\chi(g)$ taken by χ, for $g \in G$.

Suppose given an irreducible CG-module V. It gives rise to an irreducible C-representation

$$(1) \qquad\qquad T : CG \to \operatorname{End}_C(V)$$

of G (or of CG). The corresponding character χ is a function on G (or CG) with values in C. Now consider the restriction

$$(2) \qquad\qquad T : KG \to \operatorname{End}_C(V)$$

of T to the group algebra of G over K. The image $B = T(KG)$ in $\operatorname{End}_C(V)$ is a K-subalgebra of $\operatorname{End}_C(V)$. The group algebra KG is *semisimple*, so B, being a homomorphic image of KG, is isomorphic to a direct product of certain simple components of KG. We claim that B is in fact *simple*. To prove this we need only show that the center F of B is a *field*. More precisely, we wish to prove that $F = K(\chi)$, where we have identified elements $a \in C$ with the respective homotheties $a \operatorname{id}_V \in \operatorname{End}_C(V)$. Since T is (absolutely) irreducible, the map (1) is surjective, so $\operatorname{End}_C(V)$ is generated by B and C. Therefore the center of B lies in the center of $\operatorname{End}_C(V)$; that is, $F \subseteq C$.

It follows that, if an element $x = \sum a_g g$ of KG is such that $T(x)$ is central in B, we have $\chi(x) = dT(x)$, where $d = \dim_C V$. On the other hand, the value $\chi(x) = \sum a_g \chi(g)$ lies in $K(\chi)$, so $T(x) \in K(\chi)$. Thus $F \subseteq K(\chi)$. Conversely, if g is an element of G, we can take the class sum k_g of g, which lies in the center of KG. Hence $T(k_g)$ is contained in the center F of B, and $\chi(k_g) = dT(k_g)$. But $\chi(k_g)$ equals $\chi(g)$ times the number of conjugates of g, so $\chi(g)$ also lies in F. Thus $K(\chi) \subseteq F$.

In sum:

F1. *Associated to any irreducible C-character χ of G there is a unique component $B(\chi, K)$ of KG for which $\chi(B(\chi, K))$ is nonzero. The center of $B(\chi, K)$ is K-isomorphic to $K(\chi)$. Setting $F := K(\chi)$ we have an isomorphism of F-algebras $B(\chi, F) \simeq B(\chi, K)$.*

Proof of the last statement. We have $B(\chi, K) \simeq B = FB = FT(KG) = T(FG) \simeq B(\chi, F)$. In view of the K-isomorphism $B(\chi, F) \simeq B(\chi, K)$, then, we can make $B(\chi, K)$ into an F-algebra in such a way that

$$(3) \qquad B(\chi, F) \simeq B(\chi, K) \quad \text{as } F\text{-algebras} \quad (\text{where } F = K(\chi)). \qquad \square$$

In the sequel the $K(\chi)$-algebra structure of $B(\chi, K)$ will always be the one so determined.

Definition 1. Suppose the central-simple $K(\chi)$-algebra $B(\chi, K)$ has Schur index s. We also say that s is the *Schur index of χ over K*, and denote it by $s_K(\chi)$. Because of (3) we have

$$(4) \qquad s_K(\chi) = s_{K(\chi)}(\chi).$$

The unique K-character of G belonging to the simple component $B(\chi, K)$ of KG will be denoted by χ_K.

F2. *Let E/F be any extension of intermediate fields of C/K, and take the simple component $B(\chi, F)$ of FG corresponding (in the sense of F1) to the character χ. Then*

$$EG = FG \otimes_F E = (B(\chi, F) \otimes_F E) \times \cdots.$$

(a) *The simple component $B(\chi, E)$ of EG can therefore be identified with a simple component of the semisimple algebra $B(\chi, F) \otimes_F E$. In particular, if $F(\chi) = F$, the F-algebra $B(\chi, F)$ is central-simple by F1, so $B(\chi, F) \otimes_F E$ is simple and hence*

$$(5) \qquad B(\chi, E) = B(\chi, F) \otimes_F E.$$

(b) By definition, $B(\chi, C)$ is the simple component of CG corresponding in the sense of Chapter 33, F11 to a C-representation with character χ. Since, by part (a), $B(\chi, C)$ occurs as a simple component in $B(\chi, K) \otimes_K C$, we conclude: *The character χ always occurs in the character χ_K (regarded as a C-character) associated with the simple component $B(\chi, K)$ of KG; in symbols,*

$$(6) \qquad \langle \chi_K, \chi \rangle \neq 0.$$

Conversely, let ψ be an irreducible K-character of G and B the associated simple component of KG. For any irreducible C-character χ of G occurring in ψ (i.e., such that $\langle \psi, \chi \rangle \neq 0$), we have $B = B(\chi, K)$ and hence $\psi = \chi_K$.

Definition 2. Let n be the order of G and ζ a primitive n-th root of unity in C. Clearly, $K(\chi)$ is a subfield of $K(\zeta)$. Hence $K(\chi)/K$ is an *abelian field extension*. Now, it is easy to the check that the composition

$$\chi^\sigma := \sigma \circ \chi$$

of any $\sigma \in G(K(\chi)/K)$ with an *irreducible C-character* χ of G is also an irreducible C-character. We call such χ^σ the characters of G *conjugate to χ over K*. Their sum

$$(7) \qquad\qquad \mathrm{Tr}_K(\chi) = \sum_\sigma \chi^\sigma$$

is called the *trace of χ over K*. Clearly the function $\mathrm{Tr}_K(\chi)$ is a character of G with values in K, but in general it is not a K-character. Notwithstanding:

Theorem 1. *If χ is an irreducible C-character of G, then*

$$(8) \qquad\qquad \chi_K = s_K(\chi)\,\mathrm{Tr}_K(\chi) = s_K(\chi)\sum_\sigma \chi^\sigma$$

is the irreducible K-character of G corresponding to the simple component $B(\chi, K)$ of KG. One obtains all the irreducible K-characters of G in this way.

Proof. (1) We show first that

$$(9) \qquad\qquad B(\chi, K) \otimes_K C = \prod_\sigma B(\chi^\sigma, C),$$

where σ runs over all the elements of $G(K(\chi)/K)$, as in (7) and (8). Set $\psi := \chi_K$. Then $\langle \psi, \chi \rangle \neq 0$ implies $\langle \psi, \chi^\sigma \rangle = \langle \psi, \chi \rangle \neq 0$. Hence it follows from F2 that all the $B(\chi^\sigma, C)$ occur as simple components in the left side of (9). Further, they are all distinct, because so are the χ^σ. Thus, to prove (9) it suffices to show that both sides have the same dimension over C.

Set $F = K(\chi)$. Because $B(\chi^\sigma, C):C = \chi^\sigma(1)^2 = \chi(1)^2 = B(\chi, C):C$, we see using (5) that the dimension of the right side of (9) is

$$(F:K)\,(B(\chi, C):C) = (F:K)\,(B(\chi, F):F) = B(\chi, F):K,$$

and this proves the claim in view of (3).

(2) We set $s = s_K(\chi)$ and, as before, $F = K(\chi)$. The character of the KG-module $B(\chi, K)$ has the form $r\chi_K$, with r satisfying $B(\chi, K) : F = r^2 s^2$ (see F11 in Chapter 33). Correspondingly, $n\chi^\sigma$ is the character of the CG-module $B(\chi^\sigma, C)$, where n denotes the common degree of all the χ^σ, so $B(\chi, C) : C = n^2$. Then (9) implies

$$(10) \qquad\qquad r\chi_K = n\sum_\sigma \chi^\sigma.$$

But $B(\chi, C):C = B(\chi, F):F$ as before, so $n = rs$. Thus the desired equation (8) follows from (10) by cancellation of r. The last statement of the theorem is then a consequence of F2. $\qquad\qquad\square$

F3. *Let L be an intermediate field of C/K. Then χ is realizable over L if and only if L is a splitting field of the central-simple $K(\chi)$-algebra $B(\chi, K)$.*

Proof. Clearly χ is realizable over L if and only if $\chi_L = \chi$. By Theorem 1 (applied to a field L with $L \supseteq K(\chi) =: F$), this condition is equivalent to $s_L(\chi) = 1$, that is, $B(\chi, L) \sim 1$. But this says exactly that L is a splitting field of $B(\chi, F)$, because $B(\chi, L) = B(\chi, F) \otimes_F L$. Using (3) we then obtain the conclusion. $\square$

From the known properties of the Schur index of a central-simple algebra, we immediately obtain from F3:

F4. *Let L be an intermediate field of C/K such that $L : K < \infty$.*

(a) *If χ is realizable over L, then $s_K(\chi)$ divides $L : K(\chi)$.*

(b) *There exists L over which χ is realizable and such that $L : K(\chi) = s_K(\chi)$.*

(c) *If $s_K(\chi) = 1$, then χ is realizable over $K(\chi)$.*

We denote by $\langle \Theta, \psi, K \rangle$ the multiplicity of the irreducible K-character ψ in the (arbitrary) K-character Θ. Since $\langle \chi, \psi \rangle \neq 0$ only for $\psi = \chi_K$ (by F2), we see that $\langle \Theta, \chi \rangle = \langle \Theta, \chi_K, K \rangle \cdot \langle \chi_K, \chi \rangle$. Hence, by (8),

$$(11) \qquad\qquad \langle \Theta, \chi \rangle = s_K(\chi) \langle \Theta, \chi_K, K \rangle.$$

F5. (a) *For any K-character Θ of G, the index $s_K(\chi)$ divides the multiplicity of χ in Θ.*

(b) *$s_K(\chi)$ is smallest natural number m such that $m \operatorname{Tr}_K(\chi)$ is a K-character of G.*

(c) *$s_K(\chi)$ divides the degree of χ.*

Proof. Part (a) is already contained in (11). Let $\Theta = m \operatorname{Tr}_K(\chi)$ be a K-character of G. By (a), $s_K(\chi)$ divides $\langle \Theta, \chi \rangle = m \langle \operatorname{Tr}_K(\chi), \chi \rangle = m$. Hence (b) follows, thanks to Theorem 1. To prove (c), let Θ be the character of the regular representation of G, which is obviously a K-representation. The absolutely irreducible character χ occurs in Θ with multiplicity exactly $d = \chi(1)$. Thus, by (a), $s_K(\chi)$ divides $\chi(1)$. $\square$

Definition 3. We call a *central-simple K-algebra B a Schur algebra over K* if there exists a finite group G such that B is isomorphic to a simple component of the group algebra KG. By F1 and F2, the Schur algebras over K are, up to isomorphism, exactly those of the form

$$(12) \qquad\qquad B(\chi, K) \quad \text{with} \quad K(\chi) = K,$$

where χ runs over all irreducible C-characters of finite groups with values already in K. The totality of elements of the Brauer group $\operatorname{Br} K$ that can be represented by Schur algebras over K is called the *Schur group* of the field K and is written $S(K)$.

We must show that $S(K)$ really is a subgroup of Br K. If B_1, B_2 are Schur algebras over K, we can assume without loss of generality that $B_i = B(\chi_i, K)$, for two irreducible C-characters χ_i of finite groups G_i. Denote by $\chi_1 \otimes \chi_2$ the function on $G_1 \times G_2$ defined by $(\chi_1 \otimes \chi_2)(g_1, g_2) = \chi_1(g_1)\chi_2(g_2)$. If V_i denotes the representation module of CG_i associated to χ_i, we easily see that $V_1 \otimes V_2$ can be naturally made into a representation module of $C(G_1 \times G_2)$, and as such it has the character $\chi = \chi_1 \otimes \chi_2$. One also easily checks by computation that $\langle \chi, \chi \rangle = 1$, so χ is irreducible. Since $K(G_1 \times G_2) \simeq KG_1 \otimes KG_2$ it now follows that $B(\chi_1, K) \otimes B(\chi_2, K) \simeq B(\chi, K)$.

Obviously K is a Schur algebra over K, so we will be done if we show that when $B(\chi, K)$ is a Schur algebra over K, so is $B(\chi, K)^\circ$. Now, if χ is an irreducible C-character of G, so is the function $\bar{\chi}$ defined by $\bar{\chi}(g) = \chi(g^{-1})$ (see F8 and F17 in the previous chapter). And one easily sees that $B(\bar{\chi}, K) \simeq B(\chi, K)^\circ$.

Definition 4. A *crossed product* $\Gamma = (L, \mathfrak{g}, c)$ is called a *cyclotomic algebra over* K if the following conditions are satisfied:

(i) $L = K(\eta)$, where η is a root of unity.

(ii) All the values of the cocycle c are roots of unity.

(iii) The center of Γ is K; equivalently, $\mathfrak{g} = G(L/K)$ is the Galois group of L/K.

A cyclotomic algebra Γ as above is also denoted by $(K(\eta)/K, c)$, and its class in Br K is written

(13) $$[K(\eta)/K, c].$$

We denote by $S_0(K)$ the set of all such classes:

(14) $$S_0(K) = \{[B] \in \text{Br } K \mid B \text{ is a cyclotomic algebra over } K\}.$$

Remarks. (a) Let $\Gamma = (K(\eta)/K, c)$ be a cyclotomic algebra. If ζ is a root of unity in C such that $K(\eta) \subseteq K(\zeta)$, then Γ gives rise, by *inflation*, to a cyclotomic algebra $(K(\zeta)/K, \text{inf}(c))$.

(b) For a cyclotomic algebra $\Gamma = (K(\eta)/K, c)$, the values of c are roots of unity lying in $K(\eta)$, so we can choose η so that all these values are powers of η. Unless stated otherwise, we will assume from now on that this is the case.

(c) Clearly $S_0(K)$ is a *subgroup* of Br K. Indeed, to show that $[\Gamma][\Gamma']$ lies in $S_0(K)$ whenever $\Gamma = (K(\eta)/K, c)$ and $\Gamma' = (K(\eta')/K, c')$ are cyclotomic algebras, we can assume by inflation that $\eta = \eta'$; and in this case $[\Gamma][\Gamma'] = [K(\eta)/K, cc']$.

F6. *Every cyclotomic algebra* $\Gamma = (K(\eta)/K, c)$ *over* K *is a Schur algebra over* K. *Hence* $S_0(K)$ *is a subgroup of* $S(K)$.

Proof. By definition, the K-algebra Γ is generated by $K(\eta)$ and the elements u_σ for $\sigma \in \mathfrak{g}$, with relations $u_\sigma^{-1}\lambda u_\sigma = \lambda^\sigma$ for $\lambda \in K(\eta)$ and $u_\sigma u_\tau = u_{\sigma\tau}c_{\sigma,\tau}$. In the group $\Gamma^\times$ of invertible elements of Γ, take the subgroup G generated by the elements of $A := \langle \eta \rangle$ together with the u_σ. Then A is a normal subgroup of G and $G/A \simeq \mathfrak{g}$, showing that G is a *finite* group. The inclusion $G \subseteq \Gamma^\times$ gives rise to a K-algebra homomorphism $KG \to \Gamma$, which is obviously *surjective*. Hence the simple algebra Γ is isomorphic to a simple component of the semisimple algebra KG. This concludes the proof, since the center of Γ is K.   □

Theorem 2. *Let* $\Gamma = (K(\eta)/K, c)$ *be a cyclotomic algebra over* K, *and let* m *be the order of* $[\Gamma]$ *in* Br K. *Then* K *contains a primitive* m-*th root of unity*.

Proof. The values of c are all roots of unity; let n be the order of the group they generate. Then $[\Gamma]^n = [K(\eta)/K, c^n] = 1$, so m divides n. Hence the field $E = K(\eta)$ contains a primitive m-th root of unity ζ for sure. To prove the theorem we must still show that ζ is invariant under all the $\rho \in G(K(\eta)/K)$. A fixed ρ acts on all n-th roots of unity as a power map:

$$(15) \qquad\qquad x^\rho = x^r \quad \text{for some } r = r(\rho) \in \mathbb{N},$$

and moreover r is relatively prime to n and unique mod n.

Consider, next to $\Gamma = \sum_\sigma u_\sigma L$, the cyclotomic algebra

$$(16) \qquad\qquad \Gamma_r := (K(\eta)/K), c^r) = \sum_\sigma v_\sigma L.$$

We claim that the well defined K-linear map $\sum_\sigma u_\sigma \lambda_\sigma \mapsto \sum_\sigma v_\sigma \lambda_\sigma^\rho$ from Γ onto Γ_r is a ring homomorphism, that is, it preserves the cross product defining relations stated at the beginning of the proof of F6. The verification of this fact is left to the reader, with the hint that $G(L/K)$ is *abelian*. It follows that $\Gamma \simeq \Gamma_r$ and so $[\Gamma] = [\Gamma_r] = [\Gamma]^r$. This implies that $r \equiv 1 \bmod m$, showing that $\zeta = \zeta^r = \zeta^\rho$ is in fact ρ-invariant.   □

2. We continue with the notation of the previous section — in particular, K is any field of characteristic 0, with algebraic closure C. All extensions of K will be regarded as subfields of C. Further,

$$(17) \qquad\qquad \text{let } q \text{ be any prime number.}$$

As a stepping stone toward our goal of determining the Schur indices $s_K(\chi)$ of irreducible C-characters χ of finite groups, we have a reduction to the case of K-*elementary groups*, provided by the following theorem of R. *Brauer*:

Theorem 3. *Let* χ *be an irreducible* C-*character of* G *and* F *any extension of* $K(\chi)$. *For every prime* q *there is an* F-*elementary subgroup* H *of* G *and an irreducible* C-*character* ξ *of* H *such that*

$$(18) \qquad\qquad \langle \mathrm{Tr}_F(\xi), \mathrm{res}_H(\chi) \rangle \not\equiv 0 \bmod q.$$

If (18) *is satisfied for a given subgroup* H *of* G *and irreducible* C-*character* ξ *of* H, *the Schur index* $s_F(\chi)$ *of* χ *over* F *has the same* q-*component as its counterpart* $s_F(\xi)$. *Moreover,* $F(\xi):F$ *is relatively prime to* q.

Proof. By the *Brauer–Witt induction theorem* we have a representation

$$1 = \sum a_i \, \mathrm{ind}_{H_i}^{G}(\psi_i),$$

where the ψ_i are F-characters of F-elementary subgroups H_i and the a_i are integers. Multiplying by χ we obtain (by F27 in Chapter 33) the equation

$$\chi = \sum a_i \, \mathrm{ind}_{H_i}^{G}(\psi_i \, \mathrm{res}_{H_i}(\chi)).$$

As to the multiplicity with which χ occurs, we obtain

$$1 = \sum a_i \langle \psi_i \, \mathrm{res}_{H_i}(\chi), \, \mathrm{res}_{H_i}(\chi) \rangle.$$

The summands on the right cannot all be divisible by q, so there exists $H = H_i$ such that, setting $\psi = \psi_i$,

(19) $$\langle \psi \, \mathrm{res}_H(\chi), \, \mathrm{res}_H(\chi) \rangle \not\equiv 0 \bmod q.$$

The character $\psi \, \mathrm{res}_H(\chi)$ of H only takes values in F; hence in $\psi \, \mathrm{res}_H(\chi)$ the F-conjugates of any given irreducible C-character ξ of H occur with the same multiplicity as ξ. By (19), then, there must exist ξ satisfying (18). This proves the first part of the theorem.

Now suppose (18) is fulfilled. Since $F(\chi) = F$, we first have

$$\langle \mathrm{Tr}_F(\xi), \mathrm{res}_H(\chi) \rangle = (F(\xi) : F) \langle \xi, \mathrm{res}_H(\chi) \rangle,$$

whence, using (18), we immediately obtain $F(\xi) : F \not\equiv 0 \bmod q$; further,

(20) $$\langle \xi, \mathrm{res}_H(\chi) \rangle \not\equiv 0 \bmod q.$$

Now take the F-character $\xi_F = s_F(\xi) \, \mathrm{Tr}_F(\xi)$ of H. Its induction image $\mathrm{ind}_H^G(\xi_F)$ is an F-character of G, so the integer

$$\langle \mathrm{ind}_H^G(\xi_F), \chi \rangle = \langle \xi_F, \mathrm{res}_H(\chi) \rangle = s_F(\xi) \langle \mathrm{Tr}_F(\xi), \mathrm{res}_H(\chi) \rangle$$

is divisible by $s_F(\chi)$ (see F5). In view of (18), then, the q-part of $s_F(\chi)$ divides $s_F(\xi)$. To show that, conversely, the q-part of $s_F(\xi)$ divides the Schur index $s_F(\chi)$, we need only consider the F-character $s_F(\chi)\chi$. Its restriction $s_F(\chi)\,\mathrm{res}_H(\chi)$ to H is an F-character of H, in which the irreducible C-character ξ of H appears with multiplicity $s_F(\chi)\langle \mathrm{res}_H(\chi), \xi \rangle$. Hence this number is divisible by $s_F(\xi)$; but then (20) says that the q-part of $s_F(\xi)$ divides $s_F(\chi)$. $\square$

We next prepare the ground for another reduction result, first proved by *E. Witt* and *P. Roquette*.

Definition 5. Let M be the irreducible KG-module corresponding to $B(\chi, K)$ and let D be its division algebra of endomorphisms. We will regard M as a (KG, D)-bimodule. We call M *imprimitive* if it admits a nontrivial direct decomposition into right D-modules that are transitively permuted by G. If W is one of these right D-modules and H is the subgroup of $g \in G$ such that $gW = W$, the decomposition is of the form

$$(21) \qquad\qquad M = \bigoplus_{\rho \in G/H} \rho W.$$

If there is no such nontrivial decomposition, we say that the (KG, D)-bimodule M is *primitive*. The character of departure χ is then called *K-primitive*.

The KH-module W in (21) is irreducible (otherwise M would not be irreducible over KG), and it is easy to see that D represents its endomorphism ring. Thus, if we denote by $B(\xi, K)$ the simple component KH to which W belongs, we have the similarity

$$(22) \qquad\qquad B(\xi, K) \sim B(\chi, K).$$

We say that χ is *K-induced* by ξ. (Incidentally, one can check easily that ξ can be chosen so that $\chi = \operatorname{ind}_H^G(\xi)$, and then χ is also induced by ξ in the usual sense.)

F7. *Every irreducible C-character χ of G is K-induced by some K-primitive C-character ξ of some subgroup H of G, and this in such a way that $K(\chi) = K(\xi)$ and $B(\chi, K) \sim B(\xi, K)$.*

Proof. Reason by induction on the order of G. $\qquad\qquad\qquad\qquad\qquad$ □

Remarks. (1) Let χ be as above and let T be an irreducible C-representation of G in V whose character is χ. Denote by N the kernel of T and set $\widetilde{G} = G/N$. Then V is naturally a $C\widetilde{G}$-module; let $\widetilde{T}$ and $\widetilde{\chi}$ be the corresponding representation and character. Clearly V is also irreducible as a $C\widetilde{G}$-module. Since $T(KG) = \widetilde{T}(K\widetilde{G})$, we have $B(\chi, K) \simeq B(\widetilde{\chi}, K)$, and moreover, if χ is K-primitive, so is $\widetilde{\chi}$. By construction, $\widetilde{\chi}$ is a *faithful* C-character of $\widetilde{G}$ (that is, the character of a faithful C-representation). All this is to say that in dealing with elements of the Schur group over K we can always reduce the situation to that of *faithful, K-primitive characters*.

(2) In this connection the following fact is also useful: *Every subgroup and quotient group of a K-elementary group is again K-elementary.* The proof is simple and we leave it to the reader.

Lemma. *If $i : KG \to B(\chi, K)$ denotes the projection from KG onto its simple component $B = B(\chi, K)$, and if χ is faithful and K-primitive, the group algebra KA of any abelian normal subgroup A of G is taken by i onto a subfield L of B. The group A is itself mapped isomorphically under i to a group of roots of unity. In particular, A is cyclic.*

Proof. Since χ is assumed faithful, the restriction of i to G is *injective*. Let M be an irreducible summand of the KG-module B. We regard M as a KA-module, and take some *isogenous component* W thereof. Since A is normal in G, it follows as in the proof of Chapter 33, F28 that the KG-module M is induced by the KH-module W, where H is the subgroup of all $g \in G$ such that $gW = W$. Let D be the division algebra of endomorphisms of M. For every $d \in D^\times$ we obviously have $W \simeq Wd$, so $W = Wd$, so W is a right D-module. But χ was assumed K-primitive, so $W = M$, which means the KA-module M is isogenous. Thus all but one of the simple components of the semisimple algebra KA are annihilated by i. Since KA is commutative, it follows that

$$(23) \qquad L := i(KA) = K[iA]$$

is a field. Every finite subgroup of the multiplicative group of a field is cyclic, so all is proved.   □

We now investigate the situation where χ is a faithful, K-primitive character of a K-elementary group G. By assumption, G is certain to possess a cyclic normal subgroup N such that G/N is a q-group for some q. Let A be a maximal abelian normal subgroup of G containing N. Using exercise §10.6 one easily checks that A coincides with its centralizer in G, so the natural homomorphism

$$(24) \qquad G/A \to \mathrm{Aut}(A)$$

is injective. By the lemma, A is cyclic. Now choose for each $\sigma \in \mathfrak{g} := G/A$ some representative $v_\sigma \in G$ satisfying $\sigma = v_\sigma A$. For $\sigma, \tau \in \mathfrak{g}$ we have

$$(25) \qquad v_\sigma v_\tau = v_{\sigma\tau} c_{\sigma,\tau}, \quad \text{where } c_{\sigma,\tau} \in A.$$

Moreover, the elements $a \in A$ satisfy, by definition,

$$(26) \qquad a^\sigma = v_\sigma^{-1} a v_\sigma.$$

Keeping the notation of the lemma, set further $u_\sigma = i(v_\sigma)$ and consider on the algebra B the inner automorphism defined by conjugation with u_σ:

$$(27) \qquad b \mapsto u_\sigma^{-1} b u_\sigma \quad \text{for all } b \in B.$$

The elements of K are unchanged by this map. For elements $a \in A$ we have $u_\sigma^{-1} i(a) u_\sigma = i(a^\sigma)$, as we recognize by applying the map i from (26). Thus the field L maps to itself under (27), that is, (27) gives a K-automorphism $\bar\sigma$ of L. The homomorphism $\sigma \mapsto \bar\sigma$ from $\mathfrak{g}$ into the group $G(L/K)$ is injective, because (24) is injective; see (23).

To simplify the notation we now identify the group G with a subgroup of $B^\times$ via i. The cyclic subgroup A of G is then a group of roots of unity in $L \subseteq B$, and $L = K(A)$. Through the monomorphism $\sigma \to \bar\sigma$ we can view the group $\mathfrak{g}$ as a subgroup of $G(L/K)$. Since $i : KG \to B$ is surjective, the K-algebra B is generated by the elements λ of the field $L = K(A)$ and the elements $u_\sigma = i(v_\sigma) = v_\sigma$. Between these elements there hold the relations

$$u_\sigma u_\tau = u_{\sigma\tau} c_{\sigma,\tau}, \qquad u_\sigma^{-1} \lambda u_\sigma = \lambda^\sigma,$$

following from (25) and (26). Thus B is the *crossed product* of the field $L = K(A)$ with the group $\mathfrak{g}$ of automorphisms of L, with cocycle $c = (c_{\sigma,\tau})$ having values in A:

$$B = (L, \mathfrak{g}, c).$$

Let F be the fixed field of $\mathfrak{g}$ in L, so $\mathfrak{g} = G(L/F)$. Then F is the center of B (Chapter 30, Theorem 1). On the other hand, $B = B(\chi, K)$ has center $K(\chi)$, so $F = K(\chi)$. We express the results of our reasoning as follows:

F8. *Let the group H be K-elementary relative to the prime q. If the irreducible C-character ξ of H is faithful and K-primitive, then $B(\xi, K)$ is a crossed product $(K(A), \mathfrak{g}, c)$, where $K(A)$ is obtained from K by adjoining a group A of roots of unity, $\mathfrak{g} = G(K(A)/K(\xi))$ is a q-group, and the cocycle c has values in A.*

Using F7 (and the related remarks) we obtain an immediate corollary of F8:

F9. *Let the group H be K-elementary relative to the prime q. If ξ is any irreducible C-character of H, we have*

$$(28) \qquad\qquad [B(\xi, K)] = [K(A), \mathfrak{g}, c],$$

with a crossed product $\Gamma = (K(A), \mathfrak{g}, c)$ of the form specified in F8. In particular, $s_K(\xi) = s(\Gamma)$ is a q-power (since it divides the order of $\mathfrak{g}$).

Note that equation (28) should be understood as an equality in the *Brauer group* of the field $K(\xi)$.

From the considerations in this section so far, we can deduce easily at least the following result:

Theorem 4. *Let q be a fixed prime number. Every element of the q-part $S(K)_q$ of the Schur group of K has the same Schur index as a crossed product*

$$(29) \qquad\qquad \Gamma = (K(A), \mathfrak{g}, c)$$

with the following properties: (i) *A is a group of roots of unity;* (ii) *$\mathfrak{g}$ is a q-group;* (iii) *the cocycle c takes values in A;* (iv) *the fixed field K' of $\mathfrak{g}$ in $K(A)$ satisfies $K' : K \not\equiv 0 \bmod q$.*

Proof. By definition, $S(K)_q$ consists of those elements of $S(K)$ whose exponent— or, which is the same by F3 in Chapter 30, whose Schur index—is a q-power. Let $[B(\chi, K)]$ be any element of $S(K)_q$. Then $K(\chi) = K$, and by Theorem 3 (with $F = K(\chi) = K$) there exists an irreducible C-character ξ of a K-elementary group H of G such that

$$s_K(\chi) = s_K(\xi) \quad \text{and} \quad K(\xi) : K \not\equiv 0 \bmod q,$$

where we also have used that, by the last statement of F9, the Schur index $s_K(\xi)$ is a q-power. Now, by F9, $B(\xi, K)$ is similar to a crossed product Γ of the stipulated form; and as for property (iv), we only need to consider in addition that K' coincides with the center of Γ, so $K' = K(\xi)$. $\square$

Although Theorem 4 only concerns the *Schur index* of elements of $S(K)$, it is nonetheless quite a strong result. It is on this foundation that we will be able, in the next section, to determine the *Schur groups of local fields*.

But we do not wish to leave unmentioned the following general fact:

Theorem 5. *Every element $[B(\chi, K)]$ of the Schur group $S(K)$ of K can be represented by a cyclotomic algebra.*

Proof. It is just as well to assume that the given element $[B(\chi, K)]$ belongs to the q-part of $S(K)$, for some prime q. Let χ be a character of G and let n be the order of G. Denote by ζ a primitive n-th root of unity. The fixed field F of the Sylow q-group of $G(K(\zeta)/K)$ has the properties

(30) $F:K \not\equiv 0 \bmod q$ and $K(\zeta):F$ is a q-power.

Applying Theorem 3 to χ and this F yields an F-elementary subgroup H of G with irreducible C-character ξ such that

(31) $\langle \mathrm{res}_H(\chi), \xi \rangle \not\equiv 0 \bmod q$

and $F(\xi):F$ is relatively prime to q. The latter is possible only for $F(\xi) = F$, because of (30) and the inclusion $F(\xi) \subseteq K(\zeta)$. Since $K(\zeta)$ is a splitting field for ξ as well as for χ (Chapter 33, Theorem 15) and $K(\zeta):F$ is a q-power, we conclude (using also the equalities $F(\xi) = F = F(\chi)$) that the Schur indices $s_F(\xi)$ and $s_F(\chi)$, being divisors of $K(\zeta):F$, are also q-powers, so

(32) $[B(\xi, F)], [B(\chi, F)] \in S(F)_q.$

Because

$$B(\chi \otimes \bar{\chi}, F) \simeq B(\chi, F) \otimes B(\bar{\chi}, F) \sim F,$$

$\chi \otimes \bar{\chi}$ is the character of an F-representation of $G \times G$. Hence $\mathrm{res}_H(\chi) \otimes \bar{\chi}$ is the character of an F-representation of $H \times G$. One immediately checks that

$$\langle \xi \otimes \bar{\chi}, \xi \otimes \bar{\chi} \rangle = \langle \xi, \xi \rangle_H \langle \bar{\chi}, \bar{\chi} \rangle_G = 1,$$

so $\xi \otimes \bar{\chi}$ is an irreducible C-character of $H \times G$, with $F(\xi \otimes \bar{\chi}) = F(\xi, \bar{\chi}) = F$. One also verifies easily that

(33) $\langle \mathrm{res}_H(\chi) \otimes \bar{\chi}, \xi \otimes \bar{\chi} \rangle = \langle \mathrm{res}_H(\chi), \xi \rangle.$

But since $\mathrm{res}_H(\chi) \otimes \bar{\chi}$ is an F-character, equality (33) implies that $s_F(\xi \otimes \bar{\chi})$ divides $\langle \mathrm{res}_H(\chi), \xi \rangle$. But this latter number is relatively prime to q, by (31), so using (32) we conclude that

$$[B(\xi \otimes \bar{\chi}, F)] = [B(\xi, F)] \cdot [B(\chi, F)]^{-1} = 1.$$

Thus $[B(\chi, F)] = [B(\xi, F)]$ in Br F. By F9, $[B(\xi, F)]$ is represented by a cyclotomic algebra $(F(a)/F, c)$ over F, so we also have

(34) $[B(\chi, F)] = [F(a)/F, c] \in S_0(F).$

This does not yet suffice to prove the theorem, because we still need to step from F back down to K. By inflation we can assume without loss of generality that ζ lies in $\langle a \rangle$, and hence that everything takes place in the cyclotomic field $K(a)$ over K. We now make use of the *corestriction map*

$$\mathrm{cor} = \mathrm{cor}_{F/K} : \mathrm{Br}(K(a)/F) \to \mathrm{Br}(K(a)/K),$$

introduced in Definition 4* of Chapter 30. By (34) we have

$$\mathrm{cor}[B(\chi, F)] = \mathrm{cor}[K(a)/F, c] = [K(a)/K, \mathrm{cor}(c)] \in S_0(K).$$

On the other hand,

$$\mathrm{cor}[B(\chi, F)] = \mathrm{cor}(\mathrm{res}[B(\chi, K)]) = [B(\chi, K)]^{F:K},$$

by (5) above and F3* in Chapter 30. Putting everything together we conclude that $[B(\chi, K)]$ lies in $S_0(K)$, since $S_0(K)$ is a group and $F : K$ is, by (30), relatively prime to the order of $[B(\chi, K)] \in S(K)_q$. $\qquad\square$

As a consequence of Theorem 5 we immediately obtain, using Theorem 2:

Theorem 6. *If $S(K)$ contains an element of order m, there is a primitive m-th root of unity in K.*

3. In this last section we will compute the Schur group $S(K)$ when K is a *local field*.

Notation. *In the rest of this section K will be a (nonarchimedean) local field of characteristic 0. If p is the characteristic of its residue class field, we have a field extension $K/\mathbb{Q}_p$ with*

$$K : \mathbb{Q}_p < \infty.$$

For every finite extension L/K we denote by $U(L)$ the group of units of the local field L, by $U^1(L)$ the kernel of the residue class homomorphism from $U(L)$ onto the multiplicative group of the residue class field of L, and by $W(L)$ the group of roots of unity contained in L.

Further, for any natural number m we use ζ_m to denote a primitive m-th root of unity (taken, as above, in the algebraic closure C of K).

Lemma 1. *Let L/K be a Galois extension whose Galois group $\mathfrak{g}$ has order relatively prime to p. Then the natural maps*

$$H^i(\mathfrak{g}, W(L)) \to H^i(\mathfrak{g}, U(L))$$

are isomorphisms for all i.

Proof. For any arbitrary abelian torsion group M, we once again denote by M_p the p-part of M and by $M_{p'}$ the p-regular part (that is, the subgroup containing all elements whose orders are relatively prime to p). Because of Chapter 24, Theorem 4(ii), we have

$$(35) \qquad\qquad U(L) = W(L)_{p'} \times U^1(L).$$

Since all the groups involved in (35) are $\mathfrak{g}$-modules, the equality implies

$$H^i(\mathfrak{g}, U(L)) = H^i(\mathfrak{g}, W(L)_{p'}) \times H^i(\mathfrak{g}, U^1(L)).$$

Let n denote the order of $\mathfrak{g}$. By assumption, n is relatively prime to p, so the group $U^1(L)$ is uniquely n-divisible (see F8 in Chapter 25). Hence $H^i(\mathfrak{g}, U^1(L))$ is trivial for all i (see F3* in Chapter 30). Again because n and p are relatively prime, we obtain $H^i(\mathfrak{g}, W(L)) = H^i(\mathfrak{g}, W(L)_{p'})$, concluding the proof. $\qquad\square$

Lemma 2. *Let K be, as above, an extension of finite degree over $\mathbb{Q}_p$, and let $\Gamma = (K(\eta)/K, c)$ be a cyclotomic algebra over K. If $p \neq 2$ then $s(\Gamma)$ divides $p - 1$, and if $p = 2$ then $s(\Gamma) \leq 2$.*

Proof. First we recall from Theorem 5 in Chapter 31 that in this situation $s(\Gamma)$ coincides with the order of $[\Gamma]$ in Br K. Let K_0 be the fixed field of $G(K(\eta)/K)$ in $\mathbb{Q}_p(\eta)$. Clearly it suffices to prove the corresponding assertion for the cyclotomic algebra $\Gamma_0 = (\mathbb{Q}_p(\eta)/K_0, c)$ over K_0, because Γ is recoverable from Γ_0 by restriction.

To prove this case we take the *corestriction* map cor $= \mathrm{cor}_{K_0/\mathbb{Q}_p} : \mathrm{Br}\, K_0 \to \mathrm{Br}\, \mathbb{Q}_p$, which by Theorem 10 in Chapter 31 is an isomorphism, and look at the image $\mathrm{cor}(\Gamma_0) = [\mathbb{Q}_p(\eta)/\mathbb{Q}_p, \mathrm{cor}(c)]$. In view of the above, it suffices to prove the lemma's conclusion in the case $K = \mathbb{Q}_p$. Set $s = s(\Gamma)$. By Theorem 2, $K = \mathbb{Q}_p$ must contain a primitive s-th root of unity. But we know from F11 in Chapter 25 that $W(\mathbb{Q}_p)$ has order $p - 1$ if $p \neq 2$ and order 2 if $p = 2$. $\qquad\square$

If we call on Theorem 5 and look at Br $K \simeq \mathbb{Q}/\mathbb{Z}$, Lemma 2 reveals the following general fact about the Schur group of K:

F10. *Let K be an extension of finite degree over $\mathbb{Q}_p$. Then $S(K)$ is a finite, cyclic group whose order divides $p - 1$ if $p \neq 2$ and can be at most 2 if $p = 2$.*

For the time being we will settle for this much information in the case $p = 2$: $S(K)$ is either trivial or has a single nontrivial element, of Schur index 2. But in the case $p \neq 2$ we would like to determine the order of $S(K)$ exactly. And we will in fact do so without having to refer to F10 (or Theorem 5).

Let $[B]$ be any element of $S(K)$, represented by a simple component $B = B(\chi, K)$ of the group algebra KG of a finite group G, where χ denotes an irreducible $\mathbb{C}$-character of G such that $\chi(B) \neq 0$. Then $K(\chi) = K$, since $K(\chi)$ is the center of $B(\chi, K)$. The Schur index $s_K(\chi)$ equals the order of $[B]$ in Br K (again by Theorem 5 in Chapter 31). Since Br K is isomorphic to $\mathbb{Q}/\mathbb{Z}$ (Theorem 4 in Chapter 31), in order to determine $S(K)$ we need only decide which natural numbers can occur as *Schur indices* of irreducible $\mathbb{C}$-characters χ of finite groups, with the side condition $K(\chi) = K$.

We lose nothing by assuming that $[B] \in S(K)_q$ lies in the q-part of $S(K)$, where q is an arbitrary prime. By Theorem 4, $s_K(\chi)$ equals the index of a crossed product $\Gamma = (K(A), \mathfrak{g}, c)$ satisfying properties (i)–(iv) of Theorem 4.

So assume, as announced, that $p \neq 2$. First it follows from Lemma 2 (applied to $K' = \text{Fix}(\mathfrak{g})$ in lieu of K) that we are allowed to assume

$$q \neq p,$$

for otherwise $s(\Gamma) = 1$. Now let γ be the class of c in $H^2(\mathfrak{g}, L^\times)$, where we have set $L = K(A)$. Then $s_K(\chi) = s(\Gamma) = \text{ord}(\gamma)$. But γ is contained in the image of the natural map i appearing in the commutative diagram

(36)
$$
\begin{array}{ccc}
H^2(\mathfrak{g}, W(L)) & \xrightarrow{\quad i \quad} & H^2(\mathfrak{g}, L^\times) \\
 & \searrow{\scriptstyle i_1} \quad \nearrow{\scriptstyle i_2} & \\
 & H^2(\mathfrak{g}, U(L)) &
\end{array}
$$

By Lemma 1, the map i_1 is an isomorphism, and by Theorem 11 in Chapter 31, i_2 is injective and the order of its image coincides with the ramification index $e(L/K')$. Hence $\text{ord}(\gamma)$ divides $e(L/K')$, and we conclude at least that

(37) $$s_K(\chi) \quad \text{divides} \quad e(L/K).$$

Because $L = K(A_p)(A_{p'})$, the extension $L/K(A_p)$ is *unramified*, so $e(L/K) = e(K(A_p)/K)$. We can assume that ζ_p lies in A_p, otherwise $s(\Gamma) = 1$. But then $K(A_p)/K(\zeta_p)$ is a p-power, so $s(\Gamma)$, being a q-power, must divide $e(K(\zeta_p)/K)$:

(38) $$s_K(\chi) \quad \text{divides} \quad e(K(\zeta_p)/K).$$

We now claim that $e(K(\zeta_p)/K)$ is the desired order of $S(K)$:

Theorem 7. *Let K be an extension of finite degree over $\mathbb{Q}_p$. If $p \neq 2$, then $S(K)$ is a finite cyclic group whose order equals the ramification index of $K(\zeta_p)/K$.*

Proof. Set $e = e(K(\zeta_p)/K)$. Though we have proved (38), we still need to show that there exists $[\Gamma] \in S(K)$ of order $s(\Gamma) = e$. To this end we consider $L = K(\zeta_p)$ and $\mathfrak{g} = G(L/K)$. As observed just after (36), the natural map

$$H^2(\mathfrak{g}, W(L)) \to H^2(\mathfrak{g}, L^\times)$$

is injective, and its image has order e. Then let c be a cocycle of $\mathfrak{g}$ with values in $W(L)$, whose cohomology class has order e in the cyclic (!) group $H^2(\mathfrak{g}, L^\times)$. Then $\Gamma = (L, \mathfrak{g}, c) = (K(\zeta_p)/K, c)$ is a cyclotomic algebra whose class $[\Gamma]$ in $\text{Br } K$ has order $s(\Gamma) = e$. Since $[\Gamma]$ belongs to $S(K)$ (by F6), this proves the assertion. $\square$

In the language of *Hasse invariants* we can obviously rephrase Theorem 7 thus:

Theorem 7'. *Suppose $p \neq 2$ and let K be an extension of finite degree over $\mathbb{Q}_p$. Then a central-simple K-algebra A is similar to a simple component of the group algebra KG of a finite group G if and only if the Hasse invariant $\mathrm{inv}_K(A)$ of A has the form*

$$(39) \qquad \mathrm{inv}_K(A) = \frac{a}{e(K(\zeta_p)/K)} \mod 1$$

for some $a \in \mathbb{Z}$ (with no further restriction on a).

Remarks. (a) Let k be any field of characteristic 0, and denote by W the set of all roots of unity in an algebraic closure C of k. If K/k is any finite extension (in C/k), we call $K_0 = K \cap k(W)$ the *greatest cyclotomic field* of K/k. The restriction map $\mathrm{res}_{K/K_0} : \mathrm{Br}\, K_0 \to \mathrm{Br}\, K$ then satisfies

$$(40) \qquad S(K) = \mathrm{res}_{K/K_0}(S(K_0));$$

indeed, if $[B(\chi, K)] \in S(K)$, we have $K(\chi) = K$, and so K contains the cyclotomic field $k(\chi)$ over k, and it follows that $K_0(\chi) = K_0$, so $B(\chi, K) = B(\chi, K_0) \otimes_{K_0} K$.

(b) Now let K be again an extension of finite degree over $\mathbb{Q}_p$; and let the case $p = 2$ be allowed as well. Still with K_0 the greatest cyclotomic field of $K/\mathbb{Q}_p$, we consider the map $\mathrm{res}_{K/K_0} : \mathrm{Br}\, K_0 \to \mathrm{Br}\, K$. It acts simply as multiplication of Hasse invariants by the field degree $K : K_0$ (see Theorem 4 in Chapter 31), so in view of (40) we conclude that the orders of the finite cyclic groups $S(K)$ and $S(K_0)$ satisfy

$$|S(K)| = |S(K_0)|/t, \quad \text{where } t \text{ is the gcd of } |S(K_0)| \text{ and } K : K_0.$$

Now consider the case $p \neq 2$ first. As can easily be checked, $e(K(\zeta_p)/K)$ equals $e(K_0(\zeta_p)/K_0)$. But then it follows from Theorem 7 that $S(K) \simeq S(K_0)$, so that

$$t = \gcd(K : K_0, |S(K_0)|) = 1.$$

(In connection with this and subsequent statements, see F. Lorenz, "Die Schurgruppe eines lokalen Körpers", *Sitzungsber. der Math.-Naturw. Klass. Akad. Wiss. zu Erfurt* 4 (1992), 139–152.)

(c) One shows easily that *for a cyclotomic field K over $\mathbb{Q}_p$ with $p \neq 2$, the group $S(K)$ has order $(p-1)/e_z(K/\mathbb{Q}_p)$, where $e_z(K/\mathbb{Q}_p)$ denotes the tame (that is, p-regular) part of $e(K/\mathbb{Q}_p)$.*

(d) Again let K be any extension of finite degree over $\mathbb{Q}_p$ with $p \neq 2$. As a noteworthy consequence of (2) and (3) observe that *the degree $K : K_0$ is relatively prime to $(p-1)/e_z(K_0/\mathbb{Q}_p)$.*

Turning now to the case $p = 2$, we know from F10 that $S(K)$ has order 1 or 2, but deciding between the two for a particular K apparently runs into some unavoidable messiness. We omit the toilsome calculation and simply announce the result:

Theorem 8. *The Schur group of a cyclotomic field K over $\mathbb{Q}_2$ has order 2 or 1 depending on whether or not -1 is a norm of the extension $K/\mathbb{Q}_2$.*

Remark. For an arbitrary finite extension $K/\mathbb{Q}_2$, Theorem 8 does *not* hold in the stated form. In fact, -1 is a norm of $K/\mathbb{Q}_2$ if and only if it is a norm of $K_0/\mathbb{Q}_2$ (Chapter 32, F8 and Theorem 3); yet at the same time, $K:K_0$ can be *even* (in which case $S(K) \neq S(K_0)$; see Remark (b) earlier on this page). An example is afforded by $K = \mathbb{Q}_2(\sqrt[4]{2})$.

By applying *local class field theory* (Chapter 32) we can synthesize Theorems 7 and 8, as well as their accompanying remarks, as follows:

Theorem 9. *For every local field K of characteristic 0, the Schur group $S(K)$ is isomorphic to the torsion part of the Galois group of $K(W)/K$, where in the 2-adic case we must also assume that K is a cyclotomic field over $\mathbb{Q}_2$.*

The proof is left to the reader; note that the torsion part of the group

$$G(\mathbb{Q}_p(W)/\mathbb{Q}_p) = \widehat{\mathbb{Z}} \times \mathbb{Z}_p^\times$$

is generated by either $(\zeta_{p-1}, \mathbb{Q}_p(W)/\mathbb{Q}_p)$ or $(-1, \mathbb{Q}_2(W)/\mathbb{Q}_2)$, as the case may be ($p \neq 2$ or $p = 2$). Incidentally, the formulation of Theorem 9 is due to *C. Riehm* (*L'Enseignement Mathématique* **34**, 1988; but the statement there is not quite right and the justification is incomplete).

Appendix:

Problems and Remarks

References preceded by § are to this appendix.

Chapter 20: Ordered Fields and Real Fields

20.1 Let K be a *real-closed* field.

(a) Prove that a normalized polynomial q of degree at least 2 in $K[X]$ is prime if and only if it has degree 2 and is of the form $X^2 + bX + c$, where $b^2 - 4c < 0$. For each such q we have $q(x) > 0$ for every $x \in K$.

(b) Describe the polynomials $f \in K[X]$ such that $f(x) > 0$ for every $x \in K$, and those such that $f(x) \geq 0$ for every $x \in K$. Prove that each such f is a sum of squares in $K[X]$. In fact, why can one make do with *two* squares?

(c) Let $a < b$ be elements of K, and suppose $f \in K[X]$ has opposite signs at a and b, that is, $f(a) f(b) < 0$. Prove that there exists $c \in K$ such that $f(c) = 0$ and $a < c < b$ (the *intermediate value theorem*). *Hint:* Reduce to the case that f is normalized and irreducible, hence linear.

(d) Let $a < b$ be roots of $f \in K[X]$ in K. Prove that there exists $a < c < b$ such that $f'(c) = 0$ (*Rolle's Theorem*). *Hint:* Assume without loss of generality that f has no root in the interval (a, b). Write $f(X) = (X - a)^m (X - b)^n g(X)$, and apply part (c) to an appropriate divisor of f' or to g, as the case may be.

20.2 Prove that the subfield $E = \mathbb{Q}(\sqrt{2})$ of $\mathbb{R}$ admits exactly two distinct orders. (This is a special case of §20.10 below, whose proof, however, requires deeper methods.)

20.3 Let K be a *real* field and C an algebraically closed extension of K. Prove that there is a *real-closed* intermediate field R of C/K such that $C = R(\sqrt{-1})$.

20.4 Let $(K, \leq)$ be an ordered field and f an irreducible polynomial in $K[X]$ such that $f(a) f(b) < 0$ for some $a, b \in K$. Prove that if $K(\alpha)$ is an extension of K with $f(\alpha) = 0$, then $\leq$ can be extended to an order of $K(\alpha)$.

20.5 Replace the last part of the proof of Theorem 7 by the following argument: Any $\sigma \in G(E/R) \smallsetminus G(E/C)$ satisfies

$$(1) \qquad\qquad\qquad w\sigma(w) \in R.$$

(Why?) It follows that $C(w)/R$ is Galois (of degree 4). Since $R = R^2 \cup -R^2$, the Galois group is in fact *cyclic*, and if σ is one of its generators, equation (1) holds. Now apply σ to (1) to obtain $\sigma^2(w) = w$, and hence a contradiction ($w \in C$).

20.6 Let $(K, \leq)$ be an ordered field. Since K has characteristic 0, we can regard $\mathbb{Q}$ as a subfield of K. We call the order $\leq$ and the ordered field $(K, \leq)$ *archimedean* if for any $a \in K$ there exists $n \in \mathbb{N}$ such that $a < n$. Prove:

(a) $(K, \leq)$ is archimedean if and only if $\mathbb{Q}$ is *dense* in K, meaning that for every a, b in K with $a < b$ there exists $x \in \mathbb{Q}$ such that $a < x < b$).

(b) If $K : \mathbb{Q}$ is finite, $(K, \leq)$ is archimedean. (*Hint:* See F7; but the assertion can also be easily proved directly.)

(c) The field $\mathbb{Q}(X)$ of rational functions in one variable over $\mathbb{Q}$ admits both non-archimedean and archimedean orders.

20.7 Let $(K, \leq)$ be an ordered field. For $a \in K$, let $|a| \in K$ denote, as usual, the element of K such that $|a| \geq 0$ and $|a|^2 = a^2$. Then $|ab| = |a||b|$ and $|a + b| \leq |a| + |b|$. It is clear how to define in $(K, \leq)$ a *null sequence* (that is, a sequence that converges to 0), a *Cauchy sequence*, a *convergent sequence*, and the *limit point* of such. Note, however, that $|\ |$ has values in K, so it does not define a metric space in the usual sense. And in general, 0 does not have a *countable neighborhood basis*. For topological arguments in $(K, \leq)$, therefore, the usual notion of convergence of sequences is not suitable.

We cannot do justice here to the topological foundations of ordered fields in their generality. If we don't wish to abandon our habit of working with sequences, therefore, we will have to accept some restrictions. We call an ordered field ω-*complete* if it has the property that every Cauchy sequence converges. In an *archimedean* ordered field there are countable neighborhood bases and in this case we use the term *complete* instead of ω-complete.

Prove that $(K, \leq)$ *is archimedean and complete if and only if every nonempty subset of K bounded from above has a least upper bound.* (The field $\mathbb{R}$ of real numbers has both properties, as is well known; but do not resort to the fact of the existence and uniqueness of $\mathbb{R}$ in this problem and the next two, since one point is precisely to prove it.)

Now assume that $(K, \leq)$ is archimedean and complete. Prove that an arbitrary *archimedean* ordered field $(F, \leq)$ admits an order-preserving embedding in $(K, \leq)$; that is, there is an order-preserving homomorphism $\sigma : F \to K$. This is an isomorphism if and only if $(F, \leq)$ is complete. It follows that, up to an order-preserving isomorphism, there is at most one complete archimedean ordered field.

20.8 If $(K, \leq)$ is any ordered field, *there exists an ω-complete ordered extension $(\tilde{K}, \leq)$ of $(K, \leq)$ with the property that every $\alpha \in \tilde{K}$ is a limit of a sequence $(a_n)_n$*

of elements of K. Up to an order-preserving isomorphism, $(\tilde{K}, \leq)$ is uniquely determined. This extension is called the *ω-completion* of $(K, \leq)$.

Here is how the proof goes. In the ring R of all Cauchy sequences in $(K, \leq)$, consider the ideal I of all null sequences. Take as $\tilde{K}$ the quotient ring R/I, and let $\iota : K \to \tilde{K}$ be the map taking each $a \in K$ to the residue class of the constant sequence $(a, a, \dots)$. Clearly ι is injective: identify K with its image $\iota(K)$ in $\tilde{K}$.

(a) I is a maximal ideal of R; that is, $\tilde{K}$ is a field.

(b) An element $(a_n)_n$ of R is called *strictly positive* if there exists a strictly positive $\varepsilon \in K$ and a natural number N such that $a_n \geq \varepsilon$ for all $n \geq N$. Prove that an element $(a_n)_n$ of R such that neither $(a_n)_n$ nor $(-a_n)_n$ is strictly positive must be a null sequence.

It follows easily that the order $\leq$ of K can be extended to an order in $\tilde{K}$ (also denoted by $\leq$) having the following property: An element $\alpha \in \tilde{K}$ is strictly positive if and only if some (hence every) Cauchy sequence representing α is strictly positive.

Since every Cauchy sequence is bounded, it is not hard to see that for each $\alpha \in \tilde{K}$ there is a strictly positive $a \in K$ such that $\alpha \leq a$.

(c) Take $\alpha \in \tilde{K}$ and let $(a_n)_n$ be any Cauchy sequence belonging to the residue class α. Prove that $(a_n)_n$, regarded as a sequence in $\tilde{K}$, converges in $(\tilde{K}, \leq)$ to the element α:

$$\alpha = \lim a_n.$$

In particular, K is *dense* in $\tilde{K}$. *Hint:* Let $\varepsilon > 0$ be given — in fact one can require that $\varepsilon \in K$, by (b). Since $(a_n)_n$ is a Cauchy sequence, we have

$$a_n - a_m + \varepsilon > \frac{\varepsilon}{2} \quad \text{for almost all } m, n \in \mathbb{N}.$$

By the definition of the order $\leq$ on $\tilde{K}$, this implies that

$$a_n - \alpha + \varepsilon > 0 \quad \text{for almost all } n \in \mathbb{N}.$$

(d) Now let $(\alpha_n)_n$ be any Cauchy sequence in $(\tilde{K}, \leq)$. We wish to show that it converges in $(\tilde{K}, \leq)$. First consider that one can assume, without loss of generality, that $\alpha_{n+1} \neq \alpha_n$ for each n. Then, by (c), there is for each n an $a_n \in K$ such that

$$|\alpha_n - a_n| \leq |\alpha_{n+1} - \alpha_n|.$$

Thus $(\alpha_n - a_n)_n$ is a null sequence in $(\tilde{K}, \leq)$. This implies that $(a_n)_n$ is also a Cauchy sequence (with values in K), since $a_n = (a_n - \alpha_n) + \alpha_n$. Let α be the element of $\tilde{K}$ defined by $(a_n)_n$. The convergence of $(\alpha_n)_n$ now follows from (c), thanks to the equality $\alpha - \alpha_n = (\alpha - a_n) + (a_n - \alpha_n)$.

(e) Now it is straightforward to prove the desired uniqueness property of $(\tilde{K}, \leq)$.

20.9 Let $(K, \leq)$ be an ordered field and $\tilde{K}$ its ω-completion (§20.8). Clearly, if $(K, \leq)$ is archimedean, so is $(\tilde{K}, \leq)$. The completion of $\mathbb{Q}$ (with respect to the unique order on $\mathbb{Q}$) is usually denoted by $\mathbb{R}$. By §20.7, *every **archimedean** ordered* *field* $(K, \leq)$ *is order-isomorphic to a subfield of* $\mathbb{R}$. (The converse also holds, of course: every subfield of $\mathbb{R}$ has an archimedean order.)

On the other hand, if $(K, \leq)$ is nonarchimedean, the ω-completion $\tilde{K}$ of K does not in general amount to a very helpful notion, for the reasons given in §20.7. It can be replaced by a better one:

Given $(K, \leq)$, *there exists an ordered extension field* $(\hat{K}, \leq)$ *of* $(K, \leq)$ *with these properties:*

(a) K *is dense in* $\hat{K}$.

(b) *If* $(L, \leq)$ *is an ordered extension field of* $(K, \leq)$ *and K is dense in L, there is a unique order-preserving K-homomorphism of L into $\hat{K}$.*

(c) $\hat{K}$ *is complete with respect to the uniform structure defined by* $\leq$.

Such an extension is called a completion of $(K, \leq)$.

Properties (a) and (b) already imply that $\hat{K}$ is unique up to an order-preserving K-isomorphism. As for existence, one can in principle work as in §20.8, appealing to *generalized Cauchy sequences* (indexed by ordinals that lie below a fixed sufficiently large ordinal). One can also work with *Cauchy filters* (see Bourbaki, *General Topology*, Chapter 3) or use the method of *Dedekind cuts*.

Assuming the existence of the completion $\hat{K}$ of $(K, \leq)$, prove that *if K is real-closed, so is $\hat{K}$*. The proof may be require some effort. Once it is done, however, the statement can be easily generalized as follows: *An ordered field $(K, \leq)$ is dense in its real closure if an only if $\hat{K}$ is real-closed.*

20.10 Let (k, P) be an ordered field and R a real closure of (k, P). Given an algebraic extension K/k, set $M := G(K/k, R/k)$ and prove:

(a) The map $\sigma \mapsto \sigma^{-1}(R^2)$ is a bijection between M and the set of all orders Q of K that extend P.

(b) If K/k is finite and we set $s_{K/k}(x, y) = S_{K/k}(xy)$, then

$$\mathrm{sgn}_P(s_{K/k}) = \mathrm{Card}\{Q \mid Q \text{ is an order on } K \text{ and } Q \supseteq P\}.$$

Consequently, an *algebraic number field* K (that is, an extension of $\mathbb{Q}$ such that $[K : \mathbb{Q}] < \infty$) *admits at most $K : \mathbb{Q}$ distinct orders, and the signature of its trace form $s_{K/\mathbb{Q}}$ is always ≥ 0.*

(c) If K/k is finite and P can be extended to an order of K in a unique way, then $K : k$ is *odd* and therefore every order of k can be extended to K.

20.11 Let $f \in K[X]$ be a polynomial of degree $n > 1$ over a field K, and let e be a natural number. Prove that the trace form $s_{f^e/K}$ of the algebra $A := K[X]/f^e$ is equivalent to the orthogonal sum of a zero form of the appropriate dimension with the multiple $es_{f/K}$ of the trace form $s_{f/K}$ of $\bar{A} := K[X]/f$.

Hint: We have $A = K[x] = (Kx^0 + Kx^1 + \cdots + Kx^{n-1}) \oplus f(x)K[x]$, and every element of the ideal $f(x)K[x]$ of A is *nilpotent*. Moreover $S_{A/K}(a)$ is isomorphic to $eS_{\bar{A}/K}(\bar{a})$, as can be seen from the chain

$$A \supseteq f(x)A \supseteq f(x)^2 A \supseteq \cdots \supseteq f(x)^e A = 0$$

of A-modules, since each factor $f(x)^i A / f(x)^{i+1} A$ is isomorphic to $K[X]/f = \bar{A}$.

20.12 Let K/k be a finite field extension. If q is a symmetric bilinear over K, then $s_{K/k}(q) := S_{K/k} \circ q$ is a symmetric bilinear form over k. Prove that, if (k, P) is an ordered field, the following formula of *M. Knebusch* holds and represents a sharpening of §20.10(b):

$$\mathrm{sgn}_P(s_{K/k}(q)) = \sum_{Q \supseteq P} \mathrm{sgn}_Q(q),$$

where the sum is over all orders Q of K that extend P.

Hint: First reduce to the case where q is one-dimensional: $q = [\alpha]$, with $\alpha \in K^\times$. By induction and using 20.10(b), find that K can be assumed to be $k(\alpha)$. Then K is isomorphic to $k[X]/f$, with $f = \mathrm{MiPo}_k(\alpha)$. Now work just as in the proof of Theorem 9: show that $\mathrm{sgn}_P(s_{K/k}(q)) = \sum_{i=1}^r \mathrm{sgn}\,\alpha_i$, if the $\alpha_1, \alpha_2, \ldots, \alpha_r$ are all the roots of f in R.

Chapter 21: Hilbert's Seventeenth Problem and the Real Nullstellensatz

21.1 Let (K, P) be an ordered field and R its real closure. *Artin's Theorem* implies (see Remark 2 after Theorem 1): If a polynomial $f \in K[X_1, \ldots, X_n]$ is *positive definite* on R^n, it has the form $f = c_1 f_1^2 + \cdots + c_r f_r^2$, where the c_i lie in P and the f_i are rational functions in $K(X_1, \ldots, X_n)$. In many cases (including $K = \mathbb{Q}$; see Remark 1 after Theorem 1) it suffices to assume only that f takes positive values *on K^n*. In general, however, this is not so, because K need not be dense in R. A polynomial f that is positive on K may nonetheless take negative values in the "gaps" left by K in R. The next exercise gives an example.

21.2 Let $K = \mathbb{Q}(t)$ be the field of rational functions in the variable t over $\mathbb{Q}$. Prove:

(a) There exists exactly one order $\leq$ on K for which t is positive but *infinitesimal* relative to all nonzero constants; that is, $0 < t < |a|$ for all $a \in \mathbb{Q}^\times$. The strictly positive elements of K are then those of the form $t^n f(t)$, with $n \in \mathbb{Z}$ and $f(0) > 0$ (where f is understood not to have a pole at 0).

Now let R be the real closure of $(K, \leq)$.

(b) K is not dense in R: the interval $(\sqrt{t}, 2\sqrt{t})$ contains no point of K.

(c) The polynomial $f(X) = X^4 - 5tX^2 + 4t^2$ over $K = \mathbb{Q}(t)$ is positive definite on K, but takes strictly negative values on R.

21.3 Let A be a commutative ring with unity, and let $\mathfrak{a}$ be an ideal of A. Define the *real radical* $r(\mathfrak{a})$ of $\mathfrak{a}$ (as in Theorem 6) as the set of $f \in A$ for which there is some $m \in \mathbb{N}$ and some $s \in \mathrm{SQ}(A)$ such that $f^{2m} + s \in \mathfrak{a}$. Check that the same set is obtained if $m = 0$ is also allowed in the definition. Show also that $r(\mathfrak{a})$ coincides with the set of $f \in A$ for which there exists $n \in \mathbb{N}$ and $s \in \mathrm{SQ}(A)$ such that

$$f^{2^n} + s \in \mathfrak{a}.$$

21.4 In the situation of §21.3, let $r_0(\mathfrak{a})$ denote the *intersection of all real prime ideals of A that contain* $\mathfrak{a}$. It is easy to see that

(1) $$r(\mathfrak{a}) \subseteq r_0(\mathfrak{a}).$$

Now suppose we are in the situation of Theorem 6. Set $V := \mathcal{N}_R(\mathfrak{a})$ and $A := K[X_1, \ldots, X_n]$. From the definitions it can be checked without difficulty that

(2) $$r_0(\mathfrak{a}) \subseteq i(V).$$

By applying Theorem 10 of Chapter 20, show that

(3) $$i(V) \subseteq r_0(\mathfrak{a})$$

as well. Now if we take as known the conclusion (15) of the Dubois Nullstellensatz, we immediately get from the simple relations (1) and (2) the equality

(4) $$r(\mathfrak{a}) = r_0(\mathfrak{a}).$$

21.5 Give a second proof of the *Dubois real Nullstellensatz* (Theorem 6) based on relation (3) of §21.4, by showing that every ideal $\mathfrak{a}$ of a commutative ring A with unity satisfies

$$r_0(\mathfrak{a}) \subseteq r(\mathfrak{a}).$$

Hint: Suppose $f \notin r(\mathfrak{a})$. Then the set

$$S = \{f^{2m} + t \mid m \in \mathbb{N}_0, \, t \in SQ(A)\}$$

is disjoint from $\mathfrak{a}$. Take a maximal element $\mathfrak{p}$ of the set of ideals that contain $\mathfrak{a}$ but are disjoint from S. Then $\mathfrak{p}$ is a *prime ideal* (see problem §4.12 of vol. I). Clearly $f \notin \mathfrak{p}$.

We will be done if we can also prove that $\mathfrak{p}$ is *real*. Suppose there exist $f_1, \ldots, f_r$ in A such that

$$f_1^2 + f_2^2 + \cdots + f_r^2 \in \mathfrak{p}, \quad \text{but } f_1 \notin \mathfrak{p}.$$

Since $(\mathfrak{p}, f_1) \cap S \neq \varnothing$ we obtain $f^{2m} + t = g f_1 + p$; upon squaring this becomes

$$f^{4m} + \tilde{t} = g^2 f_1^2 + \tilde{p}.$$

Addition of $g^2(f_2^2 + \cdots + f_r^2)$ leads to the contradiction $S \cap \mathfrak{p} \neq \varnothing$.

21.6 Consider over $K = \mathbb{R}$ the polynomial

$$f(X, Y, Z) = X^4 - (Z^2 - 1)(X^2 + Y^2)$$

and the algebraic K-set V in K^3 defined by f. Prove:

(a) f is *irreducible* in $K[X, Y, Z]$.

(b) The ideal of V is (f); thus V is a K-*variety*.

Hint: The quickest way to get there is to draw on equality (4) of §21.4. Then what needs to be shown is that the prime ideal (f) is *real*; that is, the fraction field of

$$K[X, Y, Z]/f = K[x, y, z]$$

is formally real. This can be done by using the relation

$$z^2 - 1 = \frac{x^4}{x^2 + y^2}$$

to prove that $K(x, y, z)$ admits an order.

(c) The function $z^2 - 1 \in K[V]$ is a sum of squares in $K(V)$, but at the point $(0, 0, 0)$ of V it takes on the value $-1 < 0$.

Incidentally, this can be contrasted with the example discussed in Remark (ii) after Theorem 2. There the statement corresponding to part (b) could be established by an explicit calculation.

Chapter 23: Absolute Values on Fields

23.1 Prove that, if $|\ |$ is an absolute value on K, so is $|\ |^\rho$, where ρ is any real number such that $0 < \rho \le 1$.

Hint: If x, y are positive real numbers, then $(x + y)^\rho \le x^\rho + y^\rho$ (divide by the left-hand side).

23.2 A *quasi-absolute value* on a field K is a map $|\ |$ of K into $\mathbb{R}_{\ge 0}$ that satisfies properties (i) and (ii) of an absolute value and also

(1) $$|a + b| \le C \max(|a|, |b|),$$

where $C > 0$ is a real constant. Obviously, in this case $|\ |^\rho$ is also a quasi-absolute value, for any real $\rho > 0$. A quasi-absolute value of special importance is the function $|\ |^2_\infty$ of $\mathbb{C}$.

Prove that, if $|\ |$ is a quasi-absolute value on K, there is equivalence between:

(i) $|\ |$ is an absolute value.

(ii) Equation (1) holds with $C = 2$.

(iii) For any $n \in \mathbb{N}$ we have $|n| \le n$.

(iv) The restriction of $|\ |$ to the prime field of K is an absolute value.

Hint: Write $C = 2^\lambda$ in (1), with $\lambda \ge 0$. Prove that

(2) $$|a_1 + \cdots + a_n| \le (2n)^\lambda \max(|a_1|, \ldots, |a_n|),$$

as follows: When n is a power of 2, check the stronger inequality $|a_1 + \cdots + a_n| \le n^\lambda \max(|a_1|, \ldots, |a_n|)$; for $2^m \le n < 2^{m+1}$, supply $a_{n+1} = \cdots = a_{2^m+1} = 0$.

Now suppose that (iii) is satisfied. Then the binomial formula and inequality (2) yield, for all m, the inequality

$$|a+b|^m \leq (2(m+1))^\lambda \, (|a|+|b|)^m.$$

Taking the m-th root and the limit $m \to \infty$ we get $|a+b| \leq |a|+|b|$, as needed for (i). The implications (i) $\Rightarrow$ (iv) $\Rightarrow$ (iii) and (i) $\Rightarrow$ (ii) are clear. There remains to show that (ii) implies (i). For $C = 2$, inequality (2) leads to

$$|n| \leq 2n.$$

Again from (2) and the binomial formula we get

$$|a+b|^m \leq 2(m+1) \cdot 2 \cdot (|a|+|b|)^m,$$

from which the inequality $|a+b| \leq |a|+|b|$ is derived as before.

Note the following consequence: For every quasi-absolute value $|\ \ |$ there exists $\rho > 0$ such that $|\ \ |^\rho$ is an absolute value.

23.3 Let $|\ \ |$ be a *nonarchimedean* absolute value on K, and let d denote the corresponding metric on K. For any x, y, z in K, we have

$$d(x, y) \leq \max\big(d(x,z), d(z,y)\big).$$

Thus every triangle is isosceles: If $d(x,z) < d(y,z)$, then $d(y,z) = d(x,y)$.

For any element $a \in K$ and any real number $r > 0$, we can consider the open disk $D(a,r) = \{x \mid d(x,a) < r\}$ and the sphere $S(a,r) = \{x \mid d(x,a) = r\}$. Prove that any $b \in D(a,r)$ is a center of $D(a,r)$; that is, $D(a,r) = D(b,r)$. For any $x \in S(a,r)$, we have $D(x,r) \subseteq S(a,r)$. So $S(a,r)$ is *open*! And likewise, $D(a,r)$ is *closed*.

It follows that K is *totally disconnected*; that is, if $M \subseteq K$ is nonempty and connected, M has a single element. For suppose $a, b \in M$ satisfy $r := d(a,b) > 0$. Then $M \cap D(a,r)$ is both open and closed in M, hence equal to M; but this contradicts $b \notin D(a,r)$.

23.4 Observe that the valuation ring R of the p-adic absolute value $|\ \ |_p$ of $\mathbb{Q}$ coincides, by F5, with the *localization* $\mathbb{Z}_{(p)}$ of $\mathbb{Z}$ at the prime ideal (p) (see Exercise §4.13 in vol. I). However, we will avoid the notation $\mathbb{Z}_{(p)}$ whenever possible, to prevent confusion with the ring $\mathbb{Z}_p$ of p-adic integers. $\mathbb{Z}_p$ is the closure of $\mathbb{Z}_{(p)}$ in $\mathbb{Q}_p$ (see Definition 9 and subsequent remark). Let R be any integral domain and $\mathfrak{p}$ a *prime ideal* of R. As a generalization of the last statement in F5, prove that the inclusion $R \subset R_\mathfrak{p}$ gives rise to the canonical isomorphism

$$\mathrm{Quot}(R/\mathfrak{p}) \simeq R_\mathfrak{p}/\mathfrak{p}R_\mathfrak{p}.$$

Thus, when $\mathfrak{p}$ is a *maximal ideal*, $R_\mathfrak{p}/\mathfrak{p}R_\mathfrak{p} \simeq R/\mathfrak{p}$.

23.5 Let $K = k(X)$ be the field of rational functions in one variable over a field k. Given a normalized prime polynomial π in $k[X]$, consider the map $w_\pi : K \to \mathbb{Z} \cup \{\infty\}$ defined in vol. I, Chapter 5 (just before F5). This is a *valuation* of K in the sense of Definition 5, and its valuation group is $\mathbb{Z}$. There is also a valuation w_∞ on K, defined by

$$w_\infty(f) = -\deg f,$$

where the degree of a rational function $f = g/h$, with $g, h \in k[X]$, is defined as $\deg f = \deg g - \deg h$.

Let $| \ |$ be any absolute value on $K = k(X)$ *whose restriction to k is trivial*. Since $|n| = 1$ for any $n \in \mathbb{N} \subset K$, this absolute value is *nonarchimedean*, and thus must arise from a valuation w (see remark preceding Definition 5).

Prove that *if nontrivial, w is equivalent to either w_∞ or to a unique w_π*. Compare this with Theorem 1, which classifies absolute values on the field $\mathbb{Q}$ of rationals. In spite of the seeming analogy we must watch out for two essential differences: First, we have not considered all absolute values on $K = k(X)$, only those that are trivial on k; although this distinction disappears when k is a *finite field*. Secondly, *all* these absolute values we're considering on $k(X)$ are *nonarchimedean*, contrary to the case $K = \mathbb{Q}$, where the archimedean absolute value $| \ |_\infty$ contrasts with the other, p-adic ones. No special role attaches to the valuation w_∞ of $K = k(X)$; it simply corresponds to the prime element X^{-1} of the polynomial ring $k[X^{-1}]$.

Hint: Either $w(X) \geq 0$ or $w(X) < 0$. In the former case, $k[X]$ is contained in the valuation ring of w; in the latter it's $k[X^{-1}]$ that is. In the first case there is a unique π such that $w \sim w_\pi$, and in the second it follows that $w \sim w_{X^{-1}} = w_\infty$.

Prove further that the residue field kw of w is canonically isomorphic to the extension $k[X]/\pi$ of k in the case $w = w_\pi$, and to k in the case $w = w_\infty$. Further, with $\deg w$ defined as $kw:k$, we get a counterpart to the *product formula* (15) for $\mathbb{Q}$:

$$\sum_w w(f) \deg w = 0.$$

23.6 A subring R of a field K is called a *valuation ring* of K if, for every $a \in K$, either $a \in R$ or $a^{-1} \in R$. Consider the abelian group $\Gamma := K^\times / R^\times$ and prove that it becomes an *ordered group* if (switching to additive notation) we define

(1)
$$v(a) \geq 0 \quad \Longleftrightarrow \quad a \in R.$$

Here $v : K^\times \to \Gamma$ is the quotient homomorphism. Check that, for every $a, b \in K^\times$ such that $a + b \neq 0$,

(2)
$$v(a + b) \geq \min\left(v(a), v(b)\right).$$

Conversely: Let K be a field and $v : K^\times \to \Gamma$ a surjective homomorphism from the multiplicative group of K into an ordered abelian group Γ, and assume v satisfies (2). Then v is called a *(generalized) valuation* of K. The set $R := \{a \in K \mid v(a) \geq 0\}$ is a *valuation ring* of K; it is a local ring and $\mathfrak{p} := \{a \in K \mid v(a) > 0\}$ is its maximal ideal. Two valuations $v_1 : K^\times \to \Gamma_1$ and $v_2 : K^\times \to \Gamma_2$ of K are called *equivalent* if there is an order-preserving isomorphism $\lambda : \Gamma_1 \to \Gamma_2$ such that $v_2 = \lambda \circ v_1$. It is clear that *the valuation rings R of a field K are in one-to-one correspondence with equivalence classes of valuations on K.*

Let $v : K^\times \to \Gamma$ be a valuation on K. Set $v(0) = \infty$ to obtain a map v from K to $\Gamma \cup \{\infty\}$ satisfying $v(ab) = v(a) + v(b)$ and inequality (2) for *every* $a, b \in K$, where the usual conventions for calculating with ∞ apply. When Γ is an ordered subgroup of the reals, v is a valuation in the sense of Definition 5. Check that in this case the notion of equivalence just defined agrees with that given in the text of the chapter.

Prove also that for $K = \mathbb{Q}$, the p-adic valuations w_p exhaust all equivalence classes of generalized valuations. Likewise, the generalized valuations on $K = k(X)$ that restrict to the trivial valuation on k are all equivalent to the ones discussed in the preceding exercise. But without the triviality condition one can put on $\mathbb{Q}(X)$, for example, a valuation with group $\mathbb{Z} \times \mathbb{Z}$ (do check this). Of course there are also *archimedean* absolute values on $\mathbb{Q}(X)$.

If $(K, \leq)$ is a *nonarchimedean ordered* field, the set $\mathfrak{p}$ of $a \in K$ such that $n|a| \leq 1$ for all natural numbers n is the maximal ideal of a *valuation ring* of K.

23.7 Show that, if p and q are distinct primes, $(p^n)_n$ is not a Cauchy sequence with respect to $|\ |_q$.

Hint: $x \equiv 1 \bmod q^i$ implies $x^q \equiv 1 \bmod q^{i+1}$. Or just consider the sequence $(p^{n+1} - p^n)_n$.

23.8 Let $|\ |_1, |\ |_2, \ldots, |\ |_r$ be pairwise inequivalent nontrivial absolute values on a field K. Prove by induction that there exists $z \in K$ such that

$$(1) \qquad |z|_1 > 1 \quad \text{but} \quad |z|_j < 1 \quad \text{for } 2 \leq j \leq r.$$

Hint: By definition, there exists x such that $|x|_1 < 1$ and $|x|_2 \geq 1$, and also y such that $|y|_2 < 1$ and $|y|_1 \geq 1$. Thus $z = y/x$ proves the case $r = 2$. Now let $r > 2$ and suppose that $|x|_1 > 1$ and $|x|_j < 1$ for $2 \leq j \leq r-1$. Choose y such that $|y|_1 > 1$ and $|y|_r < 1$. If $|x|_r \leq 1$, consider $z = x^n y$. If $|x|_r > 1$, consider $z = y\, x^n/(1+x^n)$.

Derive hence *Artin's independence* (or *approximation*) *theorem: If $|\ |_1, \ldots, |\ |_r$ are pairwise inequivalent, nontrivial absolute values on K and $a_1, \ldots, a_r$ are elements of K, there exists for every $\varepsilon > 0$ an element $x \in K$ such that*

$$(2) \qquad |a_i - x|_i < \varepsilon \quad \text{for every } 1 \leq i \leq r.$$

Hint: Let z be as in (1). The sequence of $z^n/(1+z^n)$ converges to 1 with respect to $|\ |_1$, but to 0 with respect to the remaining absolute values $|\ |_j$. Similarly one gets for each i an element e_i that lies near 1 for $|\ |_i$ but near 0 for $|\ |_j$, $i \neq j$. Now $x = a_1 e_1 + \cdots + a_r e_r$ solves the problem.

23.9 Let K be an extension of $\mathbb{C}$ and $|\ |$ an absolute value of K that coincides with $|\ |_\infty$ on $\mathbb{C}$. Prove that any $a \in K \setminus \mathbb{C}$ has a nearest neighbor in $\mathbb{C}$, that is, there exists $z_0 \in \mathbb{C}$ such that

$$|a - z| \geq |a - z_0| \quad \text{for every } z \in \mathbb{C}.$$

By replacing a with an appropriate natural multiple of $a - z_0$, obtain $a \in K$ such that

(1) $$|a - z| \geq |a| > 1 \quad \text{for every } z \in \mathbb{C}.$$

Now consider $a^n - 1 = \prod_{k=1}^{n}(a - z_k)$ for $n \in \mathbb{N}$. Using (1), prove that $|a - 1| \leq |a|$, and hence that a can be replaced by $a - 1$ in (1). It follows inductively that $|a - n| = |a|$, and hence the contradiction that $n = |n| \leq 2|a|$ for every n. Contradiction to what? To the existence of $a \in K \smallsetminus \mathbb{C}$. Hence $K = \mathbb{C}$.

23.10 Let K be a *complete* field with respect to an *archimedean* absolute value $|\ |$. Suppose $|\ |$ coincides with $|\ |_\infty$ on the subfield $\mathbb{Q}$ of K.

(a) Take $c \in K$ with $|c| < 1$. Define by recursion a sequence $(x_n)_n$ in K as follows:

$$x_{n+1} = \frac{c}{x_n} - 2, \quad x_0 = 1.$$

First show that $|x_n| \geq 1$ and then that

$$|x_{n+2} - x_{n+1}| \leq |c||x_{n+1} - x_n|.$$

It follows that $(x_n)_n$ is a Cauchy sequence; its limit x satisfies the equation $x^2 + 2x - c = 0$. Hence:

(b) Every element of the form $1 + c$ with $|c| < 1$ is a square in K.

(c) Suppose -1 is *not* a square in K, and let $L = K(i)$ be an extension of K with $i^2 + 1 = 0$. Define $\|a + bi\| = |a^2 + b^2|^{1/2}$; then $\|\ \|$ is an extension of $|\ |$ to an *absolute value* on L.

Hint: It suffices to prove that $\|\ \|^2$ is a *quasi-absolute value* (see §23.2), and this in turn boils down to showing that $\|z + 1\|^2 \leq 4$ for $\|z\|^2 \leq 1$. So assume that $z = a + bi$ satisfies

(1) $$|a^2 + b^2| \leq 1.$$

Since $\|z + 1\|^2 = |a^2 + 2a + 1 + b^2| \leq 2 + 2|a|$, it is enough to show that (1) implies $|a| \leq 1$. To do this one can argue that the inequality

$$|1 + x^2| < 1$$

is *never* true for $x \in K$; indeed, if it were, (b) would imply that $1 - (1 + x^2) = -x^2$ is a square in K, and so also -1.

23.11 Let K be a extension of $\mathbb{R}$ and $|\ |$ an absolute value on K that agrees with $|\ |_\infty$ on $\mathbb{R}$. Prove that, *if -1 is a square in K, then $K = \mathbb{C}$; if -1 is not a square in K, then $K = \mathbb{R}$.*

Hint: The real work has been done in F11(b), §23.9 and §23.10(c).

23.12 Because of F9, there is a *bijection*

$$\mathbb{Z}_p \to (\mathbb{Z}/p)^{\mathbb{N}}.$$

Cardinality considerations then imply that $\mathbb{Q}_p \neq \mathbb{Q}$. Obtain the same result in a more down-to-earth way by using F15 to prove that $\sqrt{-1} \in \mathbb{Q}_p$ for $p \equiv 1 \bmod 4$; that $\sqrt{2} \in \mathbb{Q}_p$ for $p \equiv 7 \bmod 8$; that $\sqrt{-2} \in \mathbb{Q}_p$ for $p \equiv 3 \bmod 8$; and finally, that $\sqrt{-7} \in \mathbb{Q}_2$.

Even better, show that the p-adic development (29) of a number $a \in \mathbb{Q}_p$ is *periodic* (possibly with a preperiodic part) if and only if a lies in $\mathbb{Q}$.

23.13 Suppose a field F is complete with respect to a nonarchimedean absolute value $|\ |$, and let $\overline{F} = R/\mathfrak{p}$ denote the corresponding residue field. Prove that *every element u of R with $u \equiv 1 \bmod \mathfrak{p}$ is an m-th power in F, for any $m \in \mathbb{N}$ not divisible by char $\overline{F}$*. *Hint:* See F13.

23.14 Let E/K be an *algebraic* field extension.

(i) It follows from F11(b) that the only extension to E of the *trivial absolute value* on K is the trivial absolute value on E. Reprove this directly, by considering the minimal equation of some $\alpha \in E$ and assuming $|\alpha| > 1$.

(ii) If E/K is *purely inseparable*, the proof of Theorem 5 shows that every absolute value on K has a unique extension to an absolute value on E. Prove this directly using that every $x \in E$ satisfies an equation of the form

$$x^{p^m} = a, \quad \text{with } a \in K.$$

23.15 Let E/K be a *finite* field extension and let $|\ |_1, \ldots, |\ |_r$ be distinct extensions to E of an absolute value $|\ |$ of K. With the same notation as in Theorem 5, consider the canonical homomorphism of $\widehat{K}$-algebras

$$(1) \qquad\qquad \hat{d} : E \otimes_K \widehat{K} \to \prod_{i=1}^{r} \widehat{E}_i.$$

In view of F11, the absolute values $|\ |_1, \ldots, |\ |_r$ are all inequivalent. Hence *Artin's independence theorem* (§23.8) says that the diagonal map

$$d : E \to \prod_{i=1}^{r} \widehat{E}_i$$

takes E to a *dense* subset of $\prod_{i=1}^{r} \widehat{E}_i$. From this it follows through the use of F10 that the map (1) is *surjective*, because its image is a $\widehat{K}$-vector subspace.

Together these steps amount to a noticeably shorter route to Theorem 5, albeit with the exception of the theorem's last statement, according to which (1) is an isomorphism if E/K is *separable* and $|\ |_1, \ldots, |\ |_r$ run over *all* extensions of the absolute value $|\ |$.

In this connection show, more generally, that if $|\ |_1, \ldots, |\ |_r$ *are all the distinct extensions of the absolute value of $|\ |$, the kernel of (1) is the nilradical of the $\widehat{K}$-algebra $E \otimes_K \widehat{K}$*.

Hint: We just need to show that $\ker \hat{d} \subseteq \sqrt{0}$. Since $E \otimes_K \widehat{K}$ is in any case a finite-dimensional $\widehat{K}$-algebra, it suffices, by §4.14 and Chapter 2, F2 in vol. I, to show that ker $\hat{d}$ is contained in every maximal ideal $\mathfrak{m}$ of $E \otimes_K \hat{K}$. To this end, note that the residue field $E \otimes_K \hat{K}/\mathfrak{m}$ can be seen as a finite extension of $\hat{K}$; hence it gives rise to an absolute value $|\ |_i$ on E extending $|\ |$. It follows that $E \otimes_K \hat{K}/\mathfrak{m} \simeq \hat{E}_i$.

Now, if E/K is *separable*, we have $E \otimes_K \hat{K} \simeq \hat{K}[X]/f$, where $f \in K[X]$ is irreducible and *separable*; thus $E \otimes_K \hat{K}$ has no nonzero nilpotents, as asserted.

23.16 Suppose $q = c/d$, where c, d are natural numbers and $c > d$. Every natural number n has a q-adic representation, that is, it can be written as

(1) $n = a_0 + a_1 q + \cdots + a_r q^r$,   where $a_i \in \{0, 1, \ldots, c-1\}$ and $a_r \neq 0$.

Proof. Division with remainder yields first

$$n = yc + a_0, \quad \text{where } 0 \leq a_0 < c \text{ and } 0 \leq y \leq n/c.$$

Hence $dn = (dy)c + a_0 d$, with $dy \leq dn/c < n$. If $y \neq 0$, we can assume by induction that $dy = a_1 + \cdots + a_r q^{r-1}$, where the a_i are as above. Therefore

$$dn = (a_1 + \cdots + a_r q^{r-1})c + a_0 d,$$

and now division by d delivers the result.   □

In fact, if c, d are relatively prime, the representation (1) is unique.

23.17 Prove that the element $b \in R$ in F13 actually satisfies

$$|b - a| = |f(a)| < 1.$$

Correspondingly, the b in F14 actually satisfies

$$|b - a| = \frac{|f(a)|}{|f'(a)|} < |f'(a)|.$$

Chapter 24: Residue Class Degree and Ramification Index

24.1 (i) Let $A \neq 0$ be a subgroup of the additive group of $\mathbb{R}$. Show the equivalence between: (a) A is *discrete* (there are no accumulation points of A in A). (b) A has no accumulation points in $\mathbb{R}$. (c) A has a *smallest* element $a > 0$. (d) $A = \mathbb{Z}a$, where $a > 0$. (e) A is *closed* in $\mathbb{R}$ and $A \neq \mathbb{R}$.

Hint: Suppose A has an accumulation point in $\mathbb{R}$. Then, for any $\varepsilon > 0$, there exist $a, b \in A$ such that $0 < b - a < \varepsilon$. By looking at integral multiples $n(b-a)$, deduce that *every* real number is an accumulation point of A.

(ii) Deduce the counterparts of (a)–(e) for the *multiplicative* group $\mathbb{R}_{>0}$ of $\mathbb{R}$, by using the properties of the log function. In particular, a *discrete* subgroup of $\mathbb{R}_{>0}$ is *cyclic*.

24.2 Prove that $\mathbb{F}_p((X))$ is not algebraic over $\mathbb{F}_p(X)$. Using this, give an example of a *purely inseparable extension E/K of degree p that becomes trivial by completion with respect to a valuation.* This shows that formula (16) in Theorem 2 does not hold unconditionally — and likewise (60) in Chapter 23.

Hint: The field $F := \mathbb{F}_p(X)$ has countably many elements, and the same is true of any algebraic extension of F; but the field $\hat{F} := \mathbb{F}_p((X))$ is uncountable (see example after F2). Now suppose $Y \in \hat{F}$ is transcendental over F. Setting $E = F(Y)$ and $K = F(Y^p)$, we readily obtain an extension E/K as desired.

As a complement, prove that $\mathbb{F}_p((X))$ contains an element α such that $\alpha^p - \alpha - X = 0$, but $\mathbb{F}_p(X)(\alpha)/\mathbb{F}_p(X)$ is Galois of degree p.

24.3 In the algebraic closure C of $\mathbb{Q}_2$, consider the sequence of elements α_n such that $\alpha_{n+1}^2 = \alpha_n$ and $\alpha_1 = 2$, and form the extension K_2 of $\mathbb{Q}_2$ generated by these elements. Let K denote the *completion* of K_2, and set $E = K(\sqrt{3})$. Prove that *the extension E/K does not satisfy the formula $n = ef$.*

Hint: Since $w_2(\alpha_{n+1}) = 2^{-n}$, the extension $E_n := \mathbb{Q}_2(\alpha_{n+1})$ has degree 2^n over $\mathbb{Q}_2$, with ramification index 2^n and residue class degree 1 (Theorem 1). Being the union of all the E_n, then, K_2 has residue field $\mathbb{F}_2 = \mathbb{Z}/2$ and valuation group $w_2(K_2^\times) = \bigcup_n 2^{-n}\mathbb{Z}$. The same is true of the completion K of K_2.

Suppose that $\sqrt{3}$ lies in K. Since K_2 is dense in K, there exists $a \in K_2$ such that $w_2(a^2 - 3) > 2$. Now, a lies in some E_n. Applying F14 of Chapter 23 we obtain $\sqrt{3} \in E_n$. But it is easy to check by induction that

(1)
$$\sqrt{3} \notin E_n;$$

for the starting case $n = 0$, make use of F15 in Chapter 23.

Setting $E = K(\sqrt{3})$ we get a degree-2 extension E/K of complete fields. Since $\frac{1}{2}w(K^\times) = w(K^\times)$, the ramification index e of E/K must be 1. We claim that the residue class degree f is also 1. Otherwise $f = 2$, so $\bar{E}$ is the degree-2 extension of the field $\bar{K} = \mathbb{Z}/2$, generated by the roots of the polynomial $h(X) = X^2 + X + 1$ over $\bar{K}$. By F13 in Chapter 23, h splits over E as well. But $K_2(\sqrt{3})$ is dense in $K(\sqrt{3}) = E$, so we derive as before (from Chapter 23, F13) that h also splits over $E_n(\sqrt{3})$, for appropriate n. But then $E_n(\sqrt{3}) = E_n(\sqrt{-3})$, that is, E_n contains $\sqrt{-1}$. This is impossible, as can be proved much like (1).

24.4 Prove that *the group $W(\mathbb{Q}_p)$ of all roots of unity in $\mathbb{Q}_p$ is finite, and has order $p-1$ or 2 depending on whether $p \neq 2$ or $p = 2$. Hint:* See Theorem 4 and the example after F3.

24.5 (a) Prove that the statement in the example following F3 also holds if the ground field $\mathbb{Q}_p$ is replaced by an *unramified* extension K of $\mathbb{Q}_p$.

(b) Starting from the statement in the example following F3, show by induction that

(1)
$$\mathbb{Q}(\zeta_m) : \mathbb{Q} = \varphi(m).$$

Hint: Let $m = p^k m'$, where $(m', p) = 1$, and set $F = \mathbb{Q}(\zeta_{p^k}) \cap \mathbb{Q}(\zeta_{m'})$. Then $F/\mathbb{Q}$ is, with respect to $|\ |_p$, at the same time purely ramified and unramified. Hence

$F = \mathbb{Q}$. Now use the *Translation Theorem* of Galois theory (vol. I, Chapter 12, Theorem 1).

The proof method used in connection with (a) shows that (1) holds also if we replace $\mathbb{Q}$ by an algebraic number field k such that every p dividing m is unramified in k.

24.6 Let E/K be a finite field extension and let $|\ |$ be a *discrete* absolute value on E, with *normalized valuation* w. Assume also that K is $|\ |$-complete. The proof of Theorem 1 has the following corollary: If Π is a prime element of w and $\alpha_1, \ldots, \alpha_f$ are representatives of the elements of a basis of $\overline{E}/\overline{K}$, then the elements

$$\alpha_i \Pi^j, \quad \text{for } 1 \le i \le f \text{ and } 0 \le j < e$$

form an R-basis of A.

Further, assume $\overline{E}/\overline{K}$ is *separable*. By the primitive element theorem, there is $\alpha \in A$ such that the elements

$$(1) \qquad \alpha^i \Pi^j, \quad \text{for } 0 \le i < f \text{ and } 0 \le j < e,$$

form an R-basis of A. Prove that *there exists $\beta \in A$ such that $A = R[\beta]$, and so the elements $1, \beta, \beta^2, \ldots, \beta^{n-1}$, where $n = E:K$, also form an R-basis of A.*

Hint: Starting from the basis (1), choose a normalized $g \in R[X]$ of degree f such that $\overline{g} = \text{MiPo}_{\overline{K}}(\overline{\alpha})$. Conclude that $w(g(\alpha)) \ge 1$ (since $\overline{g}(\overline{\alpha})$ vanishes).

If $w(g(\alpha)) = 1$, set $\Pi = g(\alpha)$, rendering the claim true with $\beta = \alpha$. So assume $w(g(\alpha)) \ge 2$. A Taylor series development yields

$$g(\alpha + \Pi) = g(\alpha) + g'(\alpha)\Pi + c\Pi^2, \quad \text{with } c \in A.$$

There follows $w(g(\alpha + \Pi)) = 1$, and now the element $\beta = \alpha + \Pi$ serves the purpose.

24.7 Let E/K be a finite extension and $|\ |$ a nonarchimedean absolute value on E, and assume K is $|\ |$-*complete*. Prove that if E/K is *normal*, then so is the extension $\overline{E}/\overline{K}$ of residue fields, and there is a canonical homomorphism

$$(1) \qquad \begin{aligned} G(E/K) &\to G(\overline{E}/\overline{K}), \\ \sigma &\mapsto \overline{\sigma}. \end{aligned}$$

Assume that E/K is *Galois*, and let T be the kernel of (1). T is called the *inertia group* of E/K. The corresponding fixed field is called the *inertia field* of E/K. Prove that *the inertia field of E/K is the largest unramified subfield L of E/K* (compare Theorem 3). Moreover, *the map* (1) *is surjective and gives rise to the following commutative diagram of canonical isomorphisms:*

$$\begin{array}{ccc} G(E/K)/T & \xrightarrow{\ \simeq\ } & G(\overline{E}/\overline{K}) \\ {\scriptstyle\simeq}\downarrow & & \downarrow{\scriptstyle\simeq} \\ G(L/K) & \xrightarrow{\ \simeq\ } & G(\overline{L}/\overline{K}) \end{array}$$

Hint: See Theorem 3.

24.8 Let $|\ |$ be a nonarchimedean absolute value of a field K, with valuation ring R and residue field $\overline{K}$. Let $f \in R[X]$ be a normalized polynomial, and form the extension $L = K(\alpha)$ of K, where $f(\alpha) = 0$. Prove that *if $\overline{f} \in \overline{K}[X]$ is separable, then L/K is unramified with respect to* $|\ |$.

Hint: One may as well assume that K is $|\ |$-complete (see Definition 3). Let E be a splitting field of f over K. It is enough to show that E/K is unramified. To do this, prove that the natural map $G(E/K) \rightarrow G(\overline{E}/\overline{K})$ is injective. The claim then follows directly from §24.7, or by consideration of the inequalities

$$\overline{E}:\overline{K} \geq |G(\overline{E}/\overline{K})| \geq |G(E/K)| \geq E:K.$$

Another path to the proof: One can assume that f is *irreducible* over K (see Theorem 4' in Chapter 23). By *Hensel's Lemma*, $\overline{f}$ is also irreducible. It follows that $\overline{L}:\overline{K} \geq \deg \overline{f} = \deg f = L:K$.

24.9 Let K be complete with respect to a nonarchimedean absolute value $|\ |$. Prove: *If ζ is a primitive m-th root of unity in the algebraic closure of K, where m is not divisible by the characteristic of the residue field of K, the extension $K(\zeta)/K$ is unramified.* (*Hint:* It is most convenient to use §24.8.)

24.10 Let E/K be a finite extension and $|\ |$ a *discrete* absolute value of E, with normalized valuation w. Assume that K is $|\ |$-*complete* and that the extension $\overline{E}/\overline{K}$ of residue fields is *separable*. Assume further that E/K is *Galois* and consider, for every integer $i \geq -1$, the subgroup G_i of $G = G(E/K)$ consisting of all $\sigma \in G$ that operate trivially on $A/\mathfrak{P}^{i+1}$; that is, those for which

(1) $$w(\sigma x - x) > i \quad \text{for every } x \in A.$$

By definition, $G_{-1} = G$ and G_0 is the *inertia group* of E/K (see §24.7). For large enough i we clearly have $G_i = 1$. We thus obtain a well defined chain

$$G = G_{-1} \supseteq G_0 \supseteq G_1 \supseteq \cdots \supseteq G_r = 1$$

of *normal subgroups* of G. Each G_i is called the *i-th ramification group* of E/K. If F is an intermediate field of E/K with $H := G(E/F)$, the intersection $G_i \cap H$ is the i-th ramification group of E/F. Thus, for L the largest unramified subfield of E/K, the i-th ramification group of E/L (where $i \geq 0$) coincides with that of E/K (see §24.7).

Further, let Π be a prime element of E, and let $U_E = A^\times$ be the group of units of E. Setting

$$U_E^{(0)} = U_E, \quad U_E^{(1)} = 1 + \Pi A, \quad U_E^{(2)} = 1 + \Pi^2 A, \quad \cdots$$

we get a decreasing chain of *subgroups* $U_E^{(i)}$ of U_E. From now on fix $i \geq 0$. Prove:

(a) An automorphism $\sigma \in G_0$ lies in G_i if and only if

$$\sigma(\Pi)/\Pi \equiv 1 \bmod \Pi^i.$$

(b) The map $\sigma \mapsto \sigma(\Pi)/\Pi$ gives rise to a canonical isomorphism (not depending on the choice of Π) between the group G_i/G_{i+1} and a subgroup of $U_E^{(i)}/U_E^{(i+1)}$.

(c) The group G_0/G_1 is cyclic and its order is not divisible by the characteristic of the residue class field $\overline{K}$.

(d) If char $\overline{K} = 0$ we have $G_1 = 1$ (so G_0 is cyclic).

(e) If char $\overline{K} = p > 0$, the groups G_i/G_{i+1}, for $i \geq 1$, are elementary abelian p-groups and G_1 is the Sylow p-group of G_0.

Hint: Without loss of generality one can assume that $G = G_0$, that is, E/K is purely ramified. Then $A = R[\Pi]$, and (a) follows without difficulty. Once (a) is proved, one checks (b) directly. Every finite subgroup of $\overline{E}^\times = U_E^{(0)}/U_E^{(1)}$ is cyclic; this yields (c). For parts (d) and (e), show that $U_E^{(i)}/U_E^{(i+1)}$, for $i \geq 1$, is isomorphic to the *additive* group of $\overline{E}$.

Observe also that if L_1 denotes the fixed field of G_1, the extension L_1/K is the largest subextension of E/K that is *tamely ramified*, that is, whose ramification index is not divisible by the characteristic of $\overline{K}$.

24.11 Let k be an *algebraically closed* field of *characteristic zero* and let $K = k((X))$ be the field of formal Laurent series with finite principal part over k. Prove: *The algebraic closure C of K is the union of the fields $K_n := K(X^{1/n}) = k((X^{1/n}))$. More precisely, for every natural number n, C/K has exactly one intermediate field of degree n over K, namely K_n. The Galois group $G(C/K)$ is isomorphic to $\hat{\mathbb{Z}}$* (see vol. I, §12.4).

Hint: Any extension E/K of degree n is purely ramified, so $E = K(\Pi)$, and therefore $\Pi^n = uX$, where u is a unit of E. But every unit of E is an n-th power in E. (One can alternatively base the proof on statement (d) of §24.10.) To show that $G(C/K) \simeq \hat{\mathbb{Z}}$, consider that $W(k) \simeq W(\mathbb{C})$ and use that the subgroups $W_n(\mathbb{C})$ of $W(\mathbb{C})$ have canonical generators.

24.12 Let $|\ |_1$ and $|\ |_2$ be nontrivial absolute values on a field K. Prove that if $|\ |_1$ is *discrete* and $|\ |_2$ is *complete*, then $|\ |_1 \sim |\ |_2$.

Anticipating the results of Chapter 25 (Definition 1 and Remark after F2), we see that, in particular, *a local field K admits, up to equivalence, only a single nontrivial absolute value* with respect to which K is locally compact.

Hint: Let π be a prime element of K with respect to $|\ |_1$. For every natural number n, the polynomial

$$f(X) = X^n - \pi$$

is irreducible over K (see F3). Now let K be $|\ |_2$-complete. Then $|\ |_2$ is certainly *nonarchimedean*; otherwise $K = \mathbb{R}$ or $\mathbb{C}$, and f cannot be irreducible for $n > 2$. Assume $|\ |_2$ and $|\ |_1$ are not equivalent. By the *independence theorem* (§23.8) there exists, for every $\varepsilon > 0$, some $d \in K$ such that

$$|\pi - d|_1 < \varepsilon \quad \text{and} \quad |1 - d|_2 < \varepsilon.$$

For small enough ε, the first inequality says that d is, like π, a prime element for $|\ |_1$, while the second inequality guarantees that $X^n - d = 0$ has a solution in K, so long as char K does not divide n.

We mention that the result of this exercise also follows easily from a general fact proved by *F. K. Schmidt*, to which we turn next.

24.13 Let K be *complete* with respect to a nontrivial absolute value $|\ |_c$ on K. Prove: If $|\ |$ is a nontrivial absolute value on K that is *not* equivalent to $|\ |_c$, the $|\ |$-completion $\hat{K}$ of K must be *algebraically closed*.

Hint: Take $f \in \hat{K}[X]$ irreducible of degree n and *separable*. Choose a separable polynomial $h \in K[X]$ of degree n that splits into linear factors over K. Since K is $|\ |$-dense in $\hat{K}$, and thanks to the *independence theorem* (§23.8), there exists for any $\varepsilon > 0$ some $g \in K[X]$ of degree n such that

$$|f - g| < \varepsilon \quad \text{and} \quad |h - g|_c < \varepsilon.$$

For ε small enough, the first inequality ensures that g, too, is *irreducible* over $\hat{K}$, while the second implies that g, like h, splits into linear factors over K. Each of these facts are justified by *Hensel's Lemma* (Chapter 23, F14) if $|\ |$ and $|\ |_c$ are assumed to be nonarchimedean; see also §24.22. If $|\ |$ is archimedean, that is, if $\hat{K} = \mathbb{R}$ or $\mathbb{C}$ with $|\ | \sim |\ |_\infty$, then $n \le 2$, so g is irreducible. If $|\ |_c$ is archimedean, so $K = \mathbb{R}$ or $\mathbb{C}$ with $|\ |_c \sim |\ |_\infty$, then in the case $K = \mathbb{R}$ one needs to show that g, like h, has only real roots. For this use Theorem 9 in Section 20, or Rouché's Theorem.

Altogether this shows that $\hat{K}$ does not admit a nontrivial *separable* algebraic extension. The rest arises from the next problem.

24.14 Let K be *complete* with respect to a nontrivial absolute value $|\ |$. Prove that, if K is *separably closed* (that is, K admits no nontrivial separable algebraic extension), then K is *algebraically closed*.

Hint: K has characteristic $p > 0$, and it suffices to show that every $a \in K^\times$ is a p-th power in K. For that, consider, for $t \in K^\times$ such that $|t| < 1$, the polynomial

$$f_n(X) = X^p + t^n X - a.$$

Since $f_n'(X) = t^n$, this polynomial is separable, and so splits into linear factors over K. We can assume that $|a| \le 1$; then at least one root α_n of f_n satisfies $|\alpha_n| \le 1$. Hence

$$|\alpha_n^p - a| \to 0 \quad \text{for} \quad n \to \infty.$$

But then $(\alpha_n)_n$ is a Cauchy sequence, since $(\alpha_{n+1} - \alpha_n)^p = \alpha_{n+1}^p - \alpha_n^p$. Its limit α satisfies $\alpha^p = a$.

24.15 Let $|\ |$ be a nonarchimedean absolute value on a field K, and denote by $\hat{K}$ the $|\ |$-completion of K. Using Theorem 3 of Chapter 25, prove: *If K is algebraically closed, so is $\hat{K}$.*

Hint: Let $f \in \hat{K}[X]$ be an *irreducible* normalized polynomial of degree n. We must show that $n = 1$. By §24.14 it suffices to do this when f is *separable*. For every $\varepsilon > 0$ there is a normalized polynomial $g \in K[X]$ of degree n such that

$$|f - g| < \varepsilon.$$

For small enough ε, the irreducibility of f implies that of g. Because K is algebraically closed we then get $n = 1$ as needed.

24.16 Let K be any field. Prove:

(a) If K has a nontrivial, complete absolute value, its cardinality satisfies

(1) $$\operatorname{Card} K = \operatorname{Card} K^{\mathbb{N}}.$$

Hint: Let $| \ |$ be a complete absolute value on K and suppose $|\pi| < 1$. If $| \ |$ is nonarchimedean, with valuation ring R, consider a family S of representatives of R/π and expand an arbitrary $a \in R$ in powers of π.

(b) If K has two inequivalent nontrivial absolute values, both complete, then K is algebraically closed. *Hint:* See §24.13.

(c) If K is algebraically closed and condition (1) is satisfied, then K admits an infinite family of pairwise inequivalent *complete* absolute values; at the same time, there are also infinitely many pairwise inequivalent absolute values on K with respect to which K is *not* complete.

Hint: Recall §24.15 and observe that algebraically closed fields having the same characteristic and the same transcendence degree (over the prime field) are isomorphic. Further: If K is any field such that $\operatorname{Card} K > \operatorname{Card} \mathbb{N}$, then $\operatorname{Card} K$ equals the transcendence degree of K.

24.17 Given an isomorphism $\sigma : K_1 \to K_2$ of fields and nontrivial absolute values $| \ |_1$ on K_1 and $| \ |_2$ on K_2, prove: If $| \ |_1$ is *discrete* and $| \ |_2$ is *complete*, then $\sigma : (K_1, | \ |_1) \to (K_2, | \ |_2)$ is bicontinuous (a homeomorphism). *Hint:* Consider on K_1 the absolute values $| \ |_1$ and $| \ |_2 \circ \sigma$; then use §24.12.

24.18 Let K be an extension of $\mathbb{Q}_p$ such that $K : \mathbb{Q}_p < \infty$. Prove: There exists no homomorphism from $\mathbb{R}$ into K, nor from K into $\mathbb{R}$, nor yet from $\mathbb{Q}_q$ into K, for $q \neq p$. On the other hand, there are (many) homomorphisms from K into $\mathbb{C}$ and of $\mathbb{C}$ into the algebraic closure of $\mathbb{Q}_p$.

Hint: If nothing else comes to mind, use §24.13.

24.19 Let K be *complete* with respect to a *discrete* absolute value $| \ |$. By choosing a prime element π and a family of representatives $S \ni 0$ for the residue field $\overline{K} = R/\mathfrak{p}$, one has for every $a \in R$ a unique representation of the form

(1) $$a = \sum_{i=0}^{\infty} a_i \pi^i, \quad \text{with } a_i \in S;$$

see F2. We thus obtain a bijective map

(2) $$R \to S^{\mathbb{N}_0}.$$

Give the set on the right the *product topology* of discrete topologies. Check that *the map* (2) *is a homeomorphism*. As a corollary (using Tychonov's Theorem): If $|\ |$ is a *discrete* and *complete* absolute value on K and if the corresponding residue group is *finite*, the valuation ring R of $|\ |$ is *compact*.

24.20 Is there a *finite* field extension E/K and a discrete absolute value on E such that E is *complete* but K is not? In other words, is it possible for the completion $\hat{K}$ of a field K with respect to a discrete absolute value to have finite degree greater than 1 over K?

Hint: Suppose $\hat{K}:K = n > 1$. Show that $\hat{K}/K$ must in any case be *purely inseparable* (see Theorem 2 and use §24.12). To arrive at an example, start with a field of the form $E = k((X))$, with char $k = p$, and choose k so that $k:k^p = \infty$. Now let $(a_i)_i$ be a sequence of elements of k linearly independent over k^p. The element

$$b = \sum_{i=0}^{\infty} a_i X^i$$

of E is not contained in the subfield

$$F := k^p((X))\, k$$

of E. Extend b to a p-basis B of E/F, that is, to a subset B of E such that $F(B) = E$ and having the property that every finite set of pairwise distinct $b_1, \ldots, b_n$ in B satisfies the condition $F(b_1, \ldots, b_n) : F = p^n$. Then the subfield $K := F(B \smallsetminus \{b\})$ satisfies

$$E : K = p \quad \text{and} \quad E = \hat{K}.$$

24.21 Let K be complete with respect to a nonarchimedean absolute value $|\ |$, and let $f \in R[X]$ be a normalized polynomial with coefficients in the valuation ring R of $|\ |$. Assume f is *separable* and denote by $D(f)$ its discriminant. Using *Krasner's Lemma* (Chapter 25, just before Theorem 3), prove: *If there exists $\beta \in R$ such that $|f(\beta)| < |D(f)|$, then f has a root α in K.*

Hint: Over an algebraic closure C of K, write $f(X) = (X - \alpha_1)(X - \alpha_2) \ldots (X - \alpha_n)$; by assumption,

$$\prod_{i=1}^{n} |\beta - \alpha_i| \ <\ \prod_{i=1}^{n} \prod_{\substack{j=1 \\ j \neq i}}^{n} |\alpha_j - \alpha_i|.$$

Hence there exists i such that

$$|\beta - \alpha_i| < \prod_{\substack{j=1 \\ j \neq i}}^{n} |\alpha_j - \alpha_i|$$

$$\leq |\alpha_j - \alpha_i| \quad \text{for every } j \neq i.$$

Now *Krasner's Lemma* implies that $\alpha_i \in K$.

Moreover, one easily sees that α_i is the *only* root α of f satisfying the condition $|\beta - \alpha| < |\alpha' - \alpha|$ for every root $\alpha' \neq \alpha$ of f.

Remark. Clearly the result in §24.21 constitutes, next to F14 in Chapter 23, a fact of fundamental interest. Incidentally, the statement can be extended to f not necessarily normalized if the discriminant of $f(X) = aX^n + a_{n-1}X^{n-1} + \cdots + a_0$ is defined as $D(f) = a^{2n-2} \prod_{i<j}(\alpha_i - \alpha_j)^2$ (see, for instance, van der Waerden, *Algebra I*). However, the proof in this case is not nearly as straightforward; it is outlined in §24.23 below.

24.22 Let K be complete with respect to a nonarchimedean absolute value $| \ |$, which we extend to the algebraic closure C of K. Let a *separable polynomial*

$$f(X) = aX^n + a_{n-1}X^{n-1} + \cdots + a_0 = a(X - \alpha_1)(X - \alpha_2)\ldots(X - \alpha_n)$$

of degree n over K be given. Show that there exists a real number $\delta > 0$ such that every $g \in K[X]$ of degree n with $|f - g| < \delta$ has the following properties:

(i) g, too, has n distinct roots $\beta_1, \ldots, \beta_n$ in C, and they can be numbered in such a way that

(1) $\qquad |\beta_i - \alpha_i| < |\alpha_k - \alpha_i|$ for every i, k with $i \neq k$,

(2) $\qquad K(\beta_i) = K(\alpha_i)$ for every i.

(ii) g has the same *decomposition type* as f; that is, the prime factorizations of f and g have the same number of irreducible factors of degree m, for each m.

Hint: (a) Multiplying by an appropriate nonzero element in R, one immediately reduces the problem to the case $f \in R[X]$. Let a be the leading coefficient of f. For every $h \in K[X]$ of degree at most n, set

$$h^*(X) = a^{n-1} h\left(\frac{X}{a}\right).$$

Clearly, $|h^*| \leq |a^{-1}| \, |h|$. In the special case $h = f$ we get $f^* \in R[X]$, and f^* is normalized. Hence we can assume that f is a *normalized* polynomial in $R[X]$. Then $|\alpha_i| \leq 1$ for each i.

(b) Because f is separable, one can choose $\delta > 0$ such that

$$\delta \leq |f'(\alpha_i)|^2 \quad \text{for every } i.$$

Then first of all $\delta \leq 1 = |f|$, and the assumption $|f - g| < \delta$ implies $|g| = |f|$, hence $g \in R[X]$. Since

$$|f'(\alpha_i) - g'(\alpha_i)| \leq |f - g| < \delta \leq |f'(\alpha_i)|,$$

moreover, we obtain $|g'(\alpha_i)| = |f'(\alpha_i)|$, and because

$$|g(\alpha_i)| = |g(\alpha_i) - f(\alpha_i)| \leq |f - g| < \delta$$

we get

$$|g(\alpha_i)| < |g'(\alpha_i)|^2.$$

Hence, by F14 in Chapter 23, the valuation ring of $K(\alpha_i)$ (for each fixed i) contains a unique root β_i of g such that

$$|\beta_i - \alpha_i| < |f'(\alpha_i)|.$$

Since $f'(\alpha_i) = \prod_{j \neq i} (\alpha_j - \alpha_i)$, this leads to (1). But once (1) is satisfied, we see that

$$|\beta_j - \alpha_i| = |\beta_j - \alpha_j + \alpha_j - \alpha_i| = |\alpha_j - \alpha_i| \quad \text{for any } j \neq i,$$

so that — again by (1) — β_j is distinct from β_i.

(c) To prove that $K(\beta_i) = K(\alpha_i)$, observe that $K(\beta_i) \subseteq K(\alpha_i)$ is already known. If $K(\beta_i) \neq K(\alpha_i)$, there exists $\sigma \in G(C/K)$ such that $\sigma\beta_i = \beta_i$ and $\sigma\alpha_i = \alpha_k$, with $k \neq i$. Then

$$|\alpha_i - \beta_i| = |\sigma(\alpha_i - \beta_i)| = |\alpha_k - \beta_i|,$$

contradicting (1).

(d) Assertion (ii) follows easily from (2), thanks to the separability of f and g.

24.23 One solution to the problem of generalizing §24.21 to the case where $f \in R[X]$ has arbitrary leading coefficient $a \neq 0$ was found by E. Erhardt, one of my students, in his Examensarbeit (Master's thesis). He first shows that, for every $1 \leq r \leq n$ and $\sigma \in S_n$,

$$(1) \qquad \left| a \prod_{i=1}^{r} \alpha_{\sigma(i)} \right| \leq 1.$$

To see this, assume $|\alpha_1| \geq |\alpha_2| \geq \cdots \geq |\alpha_n|$ and $|\alpha_1| > 1$. If $|\alpha_i| > 1$ for $i \leq k$ and $|\alpha_i| \leq 1$ for $i > k$, we have

$$\left| a \prod_{i=1}^{r} \alpha_{\sigma(i)} \right| \leq \left| a \prod_{i=1}^{k} \alpha_i \right| = |a| \, |s_k| = |a_{n-k}| \leq 1,$$

where s_k denotes the k-th elementary symmetric function in $\alpha_1, \ldots, \alpha_n$.

Now, by definition,

$$(2) \qquad D(f) = \prod_{i=1}^{n-1} \left(a^2 \prod_{j=i+1}^{n} (\alpha_i - \alpha_j)^2 \right).$$

Denote the i-th factor on the right by D_i and renumber the roots so that $|\alpha_1| \leq |\alpha_2| \leq \cdots \leq |\alpha_n|$. From (1) we obtain immediately

$$(3) \qquad |D_i| \leq |a^2| \prod_{j=i+1}^{n} |\alpha_j|^2 = |a\alpha_{i+1} \ldots \alpha_n|^2 \leq 1.$$

Among the roots of f let α_k be one whose distance to $\beta \in R$ is minimal:

(4) $$|\alpha_k - \beta| \leq |\alpha_i - \beta| \quad \text{for all } i.$$

We wish to show that k is uniquely determined by (4). *Krasner's Lemma* will then imply that $K(\alpha_k) \subseteq K(\beta) = K$, which is all we need.

Suppose, therefore, that $|\alpha_k - \beta| = |\alpha_l - \beta|$, with $k < l$. Using (1) one sees easily that

(5) $$|D_k| \leq \left| a(\alpha_k - \alpha_l)^2 \prod_{\substack{j=k+1 \\ j \neq l}}^{n} (\alpha_k - \alpha_j) \right|$$

and

(6) $$|D_i| \leq |\alpha_i - \alpha_k| \quad \text{for } i < k.$$

Since by (3) we have in addition $|D_i| \leq 1$ for $i > k$, we obtain from (2)

$$|D(f)| \leq \left| a(\alpha_k - \alpha_l)^2 \prod_{\substack{i=1 \\ i \neq k,l}}^{n} (\alpha_k - \alpha_i) \right| \leq \left| a(\alpha_k - \beta)^2 \prod_{\substack{i=1 \\ i \neq k,l}}^{n} (\beta - \alpha_i) \right| = |f(\beta)|,$$

contradicting the assumption that $|f(\beta)| < |D(f)|$ and completing the proof.

In addition we will also show that, if $n \geq 2$, the root $\alpha := \alpha_k$ as above satisfies

(7) $$|\alpha - \beta| < 1;$$

that is, $\alpha \in R$ and $\bar{\alpha} = \bar{\beta}$. Using (3) for all $i \leq n-2$ we first obtain, from (2),

(8) $$|D(f)| \leq |D_{n-1}| = |a^2 (\alpha_{n-1} - \alpha_n)^2| \leq |a\alpha_n|^2 \leq |a\alpha_n|,$$

the last inequality because $|a\alpha_n| \leq 1$, by (1). Assume $|\alpha_n| > 1$ (otherwise there is nothing to show). Then

(9) $$|f(\beta)| = |a\alpha_n| \prod_{i=1}^{n-1} |\beta - \alpha_i|.$$

The assumption $|f(\beta)| < |D(f)|$, together with (8), now implies that

$$\prod_{i=1}^{n-1} |\beta - \alpha_i| < 1.$$

But then (4) implies that $|\beta - \alpha_k| < 1$, which is (7).

Chapter 28: Fundamentals of Modules

28.1 As a complement to F9, prove that under the assumption that an A-module M is *semisimple*, we have:

$$\text{End}_A(M) \text{ is a skew field} \quad \Longrightarrow \quad M \text{ is simple.}$$

Hint: Assume $M = N \oplus T$ and let $p \in \text{End}(M)$ be the corresponding *projector*, so $px = x$ for $x \in N$ and $px = 0$ for $x \in T$. Then $p^2 = p$.

28.2 We will construct a *noetherian and artinian K-algebra such that A° is neither artinian nor noetherian*.

Take an extension L of the field $K = \mathbb{F}_p$ such that L/L^p is not finite. On the abelian group $A = L \times L$, define a K-algebra structure by $(a, b)(c, d) = (ac, ad + bc^p)$. The only nontrivial left ideal of A is $0 \times K$, whereas for every K^p-subspace V of K the module $0 \times V$ is a right ideal of A.

28.3 Let g be an endomorphism of the A-module N. Prove: *If g is surjective and N is noetherian, or if g is injective and N is artinian, then g is an automorphism of N.*

Hint: For any n we have $\ker g^n \subseteq \ker g^{n+1}$ and $\text{im } g^{n+1} \subseteq \text{im } g^n$.

28.4 Let f be an endomorphism of an artinian and noetherian A-module M. Prove the existence of a direct decomposition $M = V \oplus W$ preserved by f and such that f restricts to an *automorphism* on V and to a *nilpotent* endomorphism on W.

Hint: There exists n such that f induces automorphisms on $\text{im } f^n$ and on $M/\ker f^n$ (see §28.3). Since f^n also induces such automorphisms, it follows that

$$\text{im } f^n \cap \ker f^n = 0, \quad M = \text{im } f^n + \ker f^n.$$

28.5 If an *artinian and noetherian A-module M* is *indecomposable*, then $C := \text{End}_A(M)$ is a *local algebra*, that is, the noninvertible elements of C form an *ideal* of C. Conversely, an A-module M for which $\text{End}_A(M)$ is a local algebra is indecomposable.

Proof. The second part is easy: If C is a local algebra, there is no nontrivial decomposition $M = V \oplus W$, because $1 = p + q$ in C implies p or q is a unit.

As for the first part: Using §28.4 we deduce from the assumptions that every element f of $C = \text{End}_A(M)$ is either a *unit* or a *nilpotent*. Take $f, g \in C$, and let f be nilpotent. Then fg is also nilpotent, for otherwise fg would be a unit and so also f. Analogously, gf is seen to be nilpotent. There remains to show that if f, g are nilpotent, so is $f + g$. Otherwise there exists h such that $fh + gh = 1$. As we just saw, fh and gh are nilpotent. But if gh is nilpotent, $fh = 1 - gh$ is a unit. $\square$

28.6 A slightly generalized version of the *Krull–Remak–Schmidt Theorem* was proved by R. G. Swan; see pp. 75 and following of his *Algebraic K-Theory*, Lecture Notes in Mathematics 76, Springer, Berlin, 1968.

28.7 Fix a prime p and consider the subring A of $\mathbb{Q}$ consisting of all a/b with b relatively prime to p (compare §23.4). Being a subring of the rationals, A has no nonzero nilpotents. But on the other hand, A is a local ring and pA is its maximal ideal; hence $\mathfrak{R}(A) = pA$, and in particular $\mathfrak{R}(A) \neq 0$.

28.8 Let an R-algebra A be finitely generated as an R-module, and let α be an element of A. Here is how to show anew that α is *integral* over R (see vol. I, Chapter 16, F1). Using Hilbert's Basis Theorem (vol. I, Chapter 19, Theorem 4), prove first that without loss R can be assumed *noetherian*. But if R is noetherian, so is the R-module A (see F24). Now look at the increasing sequence of R-submodules $R + R\alpha + \cdots + R\alpha^i$.

28.9 It is clear that in the definition of the *tensor product of R-modules*, the *commutative ground ring* R cannot simply be replaced by a noncommutative algebra A. To reach an acceptable notion of tensor product over an arbitrary R-algebra A, we start out from the following situation:

Let M be an $A°$-module (written as a right A-module) and let N be an A-module. An R-bilinear map $\beta : M \times N \to X$ into some R-module X is *balanced* if

$$\beta(xa, y) = \beta(x, ay) \quad \text{for all } a \in A, \, x \in M, \, y \in N.$$

Then a *tensor product* of M and N is an R-module $M \otimes_A N$, together with a balanced R-bilinear map $\pi : M \times N \to M \otimes_A N$, written $(x, y) \mapsto x \otimes y$, such that for every balanced R-bilinear $\beta : M \times N \to X$ there is exactly one R-linear map $f : M \otimes_A N \to X$ satisfying $f(x \otimes y) = \beta(x, y)$.

To prove that $M \otimes_A N$ exists, consider the R-module $M \otimes_R N$ and form the quotient $M \otimes_A N = (M \otimes_R N)/I$ by the submodule I generated by all elements of the form $xa \otimes y - x \otimes ay$. It is clear that $M \otimes_A N$, together with the map $\pi : M \times N \to M \otimes_R N \to M \otimes_A N$, has the required properties.

Admittedly, $M \otimes_A N$ cannot in general be regarded as a A-module — although for A commutative it can, and the reader can easily check that everything is consistent with the earlier tensor product notion introduced in Definition 6.11 of vol. I (with R replaced by A). The rules stated at that point — apart of course from the missing A-module structure and the commutativity in rule (vi) — carry over accordingly to tensor products of the type defined here. Specifically, rules (iv) and (v) hold not only with respect to the first factor of $M \otimes_A N$ but also with respect to the second. As for associativity in (vi), the reader should elucidate and prove the following statement: *Given a right A-module M, an A-B-bimodule N and a B-module Y, there is a canonical isomorphism*

$$(M \otimes_A N) \otimes_B Y \simeq M \otimes_A (N \otimes_B Y).$$

Chapter 29: Wedderburn Theory

29.1 Consider the K-vector space $V = K^{(\mathbb{N})}$ and its endomorphism algebra $A := \operatorname{End}_K(V)$. Denote by I the set of all $f \in A$ such that fV has finite dimension. Obviously I is an *ideal* of A. Check that the algebra A/I is *simple but not noetherian* (hence not artinian; see F41 in Chapter 28).

Hint: Take $f \notin I$, so $\dim fV = \infty$. There is a sequence of linearly independent elements $v_i = fw_i$, where $i = 1, 2, \ldots$. Hence there exists $g \in A$ taking each v_i to e_i. Define h by setting $he_i = w_i$. Then $gfh = 1$, so A/I simple. To show that A/I is not noetherian, start with a descending sequence of subspaces V_i with $\dim V_i/V_{i+1} = \infty$. Set $N_i = \{f \in A \mid \dim fV_i < \infty\}$; this is a descending sequence of left ideals of A all of which contain I.

29.2 Let V be any D-vector space. Prove that, unless V is finite-dimensional, $\operatorname{End}_D(V)$ is not artinian. (*Hint:* Take a sequence of linear independent vectors v_i in V and consider the descending sequence of left ideals in $\operatorname{End}_D(V)$ defined by $N_k = \{f \in \operatorname{End}_D(V) \mid fv_1 = \cdots = fv_k = 0\}$.)

29.3 Let A be a finite-dimensional *commutative* K-algebra without nonzero nilpotents. Assume that char K is either 0 or larger than the dimension $A : K$. Prove that A is *semisimple* (compare F8) by showing that the symmetric bilinear form $(x, y) \mapsto \operatorname{Tr}_{A/K}(xy)$ (vol. I, p. 134, top) is *nondegenerate*. (This is sufficient, for then every A-submodule N of A_l is a direct summand: take the orthogonal complement relative to $(\cdot, \cdot)$.)

Hint: Suppose $a \in A$ is such that $\operatorname{Tr}_{A/K}(ay) = 0$ for every $y \in A$. Then $\operatorname{Tr}_{A/K}(a^i) = 0$ for all $i \in \mathbb{N}$, and the endomorphism $f = a_A$ of the K-vector space A satisfies $\operatorname{Tr}(f^i) = 0$ for all $i \in \mathbb{N}$. The reader can easily prove (or see LA II, p. 183, Aufgabe 75) that this implies that f is *nilpotent*, and hence so is a. But then $a = 0$.

29.4 Let a K-algebra A be simple and central. Then $A \otimes_K A^\circ$ is artinian if and only if $A : K$ is finite.

Proof. If $A : K < \infty$, then $A \otimes_K A^\circ \simeq M_n(K)$ by F14; hence $A \otimes_K A^\circ$ is artinian. Conversely, suppose $A \otimes_K A^\circ$ is artinian. Since A is simple, the $A \otimes_K A^\circ$-module A is simple; since A is central over K, its endomorphism algebra equals K, by (27):

$$\operatorname{End}_{A \otimes_K A^\circ}(A) = K.$$

Applying Theorem 3 to the simple artinian algebra $A \otimes_K A^\circ$ (whose simplicity follows from Theorem 7), we get $A : K < \infty$ — and also $A \otimes_K A^\circ \simeq M_n(K)$, which gives another proof for (33). $\qquad\square$

29.5 Let $\mathbb{H}$ be the division algebra of quaternions over $\mathbb{R}$, mentioned in see Remark 4 after Theorem 9. The $\mathbb{R}$-algebra $\mathbb{H} \otimes_\mathbb{R} \mathbb{H}$ is not a division algebra, because it has dimension 16.

More generally, prove: If K is any field and $[D]$ is the class in Br K of a division algebra $D \neq K$, then $D \otimes_K D$ is not a division algebra. (*Hint:* Every splitting field L of D is also a splitting field of $D \otimes_K D$.)

Show, more precisely, that if D has Schur index n (so $D \otimes_K D$ has reduced degree n^2 over K), then $D \otimes_K D$ is similar to an algebra of reduced degree $\frac{1}{2}n(n-1)$. (*Hint:* It is enough to check that the Schur index s of $D \otimes_K D$ divides $\frac{1}{2}n(n-1)$, which is nonobvious only if n is even. One can assume, if desired, that n is a power of 2: see Chapter 30, remark 3 after F3.)

29.6 Suppose $[A] \in \mathrm{Br}\, K$. Prove that A_L is a division algebra if and only if $s(A_L) = s(A)$ and A is a division algebra. (*Hint:* See (36) and F19.)

29.7 Let D be a *division algebra* and B a subalgebra of it. Prove:

 (a) The centralizer $Z_D(B)$ is likewise a division algebra.

 (b) If B is artinian, B is also a division algebra.

Hint for (b): We know that D has no nonzero nilpotents, so $\mathfrak{R}(B)$ is zero and B is *semisimple* (Chapter 28, Theorem 4 and F29). The rest follows easily using (4) and (10). Alternatively, take $x \in B$ and look at the descending sequence with term Bx^n.

29.8 Let A and B be K-algebras. Prove that if $A \otimes_K B$ is semisimple, so are A and B.

 Hint: Suppose $A \otimes_K B$ is semisimple. Then A and B are artinian; see Remark (b) after Theorem 7 and Chapter 28, F26. Hence there is $k \in \mathbb{N}$ such that $\mathfrak{R}(A)^k = 0$, by Theorem 4 in Chapter 28. It follows that $(\mathfrak{R}(A) \otimes B)^k \subseteq \mathfrak{R}(A)^k \otimes B = 0$, and this implies $\mathfrak{R}(A) \otimes B = 0$ by the semisimplicity assumption (Chapter 28, F37 and F29). Hence $\mathfrak{R}(A)$ vanishes and A is semisimple.

29.9 Let the situation be as in Theorem 14, and set $n = B:K$ and $d = L:K$. Then parts (b) and (e) of the theorem can be sharpened to say that

$$C \otimes_K M_n(K) \simeq B^\circ \otimes_K A, \tag{1}$$

$$(B \otimes_L C) \otimes_L M_d(L) \simeq A \otimes_K L. \tag{2}$$

The isomorphism in (2) implies part (f) of Theorem 14.

 (Incidentally, from part (b), or directly from isomorphism (1), one obtains again that $A^\circ \otimes_K A \simeq M_n(K)$ if $n = A:K$.)

Proof. Let N be a simple $B \otimes A^\circ$-module. The $B \otimes A^\circ$-modules A and $B \otimes A^\circ$ satisfy

$$A \simeq N^r \quad \text{and} \quad B \otimes A^\circ \simeq N^t.$$

We obtain the following isomorphisms of A°-modules: $N^t \simeq B \otimes A^\circ \simeq A^n \simeq N^{rn}$, which implies on grounds of length that $t = rn$. Hence

$$B^\circ \otimes A = \mathrm{End}_{B \otimes A^\circ}(B \otimes A^\circ) \simeq \mathrm{End}_{B \otimes A^\circ}(N^t) \simeq M_t(\mathrm{End}_{B \otimes A^\circ}(N))$$

$$\simeq M_n(M_r(\mathrm{End}_{B \otimes A^\circ}(N))) \simeq M_n(\mathrm{End}_{B \otimes A^\circ}(N^r))$$

$$\simeq M_n(\mathrm{End}_{B \otimes A^\circ}(A)) \simeq M_n(C) = C \otimes M_n(K).$$

This yields (1).

 Now tensor (1) on the left with B over L, to obtain $(B \otimes_L C) \otimes M_n(K) \simeq (B \otimes_L B^\circ) \otimes_K A \simeq M_{n/d}(L) \otimes_K A$, hence $M_n(B \otimes_L C) \simeq M_{n/d}(M_d(B \otimes_L C)) \simeq$

$M_{n/d}(L \otimes_K A)$. This implies, by F9, that $M_d(B \otimes_L C) \simeq L \otimes_K A$, and hence (2) is proved.   □

29.10 *Let A be a central-simple K-algebra. A splitting field L of A satisfying $(L:K)^2 = A:K$ can be embedded K-isomorphically in A.*

Proof. By Theorem 18 there is an algebra A' similar to A that contains L as a maximal commutative subalgebra. At the same time, our assumption implies $A:K = (L:K)^2 = A':K$ (see Theorem 15). Hence $A \simeq A'$, which proves the conclusion.   □

29.11 We now turn to questions suggested by Theorems 15, 16 and 18. Let A be a central-simple K-algebra of degree $A:K = n^2$ and E/K a finite field extension of degree $E:K = d$.

(a) Suppose E is a subalgebra of A. The centralizer C of E in A satisfies

$$C \sim A_E, \qquad Z(C) = E, \qquad C:E = m^2, \quad \text{where } m = n/d.$$

(b) Still assuming $E \subseteq A$, the field E is a *maximal subfield* of A if and only if E is a splitting field of A *and* has no (strict) extension whose degree divides m.

(c) E is a splitting field of A if and only if there is an algebra A' similar to A that contains E as a *maximal subfield*.

Proof. (a) Use the Centralizer Theorem (page 168) and the equality

$$C : K = (C : E)(E : K).$$

(c) For the forward direction apply Theorem 18. The converse is part of (b).

(b) Again by the Centralizer Theorem, $C = Z_A(E)$ is a central-simple E-algebra satisfying

(1) $$C \sim A \otimes_K E.$$

Moreover $C \simeq M_r(D)$, for some central division algebra D over E. If A splits over E, we have $D = E$ by virtue of (1). If E is a maximal subfield of A, this is also the case in C and hence in D. But since E is the center of D, we must have $D = E$, so A splits over E.

There remains to find out when E is a maximal subfield of $M_m(E)$. Suppose a subalgebra F of the E-algebra $M_m(E)$ is a field. Applying part (a) to this situation we see that $F:E$ must divide m. Conversely, given a field extension F/E of degree t, the E-algebra F is isomorphic to a subalgebra $M_t(E)$. If $m = st$, say, $M_t(E)$ is isomorphic to a subalgebra of $M_s(M_t(E)) \simeq M_m(E)$.   □

29.12 Take $[A], [B] \in \mathrm{Br}\, K$ with $s(A), s(B)$ *relatively prime*. Prove that (a) $s(A \otimes B) = s(A)s(B)$; (b) if A and B are division algebras, so is $A \otimes B$.

Hint: Statement (b) follows from (a); see the remark after Definition 7. To prove (a) see Remark 1 after Theorem 19. Notice that if L is a splitting field of $A \otimes B$, the product $A_L \otimes B_L$ is similar to 1, so A_L is similar to B_L°. But A_L and B_L° obviously have the same Schur index, and it divides both $s(A)$ and $s(B)$, by F19. From the assumption, then, we get $A_L \sim 1 \sim B_L$; that is, $L:K$ is divisible by $s(A)$ and $s(B)$.

29.13 Let K be a field of characteristic $p > 0$ and D a division algebra with center K. If σ is a nontrivial K-algebra automorphism of D satisfying $\sigma^p = 1$, there exist nonzero $x, y \in D$ such that $\sigma y = y + x$ and $\sigma x = x$, hence also some z such that

(1) $$\sigma z = z + 1.$$

Proof. In $\mathrm{End}_K(D)$ we have $(\sigma - 1)^p = \sigma^p - 1^p = 0$. Take the largest n such that $(\sigma - 1)^n \neq 0$. For some $d \in D$ we have $x := (\sigma - 1)^n d \neq 0$, yet $\sigma x = x$. Setting $y := (\sigma - 1)^{n-1} d$ we obtain $\sigma y - y = x$, and multiplying this equation by x^{-1} we see that $z := x^{-1} y$ is a solution of $\sigma z - z = 1$. $\qquad\square$

29.14 Using §29.13 prove the following generalization of Lemma 2: *Let D be a division algebra with center K, and suppose every element of D is algebraic over K. If $D \neq K$, there is some subfield E of D strictly containing K and separable over K.*

Hint: If the conclusion fails we are in characteristic $p > 0$, and there exists $u \in D$ such that $u^p \in K$ but $u \notin K$. Apply §29.13 to the inner automorphism σ of D defined by u in order to obtain a subfield $K[z]$ of D such that $K[z]/K$ is not purely inseparable.

29.15 Let D be a a division algebra with center K, and suppose there is a natural number n such that, for any $x \in D$, the algebra $K[x]$ is a field extension over K of degree at most n. Prove that D is *finite-dimensional* of dimension at most n^2 over K.

Hint: One can assume $D \neq K$ and apply §29.14 to show there exist elements $x \in D$ such that

(1) $$K[x]/K \text{ is separable and } x \notin K.$$

Since $K[x] : K \leq n$ there is among such elements x some x_0 of *maximal degree* $K[x_0] : K$. Set $L = K[x_0]$; the centralizer C of L in D is a division algebra, and L is the center of C. If C were distinct from L, problem §29.14 would lead (through the use of the Primitive Element Theorem) a contradiction with the choice of x_0. Hence $C = L$, and the Centralizer Theorem yields $D : K = (L : K)(L : K) \leq n^2$.

29.16 Let K be a field. There exist *central* division K-algebras that have *infinite dimension* over K and admit *noninner* K-automorphisms (compare Theorem 20'). The stepwise proof of this fact is the object of Problems 29.17–21.

29.17 Let L be a field and σ an automorphism of L. Denote by K the fixed field of σ in L. Consider the set A of all maps $a : \mathbb{Z} \to L$ such that $a_i = 0$ for almost all $i < 0$, where we write a_i for $a(i)$. Such maps form an L-vector space in the obvious way. We express any element $a \in A$ as a (for now purely formal) sum

(1) $$a = \sum_i a_i X^i.$$

We define a multiplication operation in A by associating with $a, b \in A$ the map $c : \mathbb{Z} \to L$ such that $c_n = \sum_{i+j=n} a_i \sigma^i(b_j)$; that is,

$$(2) \qquad \left(\sum_i a_i X^i \right) \left(\sum_j b_j X^j \right) = \sum_n \left(\sum_{i+j=n} a_i \sigma^i(b_j) \right) X^n.$$

(Note that for negative indices, only finitely many a_i and b_j are nonzero, so c is well defined and does belong to A.) It is easy to check that this product makes the L-vector space A into a K-algebra. Calculation in A with the representation (1) proceeds as with *formal Laurent series*, apart from the fact — which later will prove decisive — that right multiplication with elements of L takes on the form

$$(3) \qquad X\lambda = \sigma(\lambda)X.$$

We call A the *algebra of σ-crossed formal Laurent series over L* and write $A = L((X; \sigma))$. If σ acts trivially on L, we obviously have $L((X; \sigma)) = L((X))$. In any case, $K((X))$ is a *commutative subalgebra* of $L((X; \sigma))$.

Working as for usual Laurent series, one can show that any nonzero element of A is invertible. Therefore $L((X; \sigma))$ *is a division algebra*.

29.18 We now determine what elements of $A = L((X; \sigma))$ commute with all others. The answer depends on whether σ has finite or infinite order. Prove:

(a) *If σ has infinite order, the center of $L((X; \sigma))$ is K.*

(b) *If σ has order n, the center of $L((X; \sigma))$ is $K((X^n))$. In this case the division algebra $L((X; \sigma))$ has finite dimension n^2 over its center.*

Thus we will have constructed central division algebras of infinite dimension over a given field K if we can find an extension L of K admitting an automorphism σ of infinite order whose fixed field is K. When char $K = 0$ we simply take $L = K(t)$, where t is an indeterminate, and define the automorphism σ of L by setting $\sigma(t) = t + 1$. If K is a finite field with q elements, let L be an algebraic closure of K and take for σ the automorphism defined by $\sigma(\lambda) = \lambda^q$. The case of an infinite field K of nonzero characteristic is postponed till §29.21.

29.19 Let $A = L((X; \sigma))$ be as above. A second automorphism τ of L/K that *commutes* with σ can be extended to a K-automorphism $\hat{\tau}$ of A by setting

$$\hat{\tau}\left(\sum_i a_i X^i \right) = \sum_i \tau(a_i) X^i.$$

We claim that *unless τ is a power of σ, the map $\hat{\tau}$ is not an inner automorphism of A.* Indeed, suppose there exists $u = \sum_i u_i X^i$ satisfying $\tau(\lambda)u = u\lambda$ merely for any $\lambda \in L$. Then equation (3) in §29.17 implies that $\tau(\lambda)u_i = \sigma^i(\lambda)u_i$ for every i; but this can only happen if all but one of the u_i vanish, that is, $\tau = \sigma^i$ for some i.

29.20 From the last two problems we see that to prove the existence of division algebras having a prescribed center K and noninner K-automorphisms, it is enough to find automorphisms σ, τ of some field extension L/K satisfying the following properties:

(1) $\mathrm{Fix}(\sigma) = K, \quad \sigma\tau = \tau\sigma, \quad \tau \neq \sigma^i$ for all $i \in \mathbb{Z}$.

(Why does this automatically imply that σ has infinite order?) If K has characteristic 0, we can choose $L = K(t)$ and $\sigma(t) = t+1$ as in §29.18 and define τ by $\tau(t) = t + \frac{1}{2}$. For K finite, again take L and σ as in §29.18; since $G(L/K)$ is abelian, all that's left to check is that $G(L/K)$ is not generated by σ, for which recall Section 12.4 in vol. I. For K arbitrary see the next problem.

29.21 For any field K, there exists an extension L/K and automorphisms σ, τ of L/K satisfying condition (1) in §29.20. Namely, take $L = K(t_i \mid i \in \mathbb{Z})$, where the t_i are indeterminates, and define σ and τ by setting $\sigma t_i = t_{i+1}$ and $\tau t_i = t_i + 1$.

29.22 A *finitely generated* field extension E/K is called *separably generated* if it possesses a transcendence basis $x_1, \ldots, x_r$ such that the (finite!) extension $E/K(x_1, \ldots, x_r)$ is separable. (Such a basis is called a *separating transcendence basis*.) Prove that *if E/K is separably generated, E/K is separable in the sense of Definition 9*.

   Hint: Let F/K be any field extension and consider the field $F(X_1, \ldots, X_r)$ of rational functions in r variables over F. There is a canonical isomorphism $K[X_1, \ldots, X_r] \otimes_K F \simeq F[X_1, \ldots, X_r]$, from which there arise canonical *injections* $K(X_1, \ldots, X_r) \otimes_K F \to F(X_1, \ldots, X_r)$ and

$$E \otimes_{K(X_1, \ldots, X_r)} (K(X_1, \ldots, X_r) \otimes_K F) \to E \otimes_{K(X_1, \ldots, X_r)} F(X_1, \ldots, X_r).$$

29.23 As is apparent from Definition 9, E/K is separable if and only if every finitely generated subextension of E/K is separable. Using §29.22, show that *if* char $K = 0$, *every extension E/K is separable*.

29.24 Now suppose that K has characteristic $p > 0$. Prove that for any field extension E/K the following conditions are equivalent:

(i) $E \otimes_K K^{1/p}$ is reduced, that it, it has no nonzero nilpotents.

(ii) If $a_1, \ldots, a_n \in E$ are linearly independent over K, the p-th powers $a_1^p, \ldots, a_n^p$ too are linearly independent over K.

   Hint: If there is a nonzero nilpotent, there exists $t \neq 0$ such that $t^p = 0$ (why?). Write such a $t \in E \otimes_K K^{1/p}$ as $t = \sum_{i=1}^n a_i \otimes b_i$, with $a_1, \ldots, a_n \in E$ linearly independent. Then $t^p = \sum a_i^p \otimes b_i^p = 0$; thus, if (ii) is satisfied, we have $b_i = 0$, contradicting the assumption $t = 0$. To prove that (i) $\Rightarrow$ (ii), work as in the second part of the proof of F20.

29.25 Again assume that char $K = p > 0$. Prove: *If a finitely generated field extension E/K satisfies condition* (ii) *in §29.24, then E/K is separably generated* (and in fact *every* set of generators of E/K contains a separating transcendence basis of E/K).

Hint: Suppose $E = K(x_1, \ldots, x_n)$, where $x_1, \ldots, x_r$ form a transcendence basis of E/K. If $r < n$, let $f(X_1, \ldots, X_{r+1})$ be a nonzero polynomial of lowest degree satisfying

$$f(x_1, \ldots, x_{r+1}) = 0.$$

Not all variables can appear in f only with exponents divisible by p; otherwise using our assumption (ii) we would get (with some thought) a contradiction to the minimality of f. So suppose X_1 appears in f with an exponent not divisible by p. Then x_1 is *separable* over $K(x_2, \ldots, x_{r+1})$, hence over $K(x_2, \ldots, x_n)$. The assertion follows by induction on n.

29.26 The results reached in §29.22-25 can be summarized as follows:

A field extension E/K is separable if and only if every finitely generated subextension of E/K is separable. A finitely generated extension is separable if and only if it is separably generated. In characteristic zero, therefore, every extension is separable. If a finitely generated extension E/K is separable, every set of generators of E/K contains a separating transcendence basis.

If char $K = p > 0$, the following statements are equivalent for an arbitrary field extension E/K:

 (i) *E/K is separable, that is, for every field extension F/K, the product $E \otimes_K F$ is reduced.*

 (ii) *$E \otimes_K K^{1/p}$ is reduced.*

 (iii) *Whenever $a_1, \ldots, a_n \in E$ are linearly independent over K, so are their p-th powers $a_1^p, \ldots, a_n^p$.*

Every extension over a perfect field is separable.

29.27 Let F be an intermediate field of a field extension E/K. If E/K is separable, so is F/K, of course (although E/F need not be; take $E = K(X)$ and $F = K(X^p)$). Using §29.26 one easily shows that *if F/K and E/F are separable, so is E/K.*

Suppose char $K = p > 0$. It can easily be proved that *a finite extension E/K is separable if and only if $E = E^p K$.* From this derive that *a finitely generated extension E/K satisfies $E = E^p K$ if and only if E/K is algebraic and separable.*

29.28 As a complement to §29.26, prove *Weil's Criterion:*

An arbitrary field extension E/K is separable if and only if the following condition is satisfied:

 (iv) *Any set of elements of K that is linearly independent over K^p is also linearly independent over E^p.*

Index

Universitext

(continued from page ii)